New Security Challenges

Series Editor
George Christou
University of Warwick
Coventry, UK

The last decade has demonstrated that threats to security vary greatly in their causes and manifestations and that they invite interest and demand responses from the social sciences, civil society, and a very broad policy community. In the past, the avoidance of war was the primary objective, but with the end of the Cold War the retention of military defence as the centrepiece of international security agenda became untenable. There has been, therefore, a significant shift in emphasis away from traditional approaches to security to a new agenda that talks of the softer side of security, in terms of human security, economic security, and environmental security. The topical New Security Challenges series reflects this pressing political and research agenda.

More information about this series at
http://www.palgrave.com/gp/series/14732

Franziskus von Lucke

The Securitisation of Climate Change and the Governmentalisation of Security

palgrave
macmillan

Franziskus von Lucke
Institute of Political Science
University of Tübingen
Tübingen, Germany

New Security Challenges
ISBN 978-3-030-50905-7 ISBN 978-3-030-50906-4 (eBook)
https://doi.org/10.1007/978-3-030-50906-4

This Palgrave Macmillan imprint is published by the registered company Springer Nature Switzerland AG.
The registered company address is: Gewerbestrasse 11, 6330 Cham, Switzerland

Acknowledgements

This book would not have been possible without the continuous support of a number of people. The journey that eventually led to this book started at the University of Hamburg in the seminars of Angela Oels, which gave me a first glimpse into the rabbit hole of Foucauldian governmentality studies and climate politics. Thus, many thanks to Angela Oels for the inspiration, various interesting discussions and the fruitful collaboration in the CliSAP excellence cluster. I am also very grateful for the support of Antje Wiener and Michael Brzoska during my time at the University of Hamburg.

The journey then continued at the University of Tübingen where I want to particularly thank my PhD supervisor Thomas Diez who supported my theoretical ideas from the beginning and with whom I had countless fruitful debates on governmentality, power, securitisation and climate change. I am also obliged to my colleagues, the student assistants and my fellow PhD students in Tübingen, with whom I had great discussions, who helped to compile empirical data and who proofread the book. Thus, many thanks go to Zehra Wellmann, Schielan Babat, Sandra Dürr, Thea Güttler, Leonie Haueisen, Benno Keppner, Miriam Keppner, Katharina Krause, Hanna Spanhel and Josefa Velten. Beyond the Tübingen crowd, a special thanks to Stefan Elbe, whose ideas greatly inspired my theoretical approach, and who gave me invaluable feedback on earlier versions of this book. I am also grateful for the input at workshops, conferences and particularly in the Tübingen IR colloquium. In particular, I want to thank Ingrid Boas, Olaf Corry, Rita Floyd, Stefano Guzzini, Andreas Hasenclever, Markus Lederer, Matthias Leese, Matt McDonald,

Aysem Mert, Delf Rothe and Jürgen Scheffran. Finally, I am indebted to my interview partners in the US, Mexico, Germany and the UK who took the time to openly discuss with me the climate security debates in their respective countries.

Last but not least, a special thanks to my wife Sabrina von Lucke, to my children as well as to my family and friends who had to endure me in times of crisis, always took the time to listen to my ideas and continuously helped me to keep up my motivation to continue this journey until the end.

Tübingen
29.02.2020

Franziskus von Lucke

Praise for *The Securitisation of Climate Change and the Governmentalisation of Security*

"In this important book, Franziskus von Lucke provides a theoretically sophisticated and empirically rich account of the relationship between security and climate change. Developing a Foucauldian-inspired account of securitization, the book rejects blanket or universal claims about the climate change–security relationship, instead insisting on the need to critically examine how the securitization of climate change plays out in particular empirical contexts. Exploring the cases of the US, Germany and Mexico, von Lucke points to distinctive dynamics of securitization in these settings, with different implications for the practices these in turn encourage. Ultimately, this book constitutes an important addition to literature on the relationship between climate change and security, while developing a distinct and nuanced account of securitization that will be of interest to a wide range of scholars of security in international relations."

—Associate Professor Matt McDonald is a Reader in International Relations at the *University of Queensland, Australia*

"In 2019 a number of states and other actors (notably the European Union) have made climate emergency declarations. It is therefore more important than ever to understand what the securitization of the climate means. That is: Who can securitize? What security measures are likely/deemed legitimate by relevant audiences? How does securitization affect the population within and outside a securitizing state? And perhaps most importantly of all, will it succeed? Franziskus von Lucke's carefully researched book offers answers to all of these questions and many others besides. von Lucke proceeds by examining with the US, Mexico and Germany, three real-life empirical cases of climate securitization. Each one provides unique insights that enable a fuller understanding of climate security. Accessibly written this is a must read for scholars and practitioners alike."

—Dr Rita Floyd, *University of Birmingham, UK*, author of *The Morality of Security: A theory of just Securitization*, 2019

"With great empirical detail and conceptual clarity, the book compares discourses and practices of climate security in different contexts. An essential reading for anyone interested in international climate politics, securitization theory, governmentality and the notion of power in International Relations."

—Dr Delf Rothe, *Institute for Peace Research and Security Policy Hamburg at the University of Hamburg, Germany*

Contents

About the Author

Franziskus von Lucke is a researcher in International Relations at the University of Tübingen. His research focuses on critical security studies, climate politics and climate justice, and he has worked extensively on the securitisation of climate change. His works have appeared in *Geopolitics*, *the Journal of International Relations and Development* and in the *Zeitschrift für Internationale Beziehungen.*

List of Abbreviations

ASP	American Security Project
BMU	Bundesministerium für Umwelt, Naturschutz und Reaktorsicherheit – Federal Ministry for the Environment, Nature Conservation and Nuclear Safety (German Environmental Ministry) (temporarily renamed BMUB in 2013)
BMVg	Bundesministerium der Verteidigung – Federal Ministry of Defence (Germany)
BMZ	Bundesministerium für wirtschaftliche Zusammenarbeit und Entwicklung – Federal Ministry for Economic Cooperation and Development (Germany)
C2ES	Center for Climate and Energy Solutions
CAN	Climate Action Network
CAP	Center for American Progress
CCC	Centro de Colaboración Cívica – Civic Cooperation Centre
CCS	Center for Climate & Security
CDM	Clean Development Mechanism
CDU	Christlich Demokratische Union Deutschlands – Christian Democratic Union of Germany
CEMDA	Centro Mexicano de Derecho Ambiental – Mexican Centre for Environmental Law
CENAPRED	Centro Nacional de Prevención de Desastres – National Centre for Disaster Prevention
CENTCOM	US Central Command
CFR	Council on Foreign Relations
CIA	Central Intelligence Agency
CICC	Comisión Intersecretarial de Cambio Climático – Inter-Ministerial Commission for Climate Change

CISEN	Centro de Investigación y Seguridad Nacional – Centre for Research and National Security
CMM	Centro Mario Molina – Mario Molina Centre
CNAS	Center for a New American Security
COP	Conference of the Parties
CPI	Climate Performance Index
CSIS	Center for Strategic & International Studies
CSS	Critical Security Studies
DHS	Department of Homeland Security
DOD	Department of Defense
ENCC	Estrategia Nacional de Cambio Climático – National Strategy on Climate Change
EPA	Environmental Protection Agency
ESS	European Security Strategy
EU	European Union
FEMA	Federal Emergency Management Agency
FONDEN	Fondo de Desastres Naturales de México – Mexican Natural Disaster Fund
FOPREDEN	Fondo para la Prevención de Desastres Naturales – Mexican Federal Fund for the Prevention of Disasters
GDP	Gross Domestic Product
GHG	Greenhouse Gas
GIZ	Deutsche Gesellschaft für Internationale Zusammenarbeit – German Federal Enterprise for International Cooperation (formerly GTZ)
GLOBE	Global Legislators Organisation for a Balanced Environment
HDI	Human Development Index
IfP-EW	Initiative for Peacebuilding – Early Warning Analysis to Action
INE	Instituto Nacional de Ecología – National Institute for Ecology
INECC	Instituto Nacional de Ecología y Cambio Climático – National Institute for Ecology and Climate Change
IPCC	Intergovernmental Panel on Climate Change
MAB	Military Advisory Board
NAFTA	North American Free Trade Agreement
NASA	National Aeronautics and Space Administration
NATO	North Atlantic Treaty Organization
NDCs	Nationally Determined Contributions
NGO	Non-Governmental Organisation
NOAA	National Oceanic and Atmospheric Administration
NSS	National Security Strategy
ODUSD-ES	Office of the Deputy Under Secretary of Defense – Environmental Security

OECD	Organisation for Economic Co-operation and Development
PAN	Partido Acción Nacional – National Action Party
PDCI	Partners for Democratic Change International
PEACC	Plan Estatal de Acción ante el Cambio Climático – State Level Plan for Climate Action
PECC	Programa Especial de Cambio Climático – Special Programme on Climate Change
PIK	Potsdam Institut für Klimafolgenforschung – Potsdam Institute for Climate Impact Research
PRI	Partido Revolucionario Institucional – Institutional Revolutionary Party
QDR	Quadrennial Defense Review
RUSI	Royal United Services Institute
SAGARPA	Secretaría de Agricultura, Ganadería, Desarrollo Rural, Pesca y Alimentación – Mexican Secretariat of Agriculture, Livestock, Rural Development, Fisheries and Food
SEGOB	Secretaría de Gobernación – Mexican Home Office
SEMARNAT	Secretaría de Medio Ambiente y Recursos Naturales – Department of the Environment and Natural Resources
SERDP	Strategic Environmental Research and Development Program
SPD	Sozialdemokratische Partei Deutschlands – Social Democratic Party of Germany
SWP	Stiftung Wissenschaft und Politik. Deutsches Institut für Internationale Politik und Sicherheit – German Institute for International and Security Affairs
UK	United Kingdom
UN	United Nations
UNAM	Universidad Nacional Autónoma de México – National Autonomous University of Mexico
UNDP	United Nations Development Programme
UNEP	United Nations Environment Programme
UNFCCC	United Nations Framework Convention on Climate Change
UNGA	United Nations General Assembly
UNSC	United Nations Security Council
US	United States
WBGU	Wissenschaftlicher Beirat der Bundesregierung für Globale Umweltveränderungen – German Advisory Council on Global Change
WWF	World Wide Fund for Nature

List of Tables

CHAPTER 1

Introduction and Theoretical Framework

Introduction

On June 22, 2018, at a European Union (EU) high-level event on 'Climate, Peace and Security: The Time for Action' High Representative for Foreign Affairs and Security Policy Federica Mogherini urged that acting on climate change was to invest 'in our own security' (EEAS 2018). Only a few weeks later, on July 11, 2018, the United Nations Security Council (UNSC) once again discussed 'climate-related security risks' and the Deputy Secretary-General Amina Mohammed made clear that 'climate change is a real threat and it is proceeding at a relentless pace' (UNSC 2018). Finally, in January 2019 at the World Economic Forum in Davos, Greta Thunberg, who has become famous for her passionate and inexorable climate activism and her role in starting the Fridays for Future movement, warned that 'our house is on fire' and urged political leaders to immediately adopt measures to stop climate change (Thunberg 2019a).

These three examples are all part of a longstanding political debate that has highlighted the catastrophic consequences of climate change and linked the issue to a range of security concerns (Brauch 2009; Rothe 2016; Dyer 2018; Lippert 2019; McDonald 2013). This 'securitisation process' (Buzan et al. 1998) already began in the 1980s when climate change first entered international politics and began to be discussed in relation to broader environmental security concerns (Floyd 2010, p. 75; Hardt 2017). Since then, the debate has expanded continuously and made

F. von Lucke, *The Securitisation of Climate Change and the Governmentalisation of Security*, New Security Challenges,
https://doi.org/10.1007/978-3-030-50906-4_1

climate change the undisputed focal point when it comes to linking changes in the environment to security concerns (Brzoska and Oels 2011, p. 51). Politically, these debates have not been without consequences. Even though the empirical and causal connection between climate change and security or conflict is contested in the academic literature (Scheffran et al. 2012b; Barnett 2000; Buhaug et al. 2014), the persistent linking of the two has nevertheless established climate change as one of the defining security problems of the twenty-first century in global politics (Chaturvedi and Doyle 2015; Rothe 2016; Lippert 2019; Dyer 2018; Dalby 2013b). Linking climate change with security thus has decisively transformed how political practitioners handle these issues and has legitimised numerous policies and practices (Floyd 2010; Diez et al. 2016; Oels 2012; UNGA 2009b; WBGU 2008; Scott and Ku 2018). However, despite the apparent consensus that climate change is not only an environmental concern, economic problem or a matter of justice but will very soon have tangible security implications, activists, political practitioners and scientists differ considerably when it comes to conceptions of security to make sense of climate change.

Some have predominately pointed to its 'national security' consequences, for example, direct threats to the territorial integrity of states and the increase in violent conflicts. As a consequence, they have urged to integrate climate change into the planning of traditional security institutions to prepare for a future ravaged by climate-induced violent conflicts (CNA 2007, p. 6; CNA Military Advisory Board 2014, p. 21; Chaturvedi and Doyle 2015; Buxton et al. 2016; Briggs 2012). In stark contrast, others have emphasised the repercussions of rising temperatures for 'human security', meaning the general deterioration of living conditions of poor populations mainly due to resource scarcity and an increase in extreme weather events (WBGU 2008, p. 1; see also GTZ 2008b, p. 8; Scheffran et al. 2012a). To handle the resulting problems, they have recommended lowering the vulnerability of affected populations by transforming problematic behaviour, to scale up adaptation efforts and to increase development aid (GTZ 2008a, p. 55; WBGU 2008, pp. 10, 115). Finally, many have refrained from concrete threat constructions and have instead depicted climate change as an overall 'risk' that will gradually affect countless variables and in turn pertain a whole range of risk groups and areas around the world (adelphi 2012, p. 31; World Bank et al. 2013, pp. xviii, xx; Corry 2012; Lippert 2019; Oels 2011; Rothe 2011b). From this point of view, the appropriate response is to develop sophisticated risk management schemes to increase the resilience of risk groups and areas in order to eventually keep the overall risk at a tolerable level (Greenpeace México 2010, p. 57; World Bank et al. 2013, p. xvii).

Thus, despite the agreement that climate change is somehow linked to security problems, the exact nature of the threat, the affected referent objects as well as the political and normative consequences of handling climate change as security issue are far from clear. This is not only true for the political debate but even more so for the academic literature, which tries to make sense of the empirical 'climate security nexus' (Scheffran et al. 2012a) and the political consequences of linking climate change to security conceptions (Brauch 2009; Diez et al. 2016; Detraz and Betsill 2009; Corry 2012; McDonald 2013; Dyer 2018; Buxton et al. 2016; Rothe 2016). The aim of this book is to contribute to these debates by exploring how specific security representations of climate change have influenced political debates, policies and practices. It thus focuses on how to theoretically make sense of the diversity of security conceptions that are associated with climate change; how different discourses of climate change as security issue have come about in diverse contexts; whether and how they make a difference in terms of political consequences and what normative implications this has.

The Evolution of the Climate Security Nexus in Academic and Political Debates

Much of the alarming political debate on climate security is based on academic literature about the nexus between the environment, climate change and security (Buhaug et al. 2014; Brauch and Scheffran 2012; Lee 2009; Raleigh and Urdal 2007; Hardt 2017). This research to a considerable extent draws on older works on environmental security and conflict originating in the 1980s and 1990s (Ullmann 1983, p. 134; Dalby 2009, p. 14; Deudney 1990; Deudney and Matthew 1999; Pirages 1991). It also stems from the theoretical debates about the 'broadening' (e.g. not only states are considered as security threats) and 'deepening' (e.g. new epistemological foundations of thinking about security and the consideration of new referent objects such as individuals) of traditional understandings of state or military security (Ullmann 1983; Booth 1991; Krause and Williams 1996, 1997; Mathews 1989). Empirically, this research focused on the questions whether and how environmental change could initiate or contribute to social, political or ultimately violent conflict (Homer-Dixon

1991, 1994) as well as lead to hundreds of millions of environmental refugees (Myers 1995). Notwithstanding the weak empirical evidence for any of these claims (Hartmann 2010, p. 235; Greenpeace 2007; Oels and Carvalho 2012), a range of different political actors eagerly adopted this argumentation to advance their political agenda.

At the beginning of these debates in the late 1980s and early 1990s, climate change was only discussed as one issue besides other environmental problems that were increasingly linked to security concerns and conflict. However, due to its global reach and overall magnitude, it soon became one of the key dangers. Thus, at the beginning of the 1990s, several environmental non-governmental organisations (NGOs) (Oels 2012, p. 186; Myers 1995) picked up the security framing to raise attention for climate change. Amongst other factors, this contributed to important breakthroughs in the international negotiations on climate change. Examples are the adoption of the United Nations Framework Convention on Climate Change (UNFCCC) in 1992, the commencement of the yearly Conferences of the Parties (COP) in 1995 and the adoption of the Kyoto Protocol in 1997, which all, to some extent, were also legitimised by referring to the threatening potential of climate change.

While environmental and climate security debates became less prevalent towards the end of the 1990s and early 2000s, the increasing scientific evidence for the far-reaching implications of global warming epitomised in the ever more detailed reports of the Intergovernmental Panel on Climate Change (IPCC) (IPCC 2001, 2007) restarted the debate in the mid-2000s (Brzoska and Oels 2011; Oels and von Lucke 2015). In contrast to the earlier discussions, other environmental problems largely ceased to play an important role, and climate change became the undisputed centre of this novel environmental security debate. Academically, this led to a renewed interest in questions about how environmental degradation and particularly climate change contributed to violent conflict or migration (Scheffran et al. 2012b; Barnett 2003; Barnett and Adger 2007; Scheffran et al. 2012a; Hsiang et al. 2013; Gleditsch 2012). While the findings of this research were mixed (Scheffran et al. 2012c; Barnett and Adger 2005, 2007; Salehyan 2008; Gleditsch 2012), this did not prevent numerous political actors from claiming that climate change indeed was one of the main security challenges of our times and necessitated urgent action, which entailed genuine climate mitigation, but also the development of military counter-strategies.

Amongst the first political actors that actively waged this debate were security policy-focused think tanks in the United States (US), which particularly since 2007 have repeatedly drawn a connection between climate change and national security (CNA 2007; Campbell et al. 2007; Campbell 2008). Beyond these, the former Vice President of the US and Democratic presidential candidate Albert 'Al' Gore at various occasions highlighted the far-reaching security implications of climate change (Gore 2007). On the other side of the Atlantic, the German Advisory Council on Global Changes (WBGU) published a widely received report on 'Climate Change as Security Risk' (WBGU 2008), and the European Union (EU) as well discussed the security implications of climate change in 2008 (Solana and EU Commission 2008) and most recently in 2018 (EEAS 2018). Moreover, several environmental and human rights NGOs began to frame climate change as security issue (Greenpeace 2007, 2013; Christian Aid 2006, 2007; Smith and Vivekananda 2007). On the global level, the UN Secretary-General has published a widely noted report on the possible security implications of climate change (UNGA 2009a), and the UNSC discussed the repercussions of climate change for 'global peace and security' in 2007, 2011, 2013 and 2018 (UNSC 2007, 2011, 2013, 2018; Scott and Ku 2018) and in 2019 held a debate on climate change as a 'threat multiplier' (UNSC 2019). Finally, even though not always directly mentioning the word 'security', Greta Thunberg, Fridays for Future, as well as other movements such as Extinction Rebellion have repeatedly emphasised the catastrophic and existentially threatening consequences of climate change. Instead of 'climate change', they hence increasingly use terms such as 'climate crisis', 'climate breakdown', 'climate apocalypse' or 'climate emergency' (Fridays for Future Austria 2019; Thunberg 2019a, b; Extinction Rebellion 2019). While aiming at conveying the far-reaching consequences of the issue and calling the public and politicians to action, they also contribute to the continued securitisation of climate change.

In general, while economic arguments (Stern 2006), justice conceptions (Caney 2006, 2010; Finley-Brook 2014) and growing scientific evidences (IPCC 2007) also mattered, constructing climate change as security issue has been central when it comes to raising attention and facilitating as well as accelerating political responses. One example is the 15th Conference of the Parties (COP) summit in Copenhagen in 2009, which received unparalleled attention in the media and public debates not least because it took place during the peak of the global climate security debate

(Oels 2012; Methmann and Rothe 2012). The widespread political and media attention for the various UNSC debates on climate change and its implication for peace and security further illustrates this argument (Sindico 2007; Goldenberg 2011; Brössler 2019; Oels and von Lucke 2015; Scott 2015; Scott and Ku 2018). Beyond the international level, linking climate change to security concerns also changed the domestic debates and influenced a range of policies and political practices in various countries (Brzoska 2012; Buxton et al. 2016; Rothe 2016). As the Chaps. 2, 3 and 4 of this book show in more detail, it helped to legitimise far-reaching climate policies in Germany, facilitated an integration of climate change into security policy in the US and transformed disaster management approaches in Mexico.

Having said that, these widespread climate security debates did not result in a consensus about what specific kind of security issue climate change is, what the appropriate countermeasures could entail and whether linking climate change to security is to be welcomed from a normative perspective. The academic literature on the climate security nexus is not of much help here because even though it casts doubt on the connection between climate change and security, it has not looked at the resulting political debates. This raises the questions why and how climate security discourses (entailing different conceptions of security) have become so prominent in the political debate notwithstanding their sometimes weak empirical foundations and what political consequences this securitisation has had exactly (see, e.g., Floyd 2010; Trombetta 2011; Oels 2012; McDonald 2005, 2008, 2013; Corry 2012; Dyer 2018; Rothe 2016; Lippert 2019).

According to the original securitisation theory, the Copenhagen School (Buzan et al. 1998), security issues are socially constructed (Buzan et al. 1998, p. 24) and a successful securitisation establishes a political platform for the legitimisation of extraordinary measures to counter a threat (Buzan et al. 1998, p. 21). While some might understand the yearly COPs or milestones of the international climate regime such as Kyoto or the latest Paris Agreement as extraordinary, most scholars agree that they do not go beyond normal politics, particularly given their more than questionable effect on the abatement of climate change (Oels and von Lucke 2015, p. 47; Gardiner 2004; Caney 2016; Buzan and Wæver 2009; Stripple 2002). The increasing criticism by social movements and climate activists such as Greta Thunberg, Fridays for Future and Extinction Rebellion, which have become much more vocal since 2016, further underline that the current handling of climate change is far from extraordinary. Theoretically, one conclusion hence could simply be that the securitisation of climate change has been unsuccessful and inconsequential, at least from the perspective of the Copenhagen School (Oswald Spring and Brauch 2011; Oels 2012). Yet, notwithstanding the absence of a successful securitisation or extraordinary responses in Copenhagen terms, many scholars have continued to analyse the climate security debate from a broader securitisation perspective. Based on the 'stubborn persistence' (Ciuta 2009, p. 312) of so many political practitioners who keep calling climate change a security threat, these scholars assume that there must be political advantages in doing so, on which the original concept of securitisation does not focus (Brzoska 2009; McDonald 2012; Detraz and Betsill 2009; Corry 2012).

Thus, instead of a clear-cut threshold between politicisation or normal politics and securitisation, this literature understands the process as a continuum (Diez et al. 2016, pp. 18–19; Oels and von Lucke 2015; Bigo and Tsoukala 2008; Stritzel 2007; Vuori 2008). In order to assess the effects of securitisation, it has largely focused on counterfactual reasoning and measuring the degree of success based on policies which without the securitisation would not have been accepted or seen as legitimate in the political debate (Trombetta 2008, p. 600). This has opened up a whole range of research avenues in relation to (different) climate security debates. A common finding is that there are multiple forms of securitisation in the case of climate change that heavily depend on the broader context in which they take place and that can have very different political consequences (Diez et al. 2016; Grauvogel and Diez 2014; Detraz and Betsill 2009; McDonald 2013; Trombetta 2011; Oels 2011; Corry 2012). These consequences are not necessarily extraordinary but nevertheless differ considerably from how the issue was handled before the security dimension was considered. The debate thus has already come a long way in overcoming some of the problems of the Copenhagen School when it comes to analysing the securitisation of climate change. However, the existing literature still has considerable blind spots, not only concerning the theoretical conception of securitisation and its political consequences but also in terms of its limited empirical focus.

The Missing Conceptualisation of Power and the Absence of Detailed Case Studies

Concerning theory, a key shortcoming of the existing research is an insufficient problematisation of the role of power in securitisation processes.

Taking a closer look at the micro dynamics and actual practices of power (Adler-Nissen and Pouliot 2014; Barnett and Duvall 2005; Burgess 2011) can help to theoretically substantiate several of the core findings of the securitisation literature and thus benefit the development of a more coherent understanding of securitisation in general. A closer look reveals that power relations are involved in enabling securitisation in the first place by forming the basis or context from which certain actors can legitimately speak security, by working as catalyst for political attention and by agenda-setting (Burgess 2011, pp. 40–41; Hansen 2000, p. 303). Moreover, they constrain the securitising actors' choices concerning the security arguments they can use (i.e. which stand a chance of resonating within a specific context) and hence lead to very different forms of securitisation entailing a diverse set of security conceptions (Balzacq 2011a, p. 26; Trombetta 2011, p. 141). Beyond that, different forms of power shape the political consequences that specific security discourses can have by transforming governance practices and making possible particular policies and ruling out others (Trombetta 2011, p. 142; Balzacq 2011a, p. 16; Elbe 2009, p. 15). Finally, understanding the underlying power dynamics can also contribute to a more thorough and nuanced discussion of the normative implications of securitisation (Elbe 2009, pp. 157–158; Floyd 2007a, 2011; Nyman and Burke 2016).

One of the crucial problems of the existing securitisation literature in relation to power is that, on the one hand, the Copenhagen School and some of its extensions have mainly operated with a state-centred top-down conception of security. In many cases, this implies a traditional and one-dimensional understanding of political power (Trombetta 2008, p. 600, 2011, p. 139; Williams 2003). This does not adequately capture the much more nuanced pathways of power in securitisation processes, as the extensive debates about different forms of climate security exemplify. On the other hand, alternative approaches to securitisation such as the Paris School of (in)securitisation around Didier Bigo and Jef Huysmans (Bigo 2002, 2008, 2009; Huysmans 2002, 2004) and the literature on risk (Kessler 2012, p. 20; Aradau and van Munster 2007; Lobo-Guerrero 2007; Neal 2004; Hameiri 2008; Hameiri and Jones 2013) have gone towards the other extreme. Here, securitisation is predominantly conceptualised as an ongoing and low-key process in which professionals of (in) security slowly expand a never-ending state of exception (Bigo 2002, p. 73; Bigo and Tsoukala 2008). In between these two more extreme poles on the power continuum are studies that have looked at different 'frames' or 'discourses' of climate security (McDonald 2013; Detraz and Betsill 2009; von Lucke et al. 2014; Grauvogel and Diez 2014). These works circumvent the problem of overestimating one particular understanding of security or power. However, they primarily derive these different climate security discourses from the existing literature. In doing so, they lack a deeper theoretically grounded problematisation of how exactly different forms of political power lie at the heart of different securitisations and enable different political consequences.

Another substantial gap in the literature on the securitisation of climate change concerns its empirical scope and depth. Apart from notable exceptions (Detraz and Betsill 2009; McDonald 2012, 2013; Diez et al. 2016; Rothe 2016; Floyd 2010), most works primarily make a theoretical point (Corry 2012; Methmann and Rothe 2012; Oels 2011, 2012; Trombetta 2011), which has led them to chiefly rely on exemplary data without substantively contributing to our understanding of the climate security debate and its political consequences in actual empirical cases. Moreover, they often focus on the global debate on climate security and its implications for the international climate negotiations (Oels 2011, 2012, 2013; Methmann and Rothe 2012; Methmann 2011, 2014; Trombetta 2008, 2011; Corry 2012; Scott and Ku 2018), on security in the Anthropocene (Dalby 2013a, 2014; Fagan 2017; Hardt 2017; Harrington and Shearing 2017), or on individual case studies (McDonald 2012), and here mainly on the US (Floyd 2010; Brzoska 2009; Hartmann 2009; Fletcher 2009; Harris 2002; Leiserowitz 2005; Nagel 2011; Richert 2009) or on specific international institutions (Scott and Ku 2018; Lippert 2019). Only very few studies compare different securitisation processes in diverse political and cultural contexts (Diez et al. 2016; Rothe 2016), which is necessary in order to understand the context-dependence and multiplicity of securitisation and its political consequences. Closely connected, there is a Western bias in the research on the securitisation of climate change, hence it has largely neglected to study securitisation processes in the Global South (Boas 2014; Bilgin 2010; von Lucke 2018). This is especially surprising in the context of climate change, as most climate security discourses as well as the estimates of the IPCC (2015, pp. 13, 50, 54) predict the first and most severe effects to take place in (poor) Southern countries, begging the question whether the political effects of securitisation differ under these circumstances.

Thus, there is considerable demand for detailed comparative empirical studies that analyse the securitisation of climate change across different political and cultural contexts. Such studies can help to move the debate beyond the development of ever more sophisticated theoretical approaches without actually applying them to empirical cases. It also strengthens our understanding of the concrete political and institutional effects of linking climate change to security conceptions.

Core Argument and Structure of the Book

The chief purpose of this book is thus to make sense of the multifaceted securitisation of climate change in different contexts (Brauch 2009; Detraz and Betsill 2009; Floyd 2010; McDonald 2013; Trombetta 2012; Oels 2012) while at the same time advancing the theoretical concept of securitisation. For this purpose, it offers a novel take on securitisation theory by focusing on the so far largely neglected role of power. The emphasis on power helps to better capture and theoretically make sense of the ambiguous and diverse variants of securitisation and the ever-changing concept of security itself (Opitz 2008, p. 206; Elbe 2009, p. 76; Hardt 2017). It also expands our understanding of the powerful political and normative consequences of constructing non-traditional issues in terms of security.

While several extensions of the Copenhagen School and entirely new approaches to securitisation have touched upon this issue, in order to fully understand the role of power in securitisation processes, we need an alternative theoretical approach. As the second part of this chapter discusses in more detail, a Foucauldian governmentality framework (Foucault 2006a, b; Dean 2010; Dillon 2006) lends itself particularly well for this endeavour (Dillon and Lobo-Guerrero 2008; Elbe 2009, 2011; Oels 2011, 2012, 2013). It primarily rests on Michel Foucault's governmentality lectures (2006a, b), which claim that political rule, in general, has undergone a decisive transformation. Thus, we can witness a 'governmentalisation of the state' (Foucault 2006b, p. 163), which means that governance does not only rest anymore on direct top-down interventions by the state (sovereign power). Instead, it consists of a power triangle, which includes 'productive' forms of power (Foucault and Faubion 2002; Foucault and Gordon 1980) such as disciplinary and governmental power that try to control and transform the behaviour of individuals or to govern the population through indirect risk management strategies (Foucault 2006b, pp. 161–165). This transformation does not only apply to governance in general, but at least since the 1980s, there is also a 'governmentalisation of security' (Elbe 2009, p. 9; Oels 2011, 2012), entailing a disentanglement of (successful) securitisation from exclusively extraordinary measures and the growing importance of disciplinary and governmental power in the realm of security. From this vantage point, securitisation becomes a specific form of 'security governance[1]' that renders issues governable resting on different forms of power, which in turn can be linked to contemporary security conceptions ranging from national security, human security to risk (Elbe 2009, pp. 12, 14–18). Eventually, the securitisation of political issues can bring about very specific and far-reaching yet not necessarily extraordinary political consequences, depending on which power forms and security conceptions prevail.

Conceptualising securitisation as part of this governmentalisation of security and hence as a specific way to exercise political power can help us to better understand at least four interrelated aspects of the securitisation process. Firstly, it establishes a theoretical framework for understanding the continuous transformation of security and the parallelism of (and connections between) different power forms. A governmentality perspective thus sets the concept of securitisation into a wider historical and cultural context, in which different security practices constantly struggle for political relevance (Opitz 2008, p. 206; Elbe 2009, p. 12; Foucault 2006b, p. 76). Secondly, it sheds light on the role of power in constituting the subjects and objects of securitisation. Based on its more nuanced understanding of political power, such a framework goes beyond the analysis of fixed securitising actors, referent objects and audiences and instead shows how these are constantly created, legitimised or discredited within different discourses of security (Burgess 2011, p. 40; Hansen 2000, p. 303). Thirdly, the multifaceted and dynamic conceptualisation of power inherent in the idea of the governmentalisation of security contributes to make sense of the diversity of securitisations and the varying political consequences. Depending on which power forms overweigh and how they are combined and enacted in different political and cultural contexts, different policies or practices seem legitimate or are discarded as irrelevant. Securitisation is hence linked to the exercise of political power by helping to put new issues onto the agenda, by acting as a catalyst for the political debate and by directly influencing key policies and practices. Finally, a governmentality perspective provides a theoretical frame of reference for discussing the normative effects and desirability of securitisation in general

[1] I understand governance in a wider sense as constituting and arranging actors around a discursively constructed 'governance-object' (Methmann 2014, p. 10; Corry 2010).

and different discourses of climate security in particular (Nyman and Burke 2016; Floyd 2011; McDonald 2011).

Beyond advancing this theoretical re-conceptualisation of securitisation, this book strengthens the detailed comparative research on the securitisation of climate change. Instead of analysing climate security discourses at the global level, it thus focuses on three in-depth country case studies, namely the US, Germany and Mexico. It thereby includes two developed countries with often diametrically opposed policies on climate change and different political cultures, in which the security implications of climate change have been discussed very prominently. Yet, it also explores the so far under-researched issue of securitisation in the Global South and in countries that are predicted to be highly affected by the adverse effects of climate change (Boas 2014; Bilgin 2010). Looking at the domestic debates of these different countries instead of the international negotiations does not only close a gap in the literature but also has methodological advantages. It makes it easier to directly trace back how different climate security discourses have legitimised and influenced specific policies and practices and hence helps to overcome one of the core problems of governmentality studies that often shy away from looking at the actual implementation of political programmes (Mckee 2009, pp. 473–474; Bröckling et al. 2012, pp. 16–17). A further advantage of the comparative case study design is that it allows studying the securitisation of climate change in diverse cultural, political and economic environments and thus to inquire how the broader contexts matter in enabling and shaping specific securitisation processes.

Concerning methodology and the empirical data, the book rests on an extensive qualitative discourse analysis of the most relevant reports, government documents and parliamentary debates on climate change (and security) in the three countries between the late 1980s and 2015. In addition to analysing documents, a range of expert interviews with key political practitioners in each country has helped to situate the findings in country-specific political debates and to uncover important political networks. The comparative focus on different country cases and the comprehensive analysis of the empirical material allow coming to a detailed and systematic understanding of how securitisation discourses emerge in specific political environments, how they function and how this eventually translates into concrete policies and practices.

In order to advance the above-introduced theoretical and empirical aims, the second part of this *introductory chapter* substantiates the theoretical claims and develops the overall analytical framework. Mainly building on Michel Foucault's governmentality lectures (2006a, b) and contemporary extensions of his work by Stefan Elbe (2009, 2011), Mitchell Dean (2010) and Angela Oels (2011, 2012), the chapter develops three distinct climate security discourses which draw on different forms of political power and are linked to different political effects. Drawing on this theoretical framework, Chaps. 2, 3 and 4 analyse the climate security debates in the US, Germany and Mexico. Each chapter briefly discusses the origins of the climate security debate in the respective country and then analyses the evolution of different climate security discourses in the period of investigation. Based on this analysis, the chapters show how each unique combination of discourses and power forms has legitimised specific political consequences. This includes the empowerment of a diverse set of actors as well as changed policies and practices in key political sectors (environment, defence and security, development and disaster management). Based on this detailed analysis, Chap. 5 draws broader comparative conclusions about how securitisation functions in general, the role of power and the normative implications of securitising climate change.

Developing an Analytical Framework to Study (Powerful) Climate Security Discourses

As argued above, a Foucauldian power and governance focused approach of securitisation can greatly add to our understanding of ongoing securitisation processes in non-traditional sectors such as climate politics. With its fine-grained analytics of political power, it helps to make sense of the divergent political effects of different threat constructions beyond and below extraordinary measures. At the same time, the original securitisation theory reminds us that invoking security conceptions and designating something as a threat, be it to national or human security or as a diffuse risk construction, transforms the political debate and makes possibly a novel treatment of the issues at stake. It may not always elevate the issue into high politics or legitimise non-democratic extraordinary measures. Yet, based on the specific threat constructions and underlying power forms, it enables new forms of governance (Dean 2010, pp. 266–267) that without the security framing would not have been accepted as legitimate or have had the same impact.

Starting out from the original ideas of Michel Foucault on governmentality, but also inspired by the works of Stefan Elbe (2009), Angela Oels (2011, 2012), Mitchell Dean (2010) and others, in this section I introduce a specific reading of the governmentality approach and discuss its relevance for securitisation theory. The origins of the governmentality approach lie in Foucault's research on political power and particularly in his lectures at the College de France (Foucault 2006a, b) where he criticises the Political Science research of that time for focusing too much on repressive forms of power and government (Foucault 1983, p. 83; Oels 2010, p. 172; Lemke 2002, p. 51). The mainstream research mainly conceptualised governance as top-down enterprise carried out by a sovereign state without much interference of non-state actors or the governed entities themselves. And despite some nuanced research on power (Lukes 2005; Bachrach and Baratz 1962), the predominant, 'juridico-political discourse' (1979, p. 88, 1983, p. 84) saw power as something that could be possessed and wielded at will. Power was directly tied to the capacity of specific actors to control the actions of others (Barnett and Duvall 2005) and hence conceptualised as something constraining, dominating and essentially bad, exercised from a top-down perspective over the governed subjects without many possibilities to resist.

Against this mainstream perspective, Foucault developed his own, much more diverse understanding of power. For Foucault, power can take very different forms, for example, strategic games, structuring fields of possibility and above all is 'productive' (Opitz 2008, p. 216). It thus enables political developments and constitutes subject positions. It mainly does so by being ingrained in what Foucault called the 'discourse' or later on the 'dispositive' (Foucault 2003; Dean 2010, 2012; Jäger and Meier 2009; Aradau et al. 2014a, p. 64). A discourse is a power-knowledge nexus that constitutes reality in a certain way, thus defining what is right or wrong and who is empowered to speak the 'truth'. It is only through discourse that humans can access reality and knowledge. Thus, all reality and truth are exposed to and shaped by power dynamics:

> Power is a relationship between actors that produce knowledges and truths that lead to individual and social practices that in turn tend to disseminate those truths. Knowledge transmits and disseminates the effects of power [...], while truth is a status given to certain knowledge by power. [...] Truth is linked in a circular relation with systems of power which produce and sustain it. (Foucault 1980, p. 133)

Power in Foucault's sense is everywhere not only in instances of direct influence or in the hands of seemingly powerful individuals. Instead, its most central functions include the constitution of subject positions, desires and truths in the first place. Moreover, by having the capacity to be productive and empowering, power is not something essentially bad from which people must be rescued. This clearly distinguishes the Foucauldian understanding from mainstream views of power as an asymmetrical force against which no resistance is possible—something which Foucault has termed 'domination' (Lemke 2002, p. 53; Dean 2010, p. 58).

Based on this more nuanced concept of power, Foucault set out to rethink the concept of governance. In his governmentality lectures, he starts out from an extensive genealogical analysis of the term 'to govern' (Dean 2010, p. 3) and its underlying power forms since the ancient times and tries to condense the dominant meaning of the term in different epochs to capture its continuous transformation. In this analysis, not the first appearance of the term is important but the point in time when people begin to consciously deal with it, enabling the development of certain tactics, strategies and modes of action (Foucault 2006b, p. 425). In order to reconceptualise contemporary debates on governance and power, Foucault takes the older, much broader meaning of the term 'to govern' as a starting point where governing is not restricted to the state, but also applies to the governing of the family, the economy or even the self (Foucault 2006b, p. 183). Thus, contrary to the mainstream understanding, Foucault claims that the term only gradually has become so closely tied to the idea of the state (Foucault 2006b, p. 135). He then combines this broader meaning with the idea of different 'mentalities' underpinning governance processes.

The basic question that Foucault tries to answer here was how it became possible that in modern societies power and governance could concentrate on the institution of the state (Lemke 2002, p. 58). In this view, governance in the form of the state is only one very specific way to exercise power and the state itself becomes a specific 'tactic' of government. Government or governmental power[2] as a specific form of power does not

[2] In Foucault's writing, he uses the term 'governmental management', which, despite some differences, sometimes is also equated with 'bio power' (Kelly 2009, p. 60; Foucault 2006b, p. 161). However, for better comparability with the other power forms and in order to delineate my approach from the existing literature, I use the term 'governmental power' throughout this book.

work in a direct, top-down fashion but tries to influence processes at the level of what Foucault called 'the population' (see Foucault 2006b, pp. 103–116, 156–159, 504–510; Dean 2010, pp. 113–115, 127). It also tries to use the potential for self-governance of every individual or other actors in society and hence seeks to only intervene if absolutely necessary (Dean 2010, p. 121). In its ideal form, the exercise of power in this way goes almost unnoticed, as it shapes the identities, the needs and desires of the governed and the knowledge of what is right and wrong.

While different meanings of the term 'governmentality' exist (Walters and Haahr 2005; Dean 2010, p. 24), in this book I focus on a historically specific one (Foucault 2006b, pp. 162–165; Walters and Haahr 2005, p. 292), which conceptualises the term as a '[…] distinctively new form of thinking about exercising power in certain societies' (Dean 2010, p. 28). It describes a new way of governing especially in Western societies since the eighteenth century, which instead of governing a territory focuses on the governance of populations through three different forms of power:

> During this era of governmentality political rule is exercised through a complex triangle of sovereignty, discipline and governmental management, which has the population as its main target and apparatuses of security as its essential mechanism. (Foucault 2006b, p. 161)

For Foucault, the concept of the population replaces the focus both on the territory (Mckee 2009, p. 466) and on the family and thus elevates governance practices to an entirely new level (Foucault 2006b, pp. 157–158; Dean 2010, p. 127). Contrasting conventional understandings, the term does not only encompass the plain number of inhabitants of a state territory. Instead, it aims at all statistical operations that have become possible with the development of sophisticated social scientific knowledge together with the fact that state bureaucracies keep track of an endless number of bio-political characteristics of the population (Foucault 2006b, pp. 74–75). Examples are mortality statistics, the age and income structure of the population, birth rates and the geographical distribution of these variables (Foucault 2006b, pp. 156–157). By using this specific knowledge, it has become possible to not only keep track of past and present developments but also to look into the future and to project the likely development of certain variables and risks within the population (Foucault 2006b, p. 396). It thus is a precondition for the transformation of political power in general.

The Governmentalisation of Security

The above-described reconceptualisation of power and governance puts a new light on the concept of securitisation. Contrary to what the original framework implies (Buzan et al. 1998), securitisation processes do not consist of given subjects that utter speech acts independently from the overall discursive framing of the situation (Hansen 2000, p. 303). In fact, the subjects of security (e.g. securitising actors) cannot be thought of as existing prior to securitisation but are a product of the power relations ingrained in the securitisation process itself (Burgess 2011, p. 49). Ultimately, securitisation can be seen as an alternative way of exercising political power, and Foucault's governmentality approach can help us to unpack this relationship.

The starting point is the idea of a 'governmentalisation of the state' since the eighteenth century, which implies that top-down, sovereign power ceases to be the dominant way of ruling societies and is accompanied by other variants such as disciplinary and governmental power (Foucault 2006b, p. 163). This refocuses the attention of the state away from securing a territory by conventional security institutions such as the army, the police or intelligence services towards fostering the well-being of the population using a triangle of three different power forms and various 'apparatuses' or 'mechanisms' of security (Dean 2010, p. 29; Foucault 2006b, p. 161; Elbe 2009, p. 62). Instead of direct force, coercion or the issuing of laws, the exercise of power rests on statistical knowledge about the population and methods to discretely influence certain variables in the background (Foucault 2006b, pp. 90–97; Dean 2010, p. 29). This also leads to a blurring of the formerly much stricter boundaries between 'normal' politics and 'extraordinary' security practices. On the one hand, security becomes a technology of government, a way of 'rendering things governable' and thus changing how populations can be governed (Oels 2011, 2013; Opitz 2008, p. 2017). On the other hand, security practices and theoretical concepts themselves undergo a transformation and become less extraordinary and more multifaceted. Examples are new security concepts such as human security, environmental security, gender security or risk (Hardt 2017, p. 43). It is exactly this double transformation that marks the connection between the governmentality approach and securitisation theory. Elbe describes this as the 'governmentalisation of security' that has its origins in the 1980s and which means that security does no longer only draw on a sovereign form of power but incorporates disciplinary and particularly governmental forms of power as well (Elbe 2009, pp. 9, 64, 71, 78).

The origins of this transformation of security lie in changing societal features and connected political and scientific debates that required new forms of government (Collier 2009). In this process, the importance of traditional security technologies declines, and they are combined with less direct and top-down forms of security based on the power triangle. Thus, the governmentalisation of security as well contributes to the blurring of (normal) and extraordinary or security politics. Security institutions and actors are progressively legitimised to help in fostering the welfare of the population outside the narrow military realm (Elbe 2009, p. 64). The field of environmental and especially climate politics illustrates this development. It shows the double movement of a 'securitisation' of non-traditional issues and a 'climatisation' (Oels 2012; Maertens 2018) of traditional concepts of security (Trombetta 2011; Corry 2012; Floyd and Matthew 2013; Hardt 2017), which I term as the 'bidirectional' qualities of securitisation.

Following the premise of the governmentalisation of security means that we cannot conceptualise securitisation as mainly resting on sovereign power and its direct, extraordinary and supposedly negative effects—as done by the Copenhagen School (Opitz 2008, pp. 219, 220). Instead, to adequately account for the transformation of security, we should broaden the analytical approach and allow for different forms of securitisation that draw on different forms of power. The process of securitising an issue in this sense is an instance of governing, a process of rendering things governable through the lens of security and the application of different forms of power. As the term 'governance' in this reading of securitisation implies, securitisation processes are not a priori considered as something extraordinary; rather they constitute a specific way of governing, which may or may not entail extraordinary effects. However, this does not mean to entirely equate normal politics or governance with security. Instead, the core premise of securitisation theory still holds: designating something as an existential threat is different from a mere politicisation because it generates a sense of urgency, increases attention and eventually can narrow down the deliberative process. Security can act as a catalyst for the exercise of any form of political power and eventually can render the situation in a certain way and thus constitute a powerful 'security truth' (Burgess 2011, p. 39). Yet, based on a governmentality reading, this process becomes much more multifaceted, with different conceptions of security and power involved and different possible consequences of securitisation processes.

The Triangle of Sovereign, Disciplinary and Governmental Power

The 'power triangle' of sovereign, disciplinary and governmental power (Foucault 2006b, pp. 76, 161) is central to the governmentality approach and also at the core of my reconceptualisation of securitisation. It accrues from Foucault's genealogy of the governance concept, in which he condenses the three forms of power from different historical periods. *Sovereign power* is the oldest one and connected to the feudalistic territorial or 'judicial state' prior to the sixteenth century, which was defined by its territory and governed mainly through issuing laws and direct interventions of the sovereign (Foucault 2006b, p. 164). From the sixteenth century onwards, the gradually arising 'administrative state', the declining importance of the state territory and the increasing focus of political rule on individuals and disciplinary practices marked the advent of *disciplinary power* (Foucault 2006b, p. 19). Finally, the eighteenth century saw the emergence of the 'governmental state' and *governmental power*, which is defined by its relationship to the population and to novel 'security mechanisms' or 'apparatuses of security' (Foucault 2006b, p. 36; Dean 2010, p. 29). Providing security not for a territory or individuals but for the entire population and at the same time seeking to ensure its welfare and prosperity became the central focus of governance practices.

In the following, I elaborate on the emergence and characteristics of each of the three power forms and link them to contemporary conceptions of security. I also briefly discuss them from a normative perspective, which rests on Foucault's ethos of siding with the subjugated and oppressed (Pickett 1996, p. 465), his rejection of domination (Foucault and Gordon 1980, p. 120) and the aim to minimise the main dangers of climate-security discourses (Elbe 2009, pp. 157–158; Pickett 1996, p. 461).

Sovereign Power: Defending the National Security of Territories

Sovereign power is similar to mainstream conceptions of power, for example, the power to enforce one's will over others' (Weber 1976, p. 571) or what Stephen Lukes describes as one-dimensional view on power (Lukes 2005, p. 16). It also resembles a Machiavellian or Hobbesian notion of power that mainly seeks to control a certain territory and to sustain the reign of the 'prince' or 'leviathan' (Foucault 2006b, p. 100; Opitz 2008, pp. 207–208). Consequently, sovereign power is mostly exercised by the sovereign—often the state and its agencies such as the police or the military—in a highly visible and direct way over a defined territory, with its main target being the perpetuation of sovereignty itself (Foucault 2003, p. 149). It has a binary and law-like character defining what is permitted or forbidden, punishing those who deviate from the law (Dean 2010, p. 29; Foucault 2006b, p. 149). In general, the exercise of sovereign power is tied to hard-politics, direct interventions of the state and its agencies such as issuing and enforcing laws, collecting taxes or punishing deviant behaviour—in the most extreme cases with death (Foucault 2003, p. 240, 2006b, p. 75). Ultimately, sovereign power culminates in a focus on traditional security conceptions and actions by the defence or military sector that all aim at preserving the reign of the sovereign over its territory.

Transferred to contemporary security debates and security discourses, this form of power can be linked to a state and military-focused national security conception (Elbe 2009, p. 86). In other words, national security—or equivalent concepts such as territorial security, state security but also international security and order—are manifestation of sovereign power. Nonetheless, it is important to stress that sovereign power in general is broader and does not necessarily have to entail a military element. However, it is always a highly visible, top-down and direct form of evoking security and primarily aims at the state territory, which in political practice (especially in the realm of international relations) often intersects with national security. Moreover, in combination with the catalysing effect of designating something as existential threat, invoking sovereign power in securitisation process can go beyond the law and can legitimise a temporary (or even permanent) state of exception where the sovereign decision replaces the written law (see Agamben 1998, p. 11, 2005; Schmitt 1963). Thus, while originally being broader, for analytical purposes it is useful to narrow it down to national security. The effects of sovereign power in securitisation processes therefore resemble what the Copenhagen School describes as effects of successful securitisation: more attention for the problem at stake and a more serious handling, though also possibly a bypassing and acceleration of normal and democratic political procedures.

From a normative standpoint, sovereign power has some advantages. Most importantly, it draws attention to issues that otherwise would probably not had been handled seriously, thereby elevating those issues into the realm of high-politics. This can be useful to draw attention to so far overlooked problems and can, at least until it reaches a certain intensity, have a politicising effect because it brings issues on the political agenda. Furthermore, an acceleration of typically slow and bureaucratic procedures can sometimes be an advantage. Both the attention generating and accelerating qualities of securitisation have been referred to as legitimisation for securitising non-military issues such as climate change or HIV/AIDS (Floyd 2013, pp. 281–282; Ingram 2010; Elbe 2006). On the downside, however, sovereign power also entails problematic features. It can narrow down the focus towards predominantly direct and short-term actions such as police or military interventions. Thus, a political response in the wake of a sovereign-based securitisation often neglects the root causes of the problem. It is a reactive form of governance that tries to cure the immediate symptoms, thereby running the risk of worsening the underlying problem in the end. Finally, a securitisation focusing on sovereign power tends to concentrate power in the hands of state, security and defence actors. It thereby increases secrecy, narrows down the space for political discussions and excludes societal and non-governmental actors. Consequently, in its extreme forms, it diminishes the opportunities for democratic control and public scrutiny. The following Table 1.1 summarises the characteristics of sovereign power.

Table 1.1 Sovereign power

			Normative assessment	
Modes of action	*Actors and sectors*	*Security conceptions*	*Opportunities*	*Dangers*
Direct, visible, negative, law, ban, attention raising, acceleration of processes *Target*: Mostly state territory	State, military, security, defence and intelligence sector, hard-politics, foreign policy, state as most important actor and referent object	National security, territorial security, international order and security, classical securitisation	More attention, elevating issues to high politics, acceleration of procedures, definitive and direct state action	Panic politics, extraordinary measures, state of exception, short-cutting of (democratic) procedures, narrowing down debates, military action, short-term measures, secrecy

Partly based on Foucault (2006a, b), Elbe (2009) and Dean (2010)

Disciplinary Power: Securing Individual Human Security

Instead of controlling a territory, disciplinary power aims at transforming individual behaviour towards a predefined, ideal-typical norm. Using sophisticated surveillance technologies and control mechanisms (Foucault 2006b, pp. 89–90), it wants to discipline, classify, control and ultimately optimise individual behaviour (Foucault 1975, p. 143). Examples are military drill schemes that use detailed codes of conduct and prescribe certain body movements, for example, when handling a rifle, in order to reign in and transform individual behaviour towards a predefined purpose. In short, disciplinary power is a deductive approach that divides reality into the abnormal and normal based on an ideal-typical or optimal model of what this 'normal' is—a process which Foucault termed 'normation' (Foucault 2006b, p. 89). In contrast to sovereign power, it is not mainly a negative but productive way of exercising power that instead of taking things away tries to create and transform identities and truth regimes (Elbe 2009, p. 117; Dean 2010, p. 81). It is also very accurate, operates at the micro-level and tries to control every single part of the process (Foucault 2006b, p. 74). Accordingly, it requires many resources and continuous attention and thus involves a constant monitoring of its subjects. Eventually, its goal is to optimise processes by assessing and understanding the function of every single element in the process and then arrange these elements into the optimal order (Foucault 2006b, p. 89). Finally, while its original conception particularly connects it to (oppressive) institutions like the military, prisons or hospitals (Foucault 1975, p. 215), disciplinary power can empower a variety of actors ranging from the state to non-state actors such as NGOs, church organisations and think tanks.

Looking at existing security conceptions, disciplinary power particularly materialises in human or individual security approaches (Elbe 2009, p. 113) or in related conceptions such as human vulnerability or food security. All of these security conceptions focus on individuals and try to enhance and empower disadvantaged people towards a predefined ideal typical norm: the secure, free and thus emancipated individual that is able to fulfil its human potential (Booth 1991, p. 319). The original conception of human security that emerged in the political realm (Boutros-Ghali 1992; UNDP 1994), as well as comparable academic concepts such as individual security (Booth 1991, 2005; Wyn Jones 2005), tried to set up a new and positive concept of security. Aiming at individuals, these concepts primarily tried to come up with an alternative to the then dominant state centric national security conceptions. The goal was to empower individuals not states and to facilitate policies that benefited their well-being. However, in the course of political and academic debates, the concept of human security has undergone a discursive shift and is partly employed to control individual behaviour (Duffield 2005, 2007; Duffield and Waddell 2006; Oels and von Lucke 2015; McCormack 2010). Especially in a securitisation context, it can be transformed and (miss)used to facilitate the interests of powerful state, military and particularly Northern actors (McCormack 2010). It can function as a justification for interventions and as rationalisation of military and neo-colonial action (McCormack 2010, p. 36; Devetak 2007, p. 152; Eriksson 1999, p. 318). Hence, the originally well-intentioned concept can itself become perverted and thus normatively problematic (Floyd 2007a, b).

Despite its emancipating origins, in contemporary political and security debates, human security often entails a distinction between people of the Global North and those living in the Global South. Thus, the ideal typical norm is the well-fed, healthy, productive, democratic and wealthy citizen in advanced and highly developed Western societies. Based on this 'normal', the human security framework produces an opposing 'abnormal' that has to be disciplined, or 'secured' often by organisations from the Global North (Lie 2015; Mawuko-Yevugah 2010; Duffield and Waddell 2006, p. 2). In this example, we can observe the productive elements of disciplinary power that have a bearing on the governed as well as on the governing. Disciplinary power prescribes certain identities and instructions and hence empowers some actors to act (e.g. Western organisations) whereas it constitutes others as passive elements of 'normation' and governance (e.g. endangered individuals in countries of the Global South). Ultimately, not least due to its 'soft' image in comparison to national security, human security especially empowers human rights and development NGOs to monitor individuals that do not resemble the predefined ideal norm.

Looking at it from a normative perspective, the positive effect of a disciplinary and human security centred securitisation is that the focus shifts from the emphasis on the security of states towards the needs and insecurities of individuals. It redirects the attention towards the most vulnerable actors, possibly legitimising policies in their interest, for instance, development aid or disaster relief measures. Furthermore, it broadens the spectrum of the actors that legitimately can participate in the securitisation process and its political aftermath, thereby opening debates and avoiding the tendency to secrecy that often comes along with national security

Table 1.2 Disciplinary power

			Normative assessment	
Modes of action	*Actors and sectors*	*Security conceptions*	*Opportunities*	*Dangers*
Productive, indirect, 'normation', disciplinary, surveillance, optimise, creating identities and truths, individual referent object *Target*: Mostly individuals	Micro-perspective, individuals, broadened actor spectrum, non-state actors, development problems	Human security, individual security, food security, vulnerability	Focus on individuals, more long-term oriented, empowerment, reduced secrecy, broad spectrum of actors	Surveillance, less visible—less easy to detect, identity and truth production, paternalistic political measures, neglect of local solutions

Partly based on Foucault (2006a, b), Elbe (2009) and Dean (2010)

conceptions. The main danger of a disciplinary power focused securitisation lies in the discursive shift and in taking away agency from threatened people (often 'poor individuals') to define the situation and the means necessary to overcome the problems. Due to the focus on an artificial ideal typical norm, it entails a paternalistic element and can oppress local solutions that might have been much more effective in a specific context. A further problematic aspect is the necessity of a constant surveillance of the governed individuals, which can constrain their freedom and suppress alternative solutions. Finally, its less direct and oppressive character, for instance, exemplified by actions of seemingly 'good' grassroots or NGO actors, makes this power form less visible. This can obscure the possibly inadequate democratic legitimisation of interventions and actors. Table 1.2 summarises the characteristics of disciplinary power.

Governmental Power: Managing the Risks of Populations

The third and 'youngest' power form, *governmental power*, is directly related to the governmentalisation of the state and the emergence of the population as the main referent object. It mainly operates through risk management schemes to pre-empt problematic developments that could imperil the well-being of the population. Moreover, instead of normation, it relies on the inductive concept of 'normalisation' (Foucault 2006b, p. 98). Normalisation tries to statistically generate a normal distribution of certain values and then focuses at bringing outliers down to this average (Foucault 2006b, pp. 97–98). According to Foucault, this is more efficient than 'normation' because it uses the natural dynamics of the population instead of acting against them (Foucault 2006b, p. 74). If handled properly, 'the phenomena themselves' will eventually bring about their own curtailment (Foucault 2006b, pp. 69, 92, 98, 102). Consequently, one of the most important instruments of governmental power is social-scientific knowledge that enables to measure and discreetly influence population dynamics. In order to manage risks based on the concept of normalisation, a key instrument of governmental power is the focus on risk or outlier groups (Dean 2010, p. 119). This is more cost-efficient (Foucault 2006b, pp. 498–499; Elbe 2009, p. 67) and in the spirit of 'laissez-faire' or 'laissez-passer' (Foucault 2006b, p. 69) ensures that governmental power only intervenes as little as possible in the natural dynamics of the population. Thus, the goal is not to regulate every aspect of life—as this would soon overstretch the government's resources—but to only intervene where it is absolutely necessary to safeguard the overall population (Dean 2010, p. 121; Foucault 2006b, pp. 69, 498).

A further characteristic of governmental power is its emphasis on governing the future by trying to (moderately) influence developments in their early stages and hence avoiding more far-reaching interventions at a later stage (Foucault 2006b, p. 39). This mainly involves risk schemes and probability calculations (Foucault 2006b, pp. 94–95) that aim at preventing problems and risks from spiralling out of control. It also means that the goal is to keep risks at a tolerable level but not necessarily to eliminate them completely (Foucault 2006b, pp. 37–38, 66–69). Ultimately, governmental power seeks to control circulation dynamics by enabling 'good' and containing 'bad' circulation (Foucault 2006b, p. 37; Opitz 2008, p. 213). An example is the European Union trying to maximise the freedom of movement of its citizens but at the same time restricting immigrant circulation by strengthening border controls (see Huysmans 2008; Oels 2009; Bigo 2008). Finally, despite its name, governmental power is not exclusively tied to the state but empowers a broad range of actors. In the words of Jef Huysmans:

> The 'cold monster' breaks down into a vast range of practices and private and public institutions that enact and develop strategies of government that arrange the conduct of freedom in modern societies. (Huysmans 2008, p. 40)

Concerning contemporary security conceptions, governmental power mainly resembles risk approaches (Elbe 2009, p. 132; see also Beck 2000; Aradau and van Munster 2007; Corry 2012; Daase and Kessler 2007). The literature has discussed many different conceptions to risk,[3] most importantly the distinction between 'known unknowns' and 'unknown unknowns' (Daase and Kessler 2007, pp. 423, 426), which mainly relates to the level of calculability of risks. While both dimensions in parts can overlap with governmental power, I deem the first one more relevant, which conceptualises risks as diffuse threats, which however, are still amenable to calculation and risk management. In practice, by using statistical methods and forecasts, political approaches to risk try to identify certain risk groups or risky activities, and then instead of trying to eliminate them altogether, aim at levelling these risk down to a tolerable, average risk level (Oels 2011, p. 18; Corry 2012, p. 245). Furthermore, similar to governmental power, risk schemes focus on long-term horizons and try to control the future by manipulating the right variables in the present, hence attempt to prevent problems from spiralling out of control. Because they build on probabilistic calculations, which are never so precise as to allow for definitive propositions at the micro-level, these interventions usually do not target single individuals (Daase and Kessler 2007, p. 423). Instead, risk assessments function on an aggregate level and hence entail a more economic use of resources than would be necessary to control every single element of the process. Finally, due to their gaze into the future, risk conceptions have a tendency to foster long-term solutions and to focus on root causes instead of symptoms (see Elbe 2009, pp. 131–135).

Concerning the normative assessment, on the one hand, the focus on groups or areas particularly at risk can enable appropriate policies to tackle risks in a cost-efficient and indirect way and without directly prohibiting or controlling behaviour. Moreover, governing though risk management schemes can detect and possibly contain problematic developments before they spiral out of control and hence contribute to tackling the root causes

[3] see Daase and Kessler (2007), Aradau and van Munster (2008), Boyle and Haggerty (2012), Hameiri and Jones (2013), Hameiri (2008).

rather than symptoms. Finally, its less direct and long-term perspective might be better compatible with a gradual politicisation of an issue while avoiding short-term panic politics. On the other hand, constructing risk groups can become problematic if it leads to a stigmatisation and marginalisation of the identified groups. Ultimately, it can also construct these groups as a danger to welfare of the population (Elbe 2009, p. 140), which on the global level could mean a danger for global peace and security. These risk groups would become the 'bad' circulation that has to be avoided to ensure the unrestricted functioning of the overall system. Another problematic aspect is that governmental power and risk tend to intervene as little as possible and only aim at lowering the aggregate risk to a tolerable level for the overall population. Eventually, this could mean that outlier groups become disposable and that some groups or areas are neglected. Finally, just as disciplinary power, governmental power is not always easy to detect; therefore its possibly negative effects can materialise unnoticed over a long time making it hard to resist or argue against it. The following Table 1.3 summarises the features of governmental power.

Table 1.3 Governmental power

			Normative assessment	
Modes of action	*Actors and sectors*	*Security conceptions*	*Opportunities*	*Dangers*
Productive, indirect, 'normalisation', statistics, focus on population, *laissez-faire*, risk-management, future-orientated, cost-efficiency, circulation, long-term *Target*: population, risk groups/areas	Macro-perspective, groups, future, insurance sector, economic solutions, broad actor spectrum	Risk, risk-management, riskisation, contingency planning, scenario planning, resilience, risk-groups, uncertainty, precautionary principle	Cost-efficiency, focus on risk groups/areas, less invasive, long-term, prevention of problems, possibly tackling root causes	Stigmatisation, risk groups vs. population, disposability of certain groups once risk is tolerable for overall population, difficult to detect

Partly based on Foucault (2006a, b), Elbe (2009) and Dean (2010)

In a nutshell, Foucault describes the three forms of power as: 'the law [sovereign power] prohibits and discipline prescribes, and the essential function of security [governmental power] [...] is to respond to a reality in such a way that this response cancels out the reality to which it responds [...]' (Foucault 2006b, p. 76). Consequently, the application of each of these three power forms encompasses very different modes of action, opportunities and dangers as well as empowers different actors. Yet, although Foucault clearly distinguishes between different power forms and talks about 'older' and 'younger' ones, the latter have not simply replaced the former (Foucault 2006b, p. 161). Rather, in the governmentalised state and security practices, all three power forms play a role and there is a constant struggle for dominance, in which 'older' power forms can re-emerge as dominant (see instance Larrinaga and Doucet 2008; Butler 2004). This is an important argument, since the literature on risk too often assumes that we now live in the 'era of risk' and that other more traditional forms of security or power cease to be important (Opitz 2008, p. 204; Larrinaga and Doucet 2008, p. 38; Hameiri 2008; Hameiri and Jones 2013). Thus, I follow Foucault's original ideas and assume that we can understand the process of the governmentalisation of the state and of security not as a straightforward decline of sovereign or disciplinary power, but as a readjustment of the system of correlation between the three power forms with governmental power gradually transforming the other two (Opitz 2008, p. 216).

Developing a Framework for Empirical Analysis: Three Climate Security Discourses

Beyond the intrinsic qualities of these three power forms, the previous section has discussed how they lie at the core of different logics or conceptions of security, e.g. national security, human security and risk or synonymous concepts. Relating the power forms to these specific concepts of security functions as a theoretical bridge between the governmentality approach and securitisation theory (see also Elbe 2009, pp. 59, 86, 108, 131). On the one hand, it acknowledges the specificity of security or rather of political process that designates something as a threat, which differs from a mere politicisation of an issue. On the other hand, it opens the meaning and consequences of security beyond the extraordinary measures postulated by the Copenhagen School. It also heeds the constantly changing nature of security conceptions and hence the political effects of securitisation processes. This transformation is particularly pronounced during the securitisation process itself, which does not only 'securitise' formerly non-security issues such as climate change but also works the other way around. It thus contributes to a transformation of the political conceptions of security used in the securitisation process, which others have called 'climatisation' or 'medicalisation' (Oels 2012; Elbe 2011; Maertens 2018), and which I name the 'bidirectional qualities' of securitisation.

To structure the empirical analysis in the following country case chapters, I assume three distinct ways of securitising an issue, that is, three ideal-typical climate security discourses.[4] Each discourse mainly draws on one of these different power forms, which is linked to different problem definitions, empowers different actors and legitimises different political responses. Having said that, these discourses are ideal-typical simplifications and in the political practice, only rarely occur completely independently from one another (Foucault 2006b, p. 23). Nevertheless, in order to structure the empirical analysis, it makes sense to start out from this simplified framework. To further guide the analysis, I divide each climate security discourse into two dimensions. The first dimension—*threat construction*—focuses on how the respective discourse evokes climate change as a threat, which security concepts it uses and which referent objects it constructs as primarily threatened. The second dimension focuses on the political consequences of each specific discourse, thus on the *power effects*. It thus looks at the transformation of the overall political debate, the legitimisation of individual policies and practices, the facilitation of certain forms of governance and the empowerment of specific actors. Due to the above-mentioned bidirectional qualities of securitisation, I do not only look for these effects in the climate field but also look at other closely connected policy sectors, most importantly defence and security, foreign, development and disaster management policy.

[4] I focus on 'discourses' and not 'dispositives' due to several reasons. Most importantly, I concur with several other scholars that there is no clear-cut boundary between discourses and dispositives as both try to understand the constitution of subjects and objects of governance and emphasise the productive as well as 'truth' generating qualities of power (Bröckling and Krasmann 2010, pp. 24, 26, 29; van Dyk and Angermüller 2010; Bührmann and Schneider 2008).

Naturally, the separation of these two dimensions is a further methodological simplification. Following Diez and others, I concur that discourses and their political consequences or power effects are co-constitutive (Kurki 2008; Milliken 1999; Diez 2001, p. 13) and similar to norms, discourses possess a dual-quality (Wiener 2007, 2008). Thus, discourses can legitimise policies and practices (i.e. have 'quasi-causal' effects, Yee 1996, p. 97), but at the same time, these practices and policies strengthen or perpetuate specific discourses (Salter 2011, p. 118). Nevertheless, individual policies or speech acts do not alone constitute a discourse. They are only one of many different layers that in the end come together in forming the discourse and only become meaningful through their discursive representation. Thus, while being in constant transformation, discourses nevertheless are relatively stable structures of meaning that do not change easily (Diez 1999, p. 607). Ultimately, climate security discourses are instrumental in constituting climate change as a distinct political issue and in legitimising political practices and policies. This does not mean that there is a deterministic causal relationship but rather that they create a field of possibility in which specific practices or policies become more likely and feasible than others, which others have described a 'causality as resonance' (Connolly 2005, p. 870).

Sovereign Power and Climate Change

The sovereign discourse securitises climate change in a direct and highly visible way, using 'national security' or similar concepts such as 'regional security', 'territorial security' or 'military security'. It renders climate change governable as a traditional security issue with a focus on sovereign actions of the state and military/defence actors. Although in its general conceptualisation, sovereign power is broader, and the focus is not exclusively on state defence matters or the military, in a securitisation context, it is justified to understand it in a narrower fashion.

Regarding the *threat construction*, the sovereign discourse focuses on security threats for states and their territory, which ultimately can also have repercussions for the international system of states. The emphasis of this discourse is mostly on second-order socio-economic effects of climate change. One of the core arguments is about climate change—in combination with population growth and degrading resources—leading to violent conflict and therefore threatening the territorial integrity, stability and thus national security of states. Whereas extreme climate change could also contribute to such problems in relatively stable countries of the Global

North (especially concerning geopolitical tensions in the Arctic), for the time being, these violent conflicts are primarily projected to break out in the Global South. However, this does not prevent actors from the Global North to construct a threat to the national security of industrialised states or the international order. The key pathways in this regard are political instability, the growth of terrorism, large-scale migration and the spread of fragile as well as failed states, all of which could ultimately threaten industrialised countries or their vital interests in the affected regions. A common example for this causal chain is the prediction of water and food scarcity in Africa or Asia, which could lead to violent conflicts and undermine already instable states even further, which could eventually necessitate or endanger (military) interventions by industrialised countries.

Concerning the *power effects*, a securitisation of climate change drawing on the sovereign discourse increases the attention for the issue to a considerable extent, elevating it into high and traditional security politics. Thus, instead of the exclusive handling by for instance the environmental ministry, it facilitates the involvement of the foreign and defence ministry or other associated agencies. Up until a certain degree of intensity, this can be beneficial in terms of effective climate policies. It can (quicker) legitimise 'normal' or non-military sovereign solutions to the climate threat such as binding emission cuts, the legal ban of greenhouse gas (GHG) emitting technologies, a CO_2 tax or the increased governance through executive orders and regulations by government agencies. However, in its more extreme forms it can go beyond this legislative realm and facilitate extra-legal decisions of the sovereign similar to what the Copenhagen School has described as the main effects of securitisation. Examples are the suspension of laws, the acceleration or even bypassing of democratic procedures or the involvement of the armed forces. This could ultimately also entail military or political interventions to destroy GHG emitting industries or the states that harbour them (Trombetta 2008, p. 599; Hartmann 2010).

Furthermore, as the sovereign discourse constructs climate change as a problem of traditional national security, it has a tendency to point to solutions in the military and defence sector. This can entail the preparation of military bases and critical infrastructure towards climate change effects or the expansion of border security measures to keep out climate migrants. It can also mean to increase military planning activities for geopolitical conflicts fuelled by climate change or simply for the altered mission scenarios around the world. In this vein, increased state efforts to secure its territory against climate threats could also reinforce political boundaries between nation states and fuel nationalistic tendencies and othering dynamics. Furthermore, while binding top-down induced emission cuts would certainly address the root causes of climate change and fall within the mitigation dimension of climate abatement measures, the immediacy and scope, as well as political weight of threats to the national security of states, often legitimise short-term solutions to tackle the immediate security threat. Thus, the sovereign discourse often facilitates solutions that aim at the symptoms and hence mainly point towards adaptation. While these solutions do not have to entail a military component, military interventions in countries that are in danger of being destabilised or overwhelmed by climate change are a legitimate response within the sovereign discourse.

Finally, a sovereign power-based securitisation narrows down the spectrum of actors and focuses on the state and its agencies, particularly the security, defence and intelligence sector. While empowering these actors, it also tends to exclude non-state and civilian actors such as environmental or human rights NGOs from the debate, which is even more reinforced by the tendency to secrecy that often comes with evoking national security. The following Table 1.4 summarises the key features of the sovereign discourse.

Disciplinary Power and Climate Change

The disciplinary discourse constructs an entirely different picture of climate change as a threat. Within this discourse, the focus is on seemingly positive and 'soft' human security conceptions and similar concepts such as individual security, food security, environmental security and human vulnerability. In the disciplinary discourse, climate change is constituted as a problem to be tackled not only by the traditional security actors and associated practices but also by the development sector and non-state actors.

Looking at the *threat construction*, the securitisation proceeds in a less alert, less extraordinary and less authoritative way. Instead of focusing on threats for the state, the emphasis is on individuals and their vulnerabilities and human security. Accordingly, the threats are not primarily the second-order socio-economic effects of climate change but the direct first-order physical effects on the daily life of people, such as extreme weather events, the spread of diseases and scarcer resources. Whereas people living in countries of the Global North might also be directly threatened by climate change, the attention is on poor people in the Global South that lack the

Table 1.4 Sovereign discourse

Security conceptions	*Threat construction*	*Discursive power effects*
National security, regional security, territorial security, International security, international order, Military security, climate conflict, Resource security, resource conflicts, water wars, energy security	Climate change is securitised in a direct, highly visible way, socio-economic effects of climate change leading to conflicts, violence and large-scale migration and therefore threatening the national security of the state *Target*: mostly state territory, international order of states, specific regions	**Transformation of the debate and of governance practices** Climate change as high politics, acceleration of procedures, decisive, effective and radical measures (possibly bypassing democratic procedures). Focus on short-term and adaptation measures that tackle the immediate symptoms. Focus on binding targets and direct interventions of the government, tendency to secrecy. Most important actors and referent objects are states and their governments and particularly the security, defence and intelligence sector. **Exemplary policies** Binding emission reduction targets, CO_2 taxes, ban of GHG-producing industries and technologies, direct political and military interventions to coerce climate laggards or to defend against secondary socio-economic dangers, military planning, prepare for geopolitical and regional tensions, secure military bases and critical infrastructure against adverse climate effects, increase border security against climate migrants

Source: Author

adequate resources and state support to adapt to the changing environment. In contrast to the sovereign discourse, the focus on the vulnerability of individuals in the disciplinary discourse can also reinforce a shared humanity perspective, hence avoiding nationalistic or othering tendencies.

Concerning the *power effects*, the focus of the disciplinary discourse is primarily on solutions that tackle the problems of individuals by transforming their behaviour. Concrete examples are setting up mitigation targets, the propagation of the clean development mechanism as well as drawing up ideal scenarios for energy use, GHG emissions and climate-friendly behaviour. Beyond that, the disciplinary discourse can also legitimise adaptation measures that directly serve the human security of individuals and prepare people to cope with the threats of climate change. On the one hand, this entails the empowerment of poor people through measures that improve their living conditions or the encouragement and support of climate-friendly behaviour. Thus, it partly fulfils the emancipatory hopes of the proponents of human security. On the other hand, however, based on the logic of 'normation', the disciplinary discourses perpetuate a set of problematic identities (Elbe 2009, p. 117). It juxtaposes well-adapted people from the powerful Global North with unprepared, powerless people in the Global South. It also propagates the need for surveillance, monitoring and control of these threatened and seemingly helpless victims. To properly discipline them, their climate-related behaviour such as GHG emissions and also their general ways of living and their choice of dwelling areas and so on have to be constantly kept under control by outside (Northern) actors with supposedly superior knowledge. The aim is to transform individuals and their circumstances towards the ideal-typical norm of the healthy, affluent and climate-resilient citizen.

While this ideal type in part mirrors a typical Northern citizen, in one crucial way it does not: the per capita emission of GHGs. Here, the aim is not a normation towards a Northern lifestyle but the prevention of an increase in emissions. This could include curtailing economic growth and the dissemination of (climate-damaging) technology to make life more comfortable. Thus, while the disciplinary discourse could legitimise measures that benefit poor individuals in the Global South, it could also curtail their freedom, remove agency and push them towards actions that are not in their immediate interest. Conceivable policies entail direct interventions as part of development aid schemes, which strive to condition the behaviour of local people towards the ideal norm. Examples are the propagation of climate preparedness measures or climate education campaigns. The goal is to raise the awareness of the affected individuals for the dangers of climate change and for the global benefits of climate-friendly behaviour. At the same time, it is also about increasing their coping capacity and decreasing their direct or 'outcome vulnerability' (O'Brien et al. 2007, p. 75) towards climate change effects.

In terms of actors, the disciplinary discourse particularly empowers non-state organisations such as environmental, development and human rights NGOs to act. However, due to the changing nature of the concept of human security, which includes an ever-closer relationship between the development and the traditional security sector, constructing climate change as a threat to human security can also foster a closer integration of military and civilian operations. The following Table 1.5 summarises the key features of the disciplinary discourse.

Governmental Power and Climate Change
A governmental power-based securitisation constructs climate change as a long-term problem that could gradually increase a range of risks. It renders climate change governable as statistical problem drawing on various

Table 1.5 Disciplinary discourse

Security conceptions	*Threat construction*	*Discursive power effects*
Human security, individual security, food security, environmental security, vulnerability, coping capacity	Climate change is securitised indirectly using human security conceptions and the like. A micro-perspective on individuals and small groups prevails. Focus on direct physical effects (everyday implications) of climate change for poor individuals. *Target*: mostly individuals, global community of individuals, humankind	**Transformation of the debate and of governance practices** Refocusing the climate security debate on the direct physical effects of climate change for human security, possibly leading to support and empowerment schemes for individuals. Yet, also increased juxtaposition of North vs. South and surveillance and control of threatened individuals. 'Normation' processes: developing vulnerable humans towards ideal norm of a climate-resilient citizen. Focus on outcome vulnerability, possibly denying poor individuals technology that endangers the global climate. Broadening of actors participating in governance processes, empowering non-state actors. **Exemplary policies** Prescribing behavioural and technological norms, monitoring emissions and behaviour, voluntary emission reduction targets, climate education, adaptation measures (reducing outcome vulnerability), energy efficiency norms, clean development mechanism, CO_2 taxes, mainstreaming climate change and climate security in development aid schemes, integrating development aid and military action, spread of 'networked security concepts'.

Source: Author

instruments such as risk-assessment and -management schemes and accepting a general degree of uncertainty.

Concerning the *threat construction*, a governmental power-based securitisation is less direct and does not point to immediate threats or to clearly defined referent objects. Instead, it relies on sophisticated statistical models to calculate certain high-risk groups and areas that climate change might affect in the future. Examples are geographical regions such as coastal and arid areas, or small islands, or socio-economic groups such as poor people in developing countries. In contrast to disciplinary power, however, the main target is not to rescue these risk groups, at least not in any case and at all costs. Instead, the governmental discourse aims primarily at the welfare of the overall population, which means that the well-being of those risk groups is obsolete as long as the overall risk for the population is bearable. Drawing on the concept of 'normalisation', the focus is on bringing the risk of the outliers down to a tolerable level in order to ensure the welfare and functioning of the overall population. While in some cases humankind constitutes the target population, depending on the perspective of the ones articulating this discourse, the circle can also be drawn much narrower and can reinforce and erect boundaries between specific groups or states. Thus, according to prominent climate scenarios, most industrialised countries only run moderate climate risks, whereas many countries of the Global South fall within the high-risk groups and thus deviate from the overall population.

The governmental discourse also constructs climate change as a problem of circulation. It aims at fostering the desired circulation (e.g. international business trips, the shipping of goods, economic growth), while avoiding the undesired, meaning excessive GHG emissions as well as people migrating due to problems caused by climate change. The particular challenge in relation to climate change is to prevent circulation dynamics that cannot be easily kept from getting out of control by the careful manipulation of other natural processes (Elbe 2009, pp. 73–74). Examples are the interconnections between economic growth and emissions, 'tipping points' and self-reinforcing events such as the offsetting of additional GHGs by the meltdown of permafrost areas, or the possibility of pandemic outbreaks of disease.

Looking at the *power effects*, a governmental securitisation of climate change fosters a view into the future, which is supposed to be controlled by preventive measures and by manipulating the right variables within the population. Instead of monitoring the whole world and each individual, the aim is to use tailor-made and cost-efficient surveillance and control measures that focus mainly on high-risk groups and regions and eventually seek to normalise their risk. The focus is on interfering indirectly by trying to decrease the 'contextual vulnerability' (O'Brien et al. 2007, p. 76) of high-risk groups by improving the general economic and political circumstances and thereby preventing problems from materialising in the first place. Thus, emission trading schemes, market instruments and voluntary behavioural change to indirectly control GHG emissions play an important role. A further central instrument is the propagation of insurance schemes that are supposed to control the future in the present by hedging against undesirable future events. Concrete examples are weather and climate insurance schemes as well as disaster relief funds. To carry out these indirect pathways of governance, non-state actors such as NGOs, think tanks, corporations and insurance firms play a decisive role. Nevertheless, the governmental discourse can also push actors in the defence and security sector to increasingly rely on probabilistic scenarios that try to control possibly problematic future climate events.

On the one hand, the governmental discourse can facilitate long-term strategies and help to increase the attention for specifically endangered risk groups. On the other hand, it can also stigmatise these risk groups and construct them as a danger *vis-à-vis* the welfare of the overall population. Moreover, because of its long-term orientation and the absence of concrete threats, it does not generate the same attention as the sovereign or disciplinary discourses. Since the overall risk can also be lowered by increasing the resilience of risk groups or by propagating climate insurance schemes, fundamental changes in industrialised countries such as widespread mitigation schemes might get out of view. At the end of the day, a governmental securitisation can be less influential in its direct political effects at least if the goal is to legitimise decisive emission reductions linked to mandatory CO_2 limits. However, on a more fundamental, yet possibly less-noticed level, the governmental discourse can be much more far-reaching as it corresponds to the broader governmentalisation of the state and thus constitutes an entirely different form of governance that touches upon almost all spheres of society. The following Table 1.6 summarises the key features of the governmental discourse.

Table 1.6 Governmental discourse

Security conceptions	*Threat construction*	*Discursive power effects*
Risk, risk-management, riskisation, scenario planning, resilience, risk-groups, risk-assessment, uncertainty, contingency-planning, precautionary principle, early warning systems	Climate change is securitised indirectly, using risk conceptions and focusing on the macro-level. Statistical construction of risk groups, areas and practices, focus not on direct effects/threats of climate change but on probabilities of these effects for risk groups *vis-à-vis* the population, identified by risk assessment schemes and vulnerability assessments, acceptance of a general degree of uncertainty. Focus on circulation dynamics and future scenarios. *Target*: population, risk groups/ areas, diffuse, unspecified	**Transformation of the debate and of governance practices** Increased surveillance of risk groups, areas and activities, 'normalisation' processes, controlling circulation, acting on the future—long-term measures aimed at root causes, bringing down the risks to a tolerable level, stigmatisation of risk groups, focus on precautionary and resilience measures, reducing the vulnerability of certain risk groups, that is, critical vulnerabilities through adaptation measures, focus on contextual vulnerability, focus on cost-efficient measures that do not interfere too much, rather acting at the level of the population, insurance solutions, empowering a broad spectrum of non-state actors **Exemplary policies** Emission trading schemes, market instruments, voluntary emission cuts, insurance schemes, risk management approaches (disaster risk atlases, early warning systems), vulnerability assessments, creation of climate risk groups and areas and development of appropriate policies to prepare/insure these risk groups, resilience measures (reducing contextual vulnerability)

Source: Author

Analysing the Securitisation of Climate Change in Three Country Cases

Drawing on the above-described theoretical framework, the remainder of this book analyses the securitisation of climate change and its political effects in three country cases, namely the US, Germany and Mexico. Mainly based on an in-depth discourse analysis, the aim is to identify which securitisation discourses (or combinations) were dominant in each country and to inquire how they are linked to specific political consequences. The focus is on understanding the multitude of securitisations and the interlinkages with varying domestic contexts and hence to foster a contextual understanding of security (see also Ciuta 2009). Looking at different country cases in a comparative manner has at least two advantages compared to focusing on the global level. Firstly, it closes a gap in the studies on securitisation and governmentality that so far have predominately focused on the global level, isolated case studies and exemplary data (Oels 2011, 2012, 2013; Methmann and Rothe 2012; Methmann 2011, 2014; Rothe 2011a, p. 7). Secondly, it addresses the bias in the literature to mainly look at political programmes without analysing the actual political and institutional effects (Rothe 2011b; Mckee 2009, pp. 473–474; Bröckling et al. 2012, pp. 16–17; van Dyk and Angermüller 2010, p. 13).

Case Selection

Why look at these specific cases? A key consideration was to find cases that are similar enough to compare, but also different enough to allow for diverse forms of the securitisation of climate change to become dominant. The US and Germany are both highly industrialised, democratic countries with a similar level of overall development (UNDP 2015). At the same time, they differ when it comes to a number of cultural and political variables, especially concerning security policy (Gerhards 2000; Böckenförde and Gareis 2014), and also have shown different approaches to climate change. To address the Western or Northern bias in securitisation studies (Boas 2014; Bilgin 2010; von Lucke 2018) and to learn about the effects of a different level of development and cultural background in securitisation processes, the book also looks at Mexico. Yet, despite its differences, due to longstanding close relations to Western countries, membership in the OECD since 1994 and a considerable economic growth since the 1990s (Bertelsmann Stiftung 2012; Dussel Peters and Maihold 2007; Akerberg 2011), Mexico is similar enough to allow for a meaningful comparison. Apart from these factors, Mexico constitutes a promising case because it is expected to be directly affected by a variety of different climate effects (Ibarrarán et al. 2008; Brodziak et al. 2011; José Cárdenas 2010) and has already put forward several progressive policies to address its perceived vulnerability.

Period of Analysis

Concerning the timeframe of the analysis, the book particularly focuses on climate security debates in each country between the late 1980s and 2015. The reason for this rather broad timeframe is that it allows me to explore the influence of the broader historical context on the climate security debate but also to capture key transformations of the discourses over time. I start with the late 1980s because it was at that time that climate change became more important on the political level, that is, it ceased to be predominantly discussed by climate experts or in an academic context. Indicators are the establishment of the IPCC in 1988, the adoption of the UNFCCC in 1992 but also several more influential domestic policies on climate regulation in the US, Germany and to a lesser extent also in Mexico (Pielke 2000a, b; Schreurs 2002; Park 2000; Jänicke 2011; Weidner and Mez 2008; Mumme and Lybecker 2002; Dussel Peters and Maihold 2007). The decision to close the research in 2015, on the one hand, rests on fact that the adoption of the Paris Agreement at the end of 2015 constitutes the end of a key cycle in the negotiations. On the other hand, it roughly coincided with the election of new governments in the US and Germany, thus once again closing a specific phase.

While the analysis of the book hence will cover a broad period, a specific emphasis is on the time from the mid-2000s. The reason behind this is that the political debates on the climate-security nexus can be divided into two crucial phases (Floyd 2010, 2012; Oels 2011, 2012). The first phase that peaked around the mid-1990s was characterised by rather broad debates on 'environmental security' or 'environmental conflict' (Homer-Dixon 1994, 1999; Myers 1995; Kaplan 1994). The second phase gained momentum in the mid-2000s (with a peak between 2006 and 2011) and focused almost exclusively on the security implications of climate change. I pay particular attention to the second phase because of its exclusive focus on climate change but also because of the much better availability of empirical material for this time.

Empirical Data and Methods

The book mainly rests on an in-depth analysis of climate security debates in each of the three countries,[5] which included governmental as well as

[5] The empirical data partly stems from a research project (ClimaSec) led by Professor Thomas Diez funded by the German Science Foundation (DFG, Grand number: DI 1688/1-1).

non-governmental sources. To assemble the empirical material, I firstly consulted the secondary literature, which already had identified several documents (NGO and think tank reports as well as key policies and laws) as particularly relevant concerning the securitisation of climate change (see for instance Oels 2012; Detraz and Betsill 2009; Brauch 2009; Brzoska 2009; Hartmann 2010). Simultaneously I conducted internet searches using keywords such as 'climate change' and 'global warming' in combination with 'security', 'threat', 'risk', 'human security', and 'national security' as well and their equivalents in German and Spanish,[6] to identify further politically influential climate security articulations. Additionally, I relied on cross-references between the acquired documents, on expert interviews in each country, and on the analysis of parliamentary debates to uncover further important documents.

The second, most important, step was to reconstruct which discourses prevailed in each country and how their specific framing of climate security changed over time. For that purpose, I conducted an in-depth, qualitative discourse analysis (Balzacq 2011b; Oels 2013; Aradau et al. 2014b, p. 8) on the most influential reports, policy documents and parliamentary debates. The aim was not a quantitative assessment of the occurrence of certain keywords but the identification of key argumentative structures and power effects linked to the three ideal-typical discourses. To structure the findings, the core results of the analysis (dominant security discourses, threats, referent objects, actors, recommendations and document specific findings) were documented in semi-standardised tables for each analysed document. To specify which discourse prevailed in a document, I relied on the frequency, the internal coherence and intensity as well as the position of the argument within the document. For instance, I considered an argument particularly important when it appeared in the introduction (or in the often included 'summary for policymakers') and in the conclusion. Naturally, the analysed documents did only seldom focus on one of the three discourses exclusively but often contained articulations of all three and tended to link discourses to each other. Nonetheless, in most cases, it was possible to identify a clear order in the importance of the different discourses.

To back up the findings of this discourses analysis, in a third step, I conducted interviews (between April and September 2014) with key NGO representatives and governmental officials in each country. The aim

[6] If not already available in English, I have translated all German and Spanish quotes myself.

was to understand how certain discourses gained relevance in political debates and to learn more about the role and motivation of key actors in these debates. The interview partners were selected based on the initial findings of the discourses analysis and on a snowball system after the first round of interviews. Each interview took between 30 and 90 minutes and was structured by a semi-standardised questionnaire. Due to the in parts sensible nature of the topic, the interviews were conducted under the condition of anonymity and were not recorded.

The last step was to understand how the dominant discourses in each country contributed to the legitimisation and adoption of policies, practices and institutional transformations. To this end, I analysed parliamentary debates (based on the same scheme as the reports) and key policy documents in central policy fields, mainly: climate/environment, security/defence/intelligence, development cooperation and disaster response/civil protection. The individual chapters will provide more details on the specific empirical sources used and will give more information on the number of analysed documents.

References

adelphi. (2012). *Climate Diplomacy: Reducing Risks for Security.* Berlin.

Adler-Nissen, R., & Pouliot, V. (2014). Power in Practice: Negotiating the International Intervention in Libya. *European Journal of International Relations, 20*, 889–911. https://doi.org/10.1177/1354066113512702.

Agamben, G. (1998). *Homo Sacer: Sovereign Power and Bare Life* (D. Heller-Roazen, Trans.). Stanford: Stanford University Press.

Agamben, G. (2005). *State of Exception* (K. Attell, Trans.). Chicago: The University of Chicago Press.

Akerberg, A. Á. (2011). Länderperspektive: Mexiko. In N. Netzer & J. Gouverneur (Eds.), *Zwischen Anspruch und Wirklichkeit: Internationale Perspektiven vor der Weltklimakonferenz in Durban* (pp. 37–40). Berlin: Friedrich-Ebert-Stiftung.

Aradau, C., & van Munster, R. (2007). Governing Terrorism Through Risk: Taking Precautions, (Un)Knowing the Future. *European Journal of International Relations, 13*(1), 89–115.

Aradau, C., & van Munster, R. (2008). Insuring Terrorism, Assuring Subjects, Ensuring Normality: The Politics of Risk after 9/11. *Alternatives, 33*, 191–210.

Aradau, C., Coward, M., Herschinger, E., Thomas, O. D., & Voelkner, N. (2014a). Discourse/Materiality. In C. Aradau, A. Neal, J. Huysmans, & N. Voelkner

(Eds.), *Critical Security Methods: New Frameworks for Analysis* (pp. 57–84). Abingdon, Oxon: Routledge.
Aradau, C., Huysmans, J., Neal, A., & Voelkner, N. (2014b). Introducing Critical Security Methods. In C. Aradau, A. Neal, J. Huysmans, & N. Voelkner (Eds.), *Critical Security Methods: New Frameworks for Analysis* (pp. 1–22). Abingdon, Oxon: Routledge.
Bachrach, P., & Baratz, M. S. (1962). The Two Faces of Power. *The American Political Science Review, 56*(4), 947–952.
Balzacq, T. (2011a). A Theory of Securitization: Origins, Core Assumptions, and Variants. In T. Balzacq (Ed.), *Securitization Theory: How Security Problems Emerge and Dissolve* (PRIO New Security Studies) (pp. 1–30). London: Routledge.
Balzacq, T. (Ed.). (2011b). *Securitization Theory: How Security Problems Emerge and Dissolve* (PRIO New Security Studies). London: Routledge.
Barnett, J. (2000). Destabilizing the Environment-Conflict Thesis. *Review of International Studies, 26*(2), 271–288.
Barnett, J. (2003). Security and Climate Change. *Global Environmental Change, 13*(1), 7–17.
Barnett, J., & Adger, W. N. (2005). *Security and Climate Change: Towards an Improved Understanding*. Human Security and Climate Change: An International Workshop Holmen Fjord Hotel, Asker, near Oslo, 20–21 June 2005.
Barnett, J., & Adger, W. N. (2007). Climate Change, Human Security and Violent Conflict. *Political Geography, 26*(6), 639–655. https://doi.org/10.1016/j.polgeo.2007.03.003.
Barnett, M., & Duvall, R. (2005). Power in International Politics. *International Organization, 59*, 471. https://doi.org/10.1017/S0020818305050010.
Beck, U. (2000). *World Risk Society*. Cambridge: Cambridge University Press.
Bertelsmann Stiftung. (2012). *BTI 2012 — Mexico Country Report*. Gütersloh.
Bigo, D. (2002). Security and Immigration: Toward a Critique of the Governmentality of Unease. *Alternatives: Global, Local, Political, 27*(Suppl), 63–92.
Bigo, D. (2008). Globalized (In)Security: The Field and the Ban-Opticon. In D. Bigo & A. Tsoukala (Eds.), *Terror, Insecurity and Liberty: Illiberal Practices of Liberal Regimes after 9/11* (Routledge Studies in Liberty and Security) (pp. 10–48). London: Routledge.
Bigo, D. (2009). International Political Sociology. In P. Williams (Ed.), *Security Studies: An Introduction* (pp. 116–129). London; New York: Routledge.
Bigo, D., & Tsoukala, A. (2008). Understanding (In)Security. In D. Bigo & A. Tsoukala (Eds.), *Terror, Insecurity and Liberty: Illiberal Practices of Liberal Regimes after 9/11* (Routledge Studies in Liberty and Security) (pp. 10–48). London: Routledge.

Bilgin, P. (2010). The 'Western-Centrism' of Security Studies: 'Blind Spot' or Constitutive Practice? *Security Dialogue, 41*, 615–622. https://doi.org/10.1177/0967010610388208.

Boas, I. (2014). Where Is the South in Security Discourse on Climate Change?: An Analysis of India. *Critical Studies on Security, 2*, 148–161. https://doi.org/10.1080/21624887.2014.905295.

Böckenförde, S., & Gareis, S. B. (Eds.). (2014). *Deutsche Sicherheitspolitik: Herausforderungen, Akteure und Strategien in einer Welt im Wandel* (UTB) (Vol. 8388, 2nd ed.). Opladen: Barbara Budrich.

Booth, K. (1991). Security and Emancipation. *Review of International Relations, 17*(4), 313–326.

Booth, K. (2005). Critical Explorations. In K. Booth (Ed.), *Critical Security Studies and World Politics* (Critical Security Studies) (pp. 1–18). Boulder, CO: Lynne Rienner.

Boutros-Ghali, B. (1992). *An Agenda for Peace. Preventive Diplomacy, Peacemaking and Peace-keeping*. New York: United Nations.

Boyle, P., & Haggerty, K. D. (2012). Planning for the Worst: Risk, Uncertainty and the Olympic Games. *The British Journal of Sociology, 63*, 241–259. https://doi.org/10.1111/j.1468-4446.2012.01408.x.

Brauch, H. G. (2009). Securitizing Global Environmental Change. In H. G. Brauch, Ú. Oswald Spring, J. Grin, C. Mesjasz, P. Kameri-Mbote, N. Chadha Behera, et al. (Eds.), *Facing Global Environmental Change: Environmental, Human, Energy, Food, Health and Water Security Concepts* (Hexagon Series on Human and Environmental Security and Peace) (Vol. 4, pp. 65–104). Berlin: Springer.

Brauch, H. G., & Scheffran, J. (2012). Introduction: Climate Change, Human Security, and Violent Conflict in the Anthropocene. In J. Scheffran, M. Brzoska, H. G. Brauch, P. M. Link, & J. Schilling (Eds.), *Climate Change, Human Security and Violent Conflict: Challenges for Societal Stability* (Hexagon Series on Human and Environmental Security and Peace) (Vol. 8, pp. 3–40). Berlin; Heidelberg; New York: Springer.

Briggs, C. M. (2012). Climate Security, Risk Assessment and Military Planning. *International Affairs, 88*(5), 1049–1064.

Bröckling, U., & Krasmann, S. (2010). Ni méthode, ni approche: Zur Forschungsperspektive der Gouvernementalitätsstudien - mit einem Seitenblick auf Konvergenzen und Divergenzen zur Diskursforschung. In J. Angermüller (Ed.), *Diskursanalyse meets Gouvernementalitätsforschung: Perspektiven auf das Verhältnis von Subjekt, Sprache, Macht und Wissen* (pp. 23–42). Frankfurt am Main: Campus-Verlag.

Bröckling, U., Krasmann, S., & Lemke, T. (2012). From Foucault's Lectures at the Collège de France to Studies of Governmentality: An Introduction. In U. Bröckling, S. Krasmann, & T. Lemke (Eds.), *Governmentality: Current*

Issues and Future Challenges (Routledge Studies in Social and Political Thought) (pp. 1–33). London: Routledge.

Brodziak, F., García, A. L., & Chow, L. G. (2011). *Climate Change Impacts on Socio-Environmental Conflicts: Vulnerability in Facing Climate Change and Social Conflicts in Mexico.* Brussels. Retrieved February 21, 2013, from http://www.ifp-ew.eu/pdf/1111mexico.pdf.

Brössler, D. (2019, January 25). Der Klimawandel wird zur Gefahr für den Frieden. *Süddeutsche Zeitung.* Retrieved March 25, 2019, from https://www.sueddeutsche.de/politik/maas-un-sicherheitsrat-klimawandel-1.4304233.

Brzoska, M. (2009). The Securitization of Climate Change and the Power of Conceptions of Security. *Sicherheit und Frieden, 27*(3), 137–145.

Brzoska, M. (2012). Climate Change as a Driver of Security Policy. In J. Scheffran, M. Brzoska, H. G. Brauch, P. M. Link, & J. Schilling (Eds.), *Climate Change, Human Security and Violent Conflict: Challenges for Societal Stability* (Hexagon Series on Human and Environmental Security and Peace) (Vol. 8, pp. 165–184). Berlin; Heidelberg; New York: Springer.

Brzoska, M., & Oels, A. (2011). "Versicherheitlichung" des Klimawandels? Die Konstruktion des Klimawandels als Sicherheitsbedrohung und ihre politischen Folgen. In M. Brzoska, M. Kalinowski, V. Matthies, & B. Meyer (Eds.), *Klimawandel und Konflikte: Versicherheitlichung versus präventive Friedenspolitik?* (1st ed., pp. 51–67). Baden-Baden: Nomos.

Buhaug, H., Nordkvelle, J., Bernauer, T., Böhmelt, T., Brzoska, M., Busby, J. W., et al. (2014). One Effect to Rule them All?: A Comment on Climate and Conflict. *Climatic Change, 127*, 391–397. https://doi.org/10.1007/s10584-014-1266-1.

Bührmann, A. D., & Schneider, W. (2008). Mehr als nur diskursive Praxis? Konzeptionelle Grundlagen und methodische Aspekte der Dispositivanalyse. *Historical Social Research/Historische Sozialforschung, 33*(1), 108–141.

Burgess, J. P. (2011). *The Ethical Subject of Security: Geopolitical Reason and the Theat to Europe* (PRIO New Security Studies). London: Routledge.

Butler, J. (2004). *Precarious Life: The Powers of Mourning and Violence.* London: Verso.

Buxton, N., Hayes, B., & George, S. (Eds.). (2016). *The Secure and the Dispossessed: How the Military and Corporations are Shaping a Climate-Changed World* (Transnational Institute). London: Pluto Press.

Buzan, B., & Wæver, O. (2009). Macrosecuritisation and Security Constellations: Reconsidering Scale in Securitisation Theory. *Review of International Studies, 35*, 253–276.

Buzan, B., Wæver, O., & de Wilde, J. (1998). *Security: A New Framework for Analysis.* Boulder, CO: Lynne Rienner.

Campbell, K. M. (2008). *Climatic Cataclysm: The Foreign Policy and National Security Implications of Climate Change*. Washington, DC: Brookings Institution Press.

Campbell, K. M., Gulledge, J., McNeill, J. R., Podesta, J., Odgen, P., Fueth, L., et al. (2007). *The Age of Consequences: The Foreign Policy and National Security Implications of Global Climate Change*. Washington, DC.

Caney, S. (2006). Environmental Degradation, Reparations, and the Moral Significance of History. *Journal of Social Philosophy, 37*(3), 464–482.

Caney, S. (2010). Climate Change and the Duties of the Advantaged. *Critical Review of International Social and Political Philosophy, 13*, 203–228. https://doi.org/10.1080/13698230903326331.

Caney, S. (2016). The Struggle for Climate Justice in a Non-Ideal World. *Midwest Studies In Philosophy, 40*, 9–26. https://doi.org/10.1111/misp.12044.

Chaturvedi, S., & Doyle, T. (2015). *Climate Terror: A Critical Geopolitics of Climate Change* (New Security Challenges). London: Palgrave Macmillan.

Christian Aid. (2006). *The Climate of Poverty: Facts, Fears and Hope*. Retrieved October 17, 2010, from http://www.clacc.net/documents/Report/Climate_of_poverty.pdf.

Christian Aid. (2007). *Human Tide: The Real Migration Crisis*. London. Retrieved October 15, 2010, from http://www.christianaid.org.uk/Images/human-tide.pdf.

Ciuta, F. (2009). Security and the Problem of Context: A Hermeneutical Critique of Securitisation Theory. *Review of International Studies, 35*, 301. https://doi.org/10.1017/S0260210509008535.

CNA. (2007). *National Security and the Threat of Climate Change*. Alexandria. Retrieved October 17, 2010, from http://securityandclimate.cna.org/.

CNA Military Advisory Board. (2014). *National Security and the Accelerating Risks of Climate Change*. Alexandria, VA.

Collier, S. J. (2009). Topologies of Power: Foucault's Analysis of Political Government beyond 'Governmentality. *Theory, Culture & Society, 26*, 78–108. https://doi.org/10.1177/0263276409347694.

Connolly, W. E. (2005). The Evangelical-Capitalist Resonance Machine. *Political Theory, 33*(6), 869–886.

Corry, O. (2010). What Is a (Global) Polity? *Review of International Studies, 36*(1), 157–180.

Corry, O. (2012). Securitisation and 'Riskification': Second-order Security and the Politics of Climate Change. *Millennium, 40*, 235–258. https://doi.org/10.1177/0305829811419444.

Daase, C., & Kessler, O. (2007). Knowns and Unknowns in the 'War on Terror': Uncertainty and the Political Construction of Danger. *Security Dialogue, 38*, 411–434. https://doi.org/10.1177/0967010607084994.

Dalby, S. (2009). *Security and Environmental Change*. Cambridge: Polity.

Dalby, S. (2013a). Biopolitics and Climate Security in the Anthropocene. *Geoforum, 49*, 184–192. https://doi.org/10.1016/j.geoforum.2013.06.013.
Dalby, S. (2013b). Climate Change: New Dimensions of Environmental Security. *The RUSI Journal, 158*, 34–43. https://doi.org/10.1080/03071847.2013.807583.
Dalby, S. (2014). Rethinking Geopolitics: Climate Security in the Anthropocene. *Global Policy, 5*, 1–9. https://doi.org/10.1111/1758-5899.12074.
Dean, M. (2010). *Governmentality: Power and Rule in Modern Society* (2nd ed.). London: Sage.
Dean, M. (2012). The Signature of Power. *Journal of Political Power, 5*, 101–117. https://doi.org/10.1080/2158379X.2012.659864.
Detraz, N., & Betsill, M. M. (2009). Climate Change and Environmental Security: For Whom the Discourse Shifts. *International Studies Perspectives, 10*, 303–320. https://doi.org/10.1111/j.1528-3585.2009.00378.x.
Deudney, D. H. (1990). The Case Against Linking Environmental Degradation and National Security. *Millennium, 19*(3), 461–476.
Deudney, D. H., & Matthew, R. A. (Eds.). (1999). *Contested Grounds: Security and Conflict in the New Environmental Politics*. Albany: State University of New York Press.
Devetak, R. (2007). Between Kant and Pufendorf: Humanitarian Intervention, Statist Anti-Cosmopolitanism and Critical Theory. *Review of International Studies, 33*(S1), 151–174.
Diez, T. (1999). Speaking Europe: The Politics of Integration Discourse. *Journal of European Public Policy, 6*(4), 598–613.
Diez, T. (2001). Europe as a Discursive Battleground: Discourse Analysis and European Integration Studies. *Cooperation and Conflict, 36*, 5–38. https://doi.org/10.1177/00108360121962245.
Diez, T., von Lucke, F., & Wellmann, Z. (2016). *The Securitisation of Climate Change: Actors, Processes and Consequences*. London: Routledge.
Dillon, M. (2006). The Security of Governance. In W. Larner & W. Walters (Eds.), *Global Governmentality: Governing International Spaces* (Routledge Advances in International Relations and Global Politics) (Vol. 28, pp. 76–94). London: Routledge.
Dillon, M., & Lobo-Guerrero, L. (2008). Biopolitics of Security in the 21st Century: An Introduction. *Review of International Studies, 34*(2), 265–292.
Duffield, M. (2005). *Human Security: Linking Development and Security in an Age of Terror*. Bonn. http://eadi.org/gc2005/confweb/papers/Mark_Duffield.pdf.
Duffield, M. (2007). *Development, Security and Unending War: Governing the World of Peoples*. Cambridge: Polity Press.
Duffield, M., & Waddell, N. (2006). Securing Humans in a Dangerous World. *International Politics, 43*(1), 1–23.

Dussel Peters, E., & Maihold, G. (2007). *Die Rolle Mexikos in der globalen Strukturpolitik* (Discussion Paper 15/2007, 15/2007). Bonn: DIE.

Dyer, H. (2018). Security Politics and Climate Change: the New Security Dilemma. In O. Corry & H. Stevenson (Eds.), *Traditions and Trends in Global Environmental Politics: International Relations and the Earth* (Routledge Research in Global Environmental Governance) (pp. 154–170). London: Routledge.

van Dyk, S., & Angermüller, J. (2010). Diskursanalyse meets Gouvernementalitätsforschung: Zur Einführung. In J. Angermüller (Ed.), *Diskursanalyse meets Gouvernementalitätsforschung: Perspektiven auf das Verhältnis von Subjekt, Sprache, Macht und Wissen* (pp. 7–21). Frankfurt am Main: Campus-Verlag.

EEAS. (2018). Climate, Peace and Security: The Time for Action. Retrieved March 21, 2019, from https://eeas.europa.eu/headquarters/headquarters-homepage/47165/climate-peace-and-security-time-action_en.

Elbe, S. (2006). Should HIV/AIDS Be Securitized?: The Ethical Dilemmas of Linking HIV/AIDS and Security. *International Studies Quarterly, 50*, 119–144. https://doi.org/10.1111/j.1468-2478.2006.00395.x.

Elbe, S. (2009). *Virus Alert: Security, Governmentality, and the AIDS Pandemic*. New York, NY: Columbia University Press.

Elbe, S. (2011). Pandemics on the Radar Screen: Health Security, Infectious Disease and the Medicalisation of Insecurity. *Political Studies, 59*, 848–866. https://doi.org/10.1111/j.1467-9248.2011.00921.x.

Eriksson, J. (1999). Observers or Advocates? On the Political Role of Security Analysts. *Cooperation and Conflict, 34*(3), 311–330.

Extinction Rebellion. (2019). The Truth. Retrieved November 28, 2019, from https://rebellion.earth/the-truth/.

Fagan, M. (2017). Security in the Anthropocene: Environment, Ecology, Escape. *European Journal of International Relations, 23*, 292–314. https://doi.org/10.1177/1354066116639738.

Finley-Brook, M. (2014). Climate Justice Advocacy. *Public Diplomacy Magazine, 12*(Summer), 11–15.

Fletcher, A. L. (2009). Clearing the Air: The Contribution of Frame Analysis to Understanding Climate Policy in the United States. *Environmental Politics, 18*, 800–816. https://doi.org/10.1080/09644010903157123.

Floyd, R. (2007a). Human Security and the Copenhagen School's Securitization Approach: Conceptualizing Human Security as a Securitizing Move. *Human Security Journal, 5*, 38–49.

Floyd, R. (2007b). Towards a Consequentialist Evaluation of Security: Bringing Together the Copenhagen and the Welsh Schools of Security Studies. *Review of International Studies, 33*(2), 327–350.

Floyd, R. (2010). *Security and the Environment: Securitisation Theory and US Environmental Security Policy*. Cambridge: Cambridge University Press.
Floyd, R. (2011). Can Securitization Theory be Used in Normative Analysis? Towards a Just Securitization Theory. *Security Dialogue, 42*, 427–439. https://doi.org/10.1177/0967010611418712.
Floyd, R. (2012). Climate Change, Environmental Security Studies and the Morality of Climate Security. *E-International Relations*.
Floyd, R. (2013). Whither Environmental Security Studies?: An Afterword. In R. Floyd & R. A. Matthew (Eds.), *Environmental Security: Approaches and Issues* (1st ed., pp. 279–297). Abingdon, Oxon; New York, NY: Routledge.
Floyd, R., & Matthew, R. A. (Eds.). (2013). *Environmental Security: Approaches and Issues* (1st ed.). Abingdon, Oxon; New York, NY: Routledge.
Foucault, M. (1975). *Discipline and Punish: The Birth of the Prison*. London: Penguin Books.
Foucault, M. (1979). *The History of Sexuality.: Vol. 1, An Introduction*. London: Allen Lane.
Foucault, M. (1980). *Power/Knowledge*. Brighton: Harvester Wheatsheaf.
Foucault, M. (1983). *Sexualität und Wahrheit* (Vol. 716). Frankfurt am Main: Suhrkamp.
Foucault, M. (2003). *Society Must Be Defended: Lectures at the Collège de France, 1975–1976*. New York: Picador.
Foucault, M. (2006a). *Die Geburt der Biopolitik: Geschichte der Gouvernementalität II* (Suhrkamp-Taschenbuch Wissenschaft) (Vol. 1809). Suhrkamp: Frankfurt am Main.
Foucault, M. (2006b). *Sicherheit, Territorium, Bevölkerung: Geschichte der Gouvernementalität I* (Suhrkamp-Taschenbuch Wissenschaft) (Vol. 1808). Frankfurt am Main: Suhrkamp.
Foucault, M., & Faubion, J. D. (Eds.). (2002). *Power* (Essential Works of Foucault 1954–1984, /Michel Foucault) (Vol. 3). London: Penguin.
Foucault, M., & Gordon, C. (Eds.). (1980). *Power/Knowledge: Selected Interviews and Other Writings 1972–1977*. New York: Vintage Books.
Fridays for Future Austria. (2019). It's Our Turn to Act. Our Future Is at Stake! Retrieved March 21, 2019, from https://fridaysforfuture.at/en/events/globaler-klimastreik/2019-03-15-globaler-klimastreik.
Gardiner, S. M. (2004). The Global Warming Tragedy and the Dangerous Illusion of the Kyoto Protocol. *Ethics & International Affairs, 18*(1), 23–39.
Gerhards, J. (Ed.). (2000). *Die Vermessung kultureller Unterschiede: USA und Deutschland im Vergleich*. Wiesbaden: Westdeutscher Verlag.
Gleditsch, N. P. (2012). Whither the Weather? Climate Change and Conflict. *Journal of Peace Research, 49*(1), 3–9.

Goldenberg, S. (2011, July 20). UN Security Council to Consider Climate Change Peacekeeping. *The Guardian*. Retrieved March 25, 2019, from https://www.theguardian.com/environment/2011/jul/20/un-climate-change-peacekeeping.

Gore, A. A. (2007). *Nobel Peace Prize Lecture*. Oslo. Retrieved October 17, 2010, from http://nobelprize.org/nobel_prizes/peace/laureates/2007/gore-lecture_en.html.

Grauvogel, J., & Diez, T. (2014). Framing und Versicherheitlichung: Die diskursive Konstruktion des Klimawandels. *Zeitschrift für Friedens- und Konfliktforschung, 3*(2), 203–232.

Greenpeace. (2007). *Klimaflüchtlinge. Die verleugnete Katastrophe* (Greenpeace). Hamburg. Retrieved from https://www.greenpeace.de/sites/www.greenpeace.de/files/klimafluechtlinge_endv_0.PDF.

Greenpeace. (2013). *Point of No Return: The Massive Climate Threats We Must Avoid*. Amsterdam.

Greenpeace México. (2010). *México ante el cambio climático: Evidencias, impactos, vulnerabilidad y adaptación*. Mexico City.

GTZ. (2008a). *Climate Change and Security: Challenges for German Development Cooperation*. Eschborn.

GTZ. (2008b). *Klimawandel und Sicherheit: Herausforderungen für die deutsche Entwicklungszusammenarbeit*. Eschborn. Retrieved September 26, 2015, from http://www.engagement-weltweit.de/fileadmin/Redaktion/ENGAGEMENT_WELTWEIT/Publik_/Wasser/GTZ_Klimawandel-und-Sicherheit.pdf.

Hameiri, S. (2008). Risk Management, Neo-Liberalism and the Securitisation of the Australian Aid Program. *Australian Journal of International Affairs, 62*, 357–371. https://doi.org/10.1080/10357710802286817.

Hameiri, S., & Jones, L. (2013). The Politics and Governance of Non-Traditional Security. *International Studies Quarterly, 57*, 462–473. https://doi.org/10.1111/isqu.12014.

Hansen, L. (2000). The Little Mermaid's Silent Security Dilemma and the Absence of Gender in the Copenhagen School. *Millennium, 29*(2), 285–306.

Hardt, J. N. (2017). *Environmental Security in the Anthropocene: Assessing Theory and Practice* (Critical Security Series). London: Taylor and Francis.

Harrington, C., & Shearing, C. (2017). *Security in the Anthropocene: Reflections on Safety and Care* (Edition Politik) (Vol. 30, 1st ed.). Bielefeld: Transcipt Verlag.

Harris, P. G. (2002). Environmental Security: Will Bush Follow Clinton's Lead? *Pacifica Review: Peace, Security & Global Change, 14*, 149–157. https://doi.org/10.1080/13239100220141191.

Hartmann, B. (2009). *Lines in the Shifting Sand: The Strategic Politics of Climate Change*. Oslo: Human Security and National Defense.

Hartmann, B. (2010). Rethinking Climate Refugees and Climate Conflict: Rhetoric, Reality and the Politics of Policy Discourse. *Journal of International Development, 22*(2), 233–246.
Homer-Dixon, T. (1991). On the Threshold: Environmental Changes as Causes of Acute Conflict. *International Security, 16*(2), 76–116.
Homer-Dixon, T. (1994). Environmental Scarcities and Violent Conflict: Evidence from Cases. *International Security, 19*(1), 5–40.
Homer-Dixon, T. (1999). *Environment, Scarcity and Violence.* Princeton, NJ: Princeton University Press.
Hsiang, S. M., Burke, M., & Miguel, E. (2013). Quantifying the Influence of Climate on Human Conflict. *Science, 341*, 1235367. https://doi.org/10.1126/science.1235367.
Huysmans, J. (2002). Defining Social Constructivism in Security Studies: The Normative Dilemma of Writing Security. *Alternatives, 27*(1), 41–62.
Huysmans, J. (2004). Minding Exceptions: The Politics of Insecurity and Liberal Democracy. *Contemporary Political Theory, 3*(3), 321–341.
Huysmans, J. (2008). *The Politics of Insecurity: Fear, Migration and Asylum in the EU* (The New International Relations Series). London: Routledge.
Ibarrarán, M. E., Malone, E. L., & Brenkert, A. L. (2008). *Climate Change Vulnerability and Resilience: Current Status and Trends for Mexico.* Retrieved June 26, 2013, from http://www.cakex.org/sites/default/files/Current%20Status%20and%20Trends%20for%20Mexico.pdf.
Ingram, A. (2010). Governmentality and security in the US President's Emergency Plan for AIDS Relief (PEPFAR). *Geoforum, 41*, 607–616. https://doi.org/10.1016/j.geoforum.2010.02.002.
IPCC. (2001). *Climate Change 2001: Synthesis Report.* Cambridge: Cambridge University Press.
IPCC. (2007). *Climate Change 2007 Synthesis Report: Contributions of Working Groups I, II and III to the Fourth Assessment Report of the Intergovernmental on Climate Change.* Geneva: IPCC.
IPCC. (2015). *Climate Change 2014: Synthesis Report. Contribution of Working Groups I, II and III to the Fifth Assessment Report of the Intergovernmental Panel on Climate Change.* Geneva, Switzerland: Intergovernmental Panel on Climate Change.
Jäger, S., & Meier, F. (2009). Theoretical and Methodological Aspects of Foucauldian Critical Discourse Analysis and Dispositive Analysis. In R. Wodak & M. Meyer (Eds.), *Methods for Critical Discourse Analysis* (pp. 34–61). Sage.
Jänicke, M. (2011). German Climate Change Policy: Political and Economic Leadership. In R. Wurzel & J. Connelly (Eds.), *The European Union as a Leader in International Climate Change Politics* (Routledge/UACES Contemporary European Studies) (Vol. 15, pp. 129–146). London; New York: Routledge.

José Cárdenas, M. (2010). *Mexico Ante el Cambio Climático: Evidencias, Impactos, Vulnerabilidad y Adaptación*. México D.F. Retrieved April 21, 2013, from http://www.greenpeace.org/mexico/Global/mexico/report/2010/6/vulnerabilidad-mexico.pdf.

Kaplan, R. (1994). The Coming Anarchy: How Scarcity, Crime, Overpopulation, Tribalism, and Disease are Rapidly Destroying the Social Fabric of Our Planet. *The Atlantic Monthly, 273*(2), 44–76.

Kelly, M. G. E. (2009). *The Political Philosophy of Michel Foucault* (Routledge Studies in Social and Political Thought) (Vol. 61, 1st ed.). New York: Routledge.

Kessler, O. (2012). Risk. In J. P. Burgess (Ed.), *The Routledge Handbook of New Security Studies* (pp. 17–26). London; New York, NY: Routledge.

Krause, K., & Williams, M. C. (1996). Broadening the Agenda of Security Studies: Politics and Methods. *Mershon International Studies Review, 40*(2), 229–254.

Krause, K., & Williams, M. C. (Eds.). (1997). *Critical Security Studies: Concepts and Cases* (Borderlines) (Vol. 8). Minneapolis: University of Minnesota Press.

Kurki, M. (2008). *Causation in International Relations: Reclaiming Causal Analysis* (Cambridge Studies in International Relations) (Vol. 108). Cambridge: Cambridge University Press.

Larrinaga, M. d., & Doucet, M. G. (2008). Sovereign Power and the Biopolitics of Human Security. *Security Dialogue, 39*, 517–537. https://doi.org/10.1177/0967010608096148.

Lee, J. R. (2009). *Climate Change and Armed Conflict: Hot and Cold Wars* (Routledge Studies in Peace and Conflict Resolution). London: Routledge.

Leiserowitz, A. (2005). American Risk Perceptions: Is Climate Change Dangerous? *Risk Analysis, 25*(6), 1433–1442.

Lemke, T. (2002). Foucault, Governmentality, and Critique. *Rethinking Marxism, 14*(Fall), 49–64.

Lie, J. H. S. (2015). Developmentality: Indirect Governance in the World Bank–Uganda Partnership. *Third World Quarterly, 36*, 723–740. https://doi.org/10.1080/01436597.2015.1024435.

Lippert, T. H. (2019). *NATO, Climate Change, and International Security: A Risk Governance Approach*. Cham: Palgrave Macmillan.

Lobo-Guerrero, L. (2007). Biopolitics of Specialist Risk: Kidnap and Ransom Insurance. *Security Dialogue, 38*(3), 315–334.

von Lucke, F. (2018). Linking Climate Change and Security in Mexico: Explorations into an Attempted Securitisation in the Global South. *Journal of International Relations and Development, 21*(2), 415–441.

von Lucke, F., Diez, T., & Wellmann, Z. (2014). What's at Stake in Securitising Climate Change? Towards a Differentiated Approach. *Geopolitics, 19*, 857–884. https://doi.org/10.1080/14650045.2014.913028.

Lukes, S. (2005). *Power: A Radical View* (2nd ed.). New York: Palgrave Macmillan.

Maertens, L. (2018). The Partial Climatisation of Migration, Security and Conflict. In S. Aykut, J. Foyer, & E. Morena (Eds.), *Globalising the Climate: COP21 and the Climatisation of Global Debates* (pp. 116–134). Abingdon, Oxon: Routledge.

Mathews, J. T. (1989). Redefining Security. *Foreign Affairs, 68*(2), 162–177.

Mawuko-Yevugah. (2010). *Governing through Developmentality: The Politics of International Aid Reform and the (Re)Production of Power, Neoliberalism and Neocolonial Interventions in Ghana.* Doctoral Thesis. University of Alberta, Edmonton.

McCormack, T. (2010). *Critique, Security and Power: The Political Limits to Emancipatory Approaches.* London: Routledge.

McDonald, M. (2005). *Discourses of Security and Environmental Change.* Paper presented at the International Studies Association (ISA) Conference in Hawaii, on March, 2005.

McDonald, M. (2008). Securitization and the Construction of Security. *European Journal of International Relations, 14*(4), 563–587.

McDonald, M. (2011). Deliberation and Resecuritization: Australia, Asylum-Seekers and the Normative Limits of the Copenhagen School. *Australian Journal of Political Science, 46*, 281–295. https://doi.org/10.1080/10361146.2011.568471.

McDonald, M. (2012). The Failed Securitization of Climate Change in Australia. *Australian Journal of Political Science, 47*(4), 579–592.

McDonald, M. (2013). Discourses of Climate Security. *Political Geography, 33*, 42–51. https://doi.org/10.1016/j.polgeo.2013.01.002.

Mckee, K. (2009). Post-Foucauldian Governmentality: What Does It Offer Critical Social Policy Analysis? *Critical Social Policy, 29*, 465–486. https://doi.org/10.1177/0261018309105180.

Methmann, C. (2011). The Sky Is the Limit: Global Warming as Global Governmentality. *European Journal of International Relations*, 1–23. https://doi.org/10.1177/1354066111415300.

Methmann, C. (2014). *We Are All Green Now: Hegemony, Governmentality and Fantasy in the Global Climate Polity.* Dissertation zur Erlangung des Grades eines Doktors der Philosophie. Hamburg: Universität Hamburg.

Methmann, C., & Rothe, D. (2012). Politics for the Day after Tomorrow: The Logic of Apocalypse in Global Climate Politics. *Security Dialogue, 43*(4), 323–344.

Milliken, J. (1999). The Study of Discourse in International Relations: A Critique of Research and Methods. *European Journal of International Relations, 5*(2), 225–254.

Mumme, S. P., & Lybecker, D. (2002). Environmental Capacity in Mexico: An Assessment. In H. Weidner, M. Jänicke, & H. Jörgens (Eds.), *Capacity Building in National Environmental Policy: A Comparative Study of 17 Countries* (pp. 311–327). Berlin; New York, NY: Springer.

Myers, N. (1995). *Environmental Exodus: An Emergent Crisis in the Global Arena*. Washington, DC: The Climate Institute.

Nagel, J. (2011). Climate Change, Public Opinion and the Military Security Complex. *The Sociological Quarterly, 52*, 203–211.

Neal, A. W. (2004). Cutting off the King's Head: Foucault's Society Must Be Defended and the Problem of Sovereignty. *Alternatives, 29*, 373–398. https://doi.org/10.1177/030437540402900401.

Nyman, J., & Burke, A. (Eds.). (2016). *Ethical Security Studies: A New Research Agenda* (Routledge Critical Security Studies Series). London: Routledge.

O'Brien, K., Eriksen, S., Nygaard, L. P., & Schjolden, A. (2007). Why Different Interpretations of Vulnerability Matter in Climate Change Discourses. *Climate Policy, 7*(1), 73–88.

Oels, A. (2009). *The "Vulnerable" as "Becoming Dangerous": The Relevance of a Foucaultian Framework for Studying the "Securitization" of Climate Change*. Paper presented at the workshop 'Climate Change, Social Stress and Violent Conflict: State of the Art and Research Needs', Universität Hamburg, 19/20 November 2009.

Oels, A. (2010). Die Gouvernementalität der internationalen Klimapolitik: Biomacht oder fortgeschritten liberales Regieren? In M. Voss (Ed.), *Der Klimawandel - Sozialwissenschaftliche Perspektiven* (pp. 171–186). Wiesbaden: VS-Verlag für Sozialwissenschaften.

Oels, A. (2011). Rendering Climate Change Governable by Risk: From Probability to Contingency. *Geoforum, 45*, 1–13. https://doi.org/10.1016/j.geoforum.2011.09.007.

Oels, A. (2012). From 'Securitization' of Climate Change to 'Climatization' of the Security Field: Comparing Three Theoretical Perspectives. In J. Scheffran, M. Brzoska, H.-G. Brauch, M. P. Link, & J. P. Schilling (Eds.), *Climate Change, Human Security and Violent Conflict. Challenges for Societal Stability* (Hexagon Series) (Vol. 8, pp. 185–205). Berlin: Springer.

Oels, A. (2013). Climate Security as Governmentality: From Precaution to Preparedness. In H. Bulkeley & J. Stripple (Eds.), *Governing the Global Climate: Rationality, Practice and Power* (pp. 197–216). Cambridge: Cambridge University Press.

Oels, A., & Carvalho, A. (2012). Wer hat Angst vor 'Klimaflüchtlingen'? Wie die mediale und politische Konstruktion des Klimawandels den politischen Handlungsspielraum strukturiert. In I. Neverla & M. S. Schäfer (Eds.), *Klimawandel in den Medien: Kommunikationswissenschaftliche Perspektiven* (pp. 253–276). VS-Verlag für Sozialwissenschaften.

Oels, A., & von Lucke, F. (2015). Gescheiterte Versicherheitlichung oder Sicherheit im Wandel: Hilft uns die Kopenhagener Schule beim Thema Klimawandel? *Zeitschrift für Internationale Beziehungen, 22*(1), 43–70.

Opitz, S. (2008). Zwischen Sicherheitsdispositiven und Securitization: Zur Analytik illiberaler Gouvernementalität. In P. Purtschert, K. Meyer, & Y. Winter (Eds.), *Gouvernementalität und Sicherheit: Zeitdiagnostische Beiträge im Anschluss an Foucault* (pp. 201–228). Bielefeld: transcript.

Oswald Spring, Ú., & Brauch, H. G. (2011). Coping with Global Environmental Change – Sustainability Revolution and Sustainable Peace. In H. G. Brauch, Ú. O. Spring, C. Mesjasz, J. Grin, P. Kameri-Mbote, B. Chourou, et al. (Eds.), *Coping with Global Environmental Change, Disasters and Security – Threats, Challenges, Vulnerabilities and Risks* (Hexagon Series on Human and Environmental Security and Peace) (Vol. 5, pp. 1487–1504). Berlin; Heidelberg; New York: Springer.

Park, J. (2000). Governing Climate Change Policy. In P. G. Harris (Ed.), *Climate Change and American Foreign Policy* (pp. 73–87). New York, NY: St. Martin's Press.

Pickett, B. L. (1996). Foucault and the Politics of Resistance. *Polity, 28*, 445–466. https://doi.org/10.2307/3235341.

Pielke, R. A., Jr. (2000a). Policy History of the US Global Change Research Program: Part I. Administrative Development. *Global Environmental Change, 10*, 9–25.

Pielke, R. A., Jr. (2000b). Policy History of the US Global Change Research Program: Part II. Legislative Process. *Global Environmental Change, 10*, 133–144.

Pirages, D. (1991). Social Evolution and Ecological Security. *Bulletin of Peace Proposals, 22*(3), 329–334.

Raleigh, C., & Urdal, H. (2007). Climate Change, Environmental Degradation and Armed Conflict. *Political Geography, 26*, 674–694.

Richert, J. (2009). *Klimawandel und Sicherheit in der Amerikanischen Politik* (Diskussionspapier). Berlin. Retrieved August 14, 2012, from http://www.swp-berlin.org/fileadmin/contents/products/arbeitspapiere/2009_DisP_Richert05_ks.pdf.

Rothe, D. (2011a). *Cleaning Foucault's Glasses: Problems and Blind-Spots of a Governmentality Approach to Global Climate Governance.* Paper prepared for the Workshop: Governing the Global Climate Polity: Rationality, Practice and Power, Lund University, Sweden, 19–21 June 2011.

Rothe, D. (2011b). Managing Climate Risks or Risking a Managerial Climate: State, Security and Governance in the International Climate Regime. *International Relations, 25*, 330–345. https://doi.org/10.1177/0047117811415486.

Rothe, D. (2016). *Securitizing Global Warming: A Climate of Complexity* (Routledge Studies in Resilience). Abingdon, Oxon: Routledge.

Salehyan, I. (2008). From Climate Change to Conflict? No Consensus Yet. *Journal of Peace Research, 45*, 315–326. https://doi.org/10.1177/0022343308088812.

Salter, M. B. (2011). When Securitization Fails: The Hard Case of Counter-Terrorism Programs. In T. Balzacq (Ed.), *Securitization Theory: How Security Problems Emerge and Dissolve* (PRIO New Security Studies). London: Routledge.

Scheffran, J., Brzoska, M., Brauch, H. G., Link, P. M., & Schilling, J. (Eds.). (2012a). *Climate Change, Human Security and Violent Conflict* (Hexagon Series on Human and Environmental Security and Peace). Berlin; Heidelberg: Springer.

Scheffran, J., Brzoska, M., Kominek, J., Link, P. M., & Schilling, J. (2012b). Disentangling the Climate-Conflict Nexus: Empirical and Theoretical Assessment of Vulnerabilities and Pathways. *Review of European Studies.* https://doi.org/10.5539/res.v4n5p1.

Scheffran, J., Brzoska, M., Kominek, J., Link, P. M., & Schilling, J. (2012c). Climate Change and Violent Conflict. *Science, 336*(6083), 869–871.

Schmitt, C. (1963). *Der Begriff des Politischen: Text von 1932 mit einem Vorwort und drei Corollarien.* Berlin: Duncker und Humblot.

Schreurs, M. A. (2002). *Environmental Politics in Japan, Germany, and the United States.* Cambridge: Cambridge University Press.

Scott, S. V. (2015). Implications of Climate Change for the UN Security Council: Mapping the Range of Potential Policy Responses. *International Affairs, 91*, 1317–1333. https://doi.org/10.1111/1468-2346.12455.

Scott, S. V., & Ku, C. (Eds.). (2018). *Climate Change and the UN Security Council.* Cheltenham: Edward Elgar.

Sindico, F. (2007). Climate Change: A Security (Council) Issue? *Carbon & Climate Law Review, 29*(1), 29–34.

Smith, D., & Vivekananda, J. (2007). *A Climate of Conflict: The Links between Climate Change, Peace and War.* London: International Alert.

Solana, J., & EU Commission. (2008). *Climate Change and International Security. Paper from the High Representative and the European Commission to the European Council (S113/08).* Brussels. Retrieved September 21, 2010, from http://www.consilium.europa.eu/uedocs/cms_data/docs/pressdata/en/reports/99387.pdf.

Stern, N. (2006). *Stern Review on the Economics of Climate Change.* London. Retrieved October 17, 2010, from http://siteresources.worldbank.org/INTINDONESIA/Resources/226271-1170911056314/3428109-1174614780539/SternReviewEng.pdf.

Stripple, J. (2002). Climate Change as a Security Issue. In E. Page & M. Redclift (Eds.), *Human Security and the Environment: International Comparisons* (pp. 105–127). Cheltenham: Edward Elgar.

Stritzel, H. (2007). Towards a Theory of Securitization: Copenhagen and Beyond. *European Journal of International Relations, 13*(3), 357–383.

Thunberg, G. (2019a, January 25). 'Our House Is on Fire': Greta Thunberg, 16, Urges Leaders to Act on Climate. *The Guardian*. Retrieved February 1, 2019, from https://www.theguardian.com/environment/2019/jan/25/our-house-is-on-fire-greta-thunberg16-urges-leaders-to-act-on-climate.

Thunberg, G. (2019b). Speech by Greta Thunberg, Climate Activist: Delivered at the Civil Society for rEUnaissance event at the European Economic and Social Committee. Retrieved March 18, 2019, from https://www.eesc.europa.eu/en/avdb/video/speech-greta-thunberg-climate-activist.

Trombetta, M. J. (2008). Environmental Security and Climate Change: Analysing the Discourse. *Cambridge Review of International Affairs, 21*(4), 585–602.

Trombetta, M. J. (2011). Rethinking the Securitization of the Environment: Old Beliefs, New Insights. In T. Balzacq (Ed.), *Securitization Theory: How Security Problems Emerge and Dissolve* (PRIO New Security Studies) (pp. 135–149). Abingdon, Oxon; New York: Routledge.

Trombetta, M. J. (2012). Climate Change and the Environmental Conflict Discourse. In J. Scheffran, M. Brzoska, H. G. Brauch, P. M. Link, & J. Schilling (Eds.), *Climate Change, Human Security and Violent Conflict* (Hexagon Series on Human and Environmental Security and Peace) (pp. 151–164). Berlin; Heidelberg: Springer.

Ullmann, R. (1983). Redefining Security. *International Security, 8*(1), 129–153.

UNDP. (1994). *Human Development Report 1994: New Dimensions of Human Security*. New York; Oxford: Oxford University Press.

UNDP. (2015). *Human Development Report 2015: Work for Human Development*. New York.

UNGA. (2009a). *Climate Change and Its Possible Security Implications*. Report of the Secretary-General, A/64/350 (2009). New York, NY.

UNGA. (2009b). *Resolution Adopted by the General Assembly: Climate Change and Its Possible Security Implications, A/RES/63/281*. New York, NY.

UNSC. (2007). *5663rd Meeting, Tuesday, 17 April 2007, 10 a.m.* New York, NY.

UNSC. (2011). *6587th Meeting, Wednesday, 20 July 2011, 10 a.m.* New York, NY. Retrieved February 28, 2013, from http://www.circleofblue.org/waternews/wp-content/uploads/2012/09/UN-Security-Council_climate-change-July-2011.pdf.

UNSC. (2013). *Press Conference on Impact of Climate Change on Marshall Islands*. United Nations Security Council, Department of Public Information. Retrieved December 13, 2013, from http://www.un.org/News/briefings/docs/2013/130215_MI.doc.htm.

UNSC. (2018). Deputy Secretary-General's Remarks at Security Council Debate on "Understanding and Addressing Climate-related Security Risks". Retrieved March 22, 2019, from https://www.un.org/sg/en/content/dsg/statement/2018-07-11/deputy-secretary-generals-remarks-security-council-debate-.

UNSC. (2019). Climate Change Recognized as 'Threat Multiplier', UN Security Council Debates Its Impact on Peace. Retrieved March 21, 2019, from https://news.un.org/en/story/2019/01/1031322.

Vuori, J. A. (2008). Illocutionary Logic and Strands of Securitization: Applying the Theory of Securitization to the Study of Non-Democratic Political Orders. *European Journal of International Relations, 14*(1), 65–99.

Walters, W., & Haahr, J. H. (2005). Governmentality and Political Studies. *European Political Science, 4*, 288–300. https://doi.org/10.1057/palgrave.eps.2210038.

WBGU. (2008). *World in Transition: Climate Change as a Security Risk*. Berlin; London; Sterling: Earthscan.

Weber, M. (1976). *Wirtschaft und Gesellschaft. Grundriss der Verstehenden Soziologie*. Tübingen: Mohr.

Weidner, H., & Mez, L. (2008). German Climate Change Policy: A Success Story With Some Flaws. *The Journal of Environment & Development, 17*, 356–378. https://doi.org/10.1177/1070496508325910.

Wiener, A. (2007). The Dual Quality of Norms and Governance beyond the State: Sociological and Normative Approaches to 'Interaction'. *Critical Review of International Social and Political Philosophy, 10*(1), 47–69.

Wiener, A. (2008). *The Invisible Constitution of Politics: Contested Norms and International Encounters*. Cambridge; New York, NY: Cambridge University Press.

Williams, M. C. (2003). Words, Images, Enemies: Securitization and International Politics. *International Studies Quarterly, 47*, 511–531.

World Bank, PIK, & Climate Analytics. (2013). *Turn Down the Heat: Climate Extremes, Regional Impacts, and the Case for Resilience*. Washington, DC.

Wyn Jones, R. (2005). On Emancipation: Necessity, Capacity, and Concrete Utopias. In K. Booth (Ed.), *Critical Security Studies and World Politics* (Critical Security Studies) (pp. 215–236). Boulder, CO: Lynne Rienner.

Yee, A. S. (1996). The Causal Effects of Ideas on Policies. *International Organization, 50*(1), 69–108.

CHAPTER 2

United States: Climate Change, National Security and the Climatisation of the Defence Sector

Introduction

National security leaders now recognize that global climate change is a matter of national security and may even be a defining security challenge of the 21st century. (Rogers and Gulledge 2010, p. 7)

The representation of climate change has undergone a decisive transformation since the issue began to be more important in the US in the late 1980s. At first only being one amongst other environmental problems and discussed mostly in scientific and expert communities, the issue soon rose to become the most important environmental concern. Eventually, this has culminated in the construction of climate change as an important national security issue that necessitates far-reaching changes in the planning of the US defence sector. The following chapter traces back this development from 1989 to 2015, shows how climate change has been articulated as a security issue within three climate security discourses and discusses resulting political responses.

The US is a crucial case when it comes to political responses to climate change. It is the biggest contributor of historic CO_2 emissions (Ge et al. 2014), the second biggest contemporary emitter (Dröge 2016, p. 11) and has one of the highest per capita emissions (World Bank 2013; Dröge 2016, p. 17). This book includes the US as a case study of an industrialised country that is heavily engaged in the international negotiations and has

F. von Lucke, *The Securitisation of Climate Change and the Governmentalisation of Security*, New Security Challenges,
https://doi.org/10.1007/978-3-030-50906-4_2

undergone decisive transformations concerning its domestic and international climate policies. While it was a vanguard in the early stages of the international negotiations, this position shifted during the end of the 1990s towards a strong laggard position, which only was mediated somewhat under the second term of Barack Obamas Presidency.

The remainder of this chapter shows that the securitisation of climate change has had a considerable influence on how political practitioners in the US perceived climate change and which political practices and solutions have been legitimised. Empirically, this case rests on a discourse analysis of over 50 publications of relevant climate change actors among think tanks, and in civil society and government institutions. Furthermore, it includes an analysis of 270 parliamentary debates on climate change that contain securitising articulations and insights from 20 semi-structured informal interviews with think tanks, NGOs and state representatives in the US.

The Origins of the Climate and Security Debate in the US

Problematising environmental issues as security threats first became more widespread in the 1960s, in the beginning often in connection to nuclear weapons testing and the 'nuclear winter hypothesis' (Dalby 2009, pp. 37–39). A decade later, 'environmental security' became an increasingly important concern on the political level. Although the discussion at this time was still broad and revolved around issues such as resource scarcity, overpopulation (Ehrlich 1968) and the limits to growth (Meadows and Meadows 1972), climate change—at that time mostly termed 'global warming'—was already one of the concerns (Dalby 2009, p. 14; Brown 1977, p. 7). Based on these discussions and disturbing projections of the US National Academy of Sciences about climate change (National Academy of Sciences 1975, 1977), at the end of the 1970s, the first voices appeared that explicitly depicted environmental problems and climate change as threats to 'national security' (Brown 1977; see also Floyd 2010, p. 67). Hence, they were the very first precursors of the sovereign discourse that later would become the defining articulation of climate security in the US.

While the inclusion of climate change into this early environmental security debate certainly was not the only reason, it nevertheless falls into the same time when climate change in general began to be more important in scientific and political discussions in the US. Scientists had already discussed climate change as one of the more important problems in the

1950s (Park 2000, p. 79), but it was not until the 1970s that political actors picked up the issue on a broader basis and adopted first tangible policies on the matter. In 1978, the National Program on Climate Change was established as one of the earliest concrete policies in this sector (Pielke Jr 2000a, p. 12). In the 1980s, climate change gained more relevance on the domestic political level and several actors in Congress, particularly the Democrats Albert 'Al' Gore and Timothy Wirth and the Republican John Chaffee, brought the issue onto the agenda (Pielke Jr 2000a, p. 16; Park 2000, p. 80; Bryner 2000, p. 112). The increasing interest in climate matters eventually had further concrete policy implications such as the 1987 Global Climate Change Protection Act (US Congress 1988) and the creation of the Committee on Earth and Environmental Sciences in 1990. Finally, one of the most influential decisions was the creation of the US Global Change Program (USGCRP 2015; Pielke Jr 2000b, pp. 134–136).

When the end of the Cold War left the US security establishment without clear adversary (Dalby 2002, pp. 16–17; Dabelko et al. 1995, p. 4; Sperling 2010, pp. 174, 201) and hence opened up spaces for 'new discourses of danger' (Harris 2002, pp. 150–151; Floyd 2010, p. 65; Campbell 1992, p. 57), environmental security in general became a central concern in the US debate. Thus, at the end of the 1980s and early 1990s, climate security articulations became more common in Congressional debates. Climate change also increasingly became an important issue in the US-American public, not least due to the record summer of 1988 and a dire warning by NASA scientist James Hansen (Leiserowitz 2005, p. 1435; Grundmann and Scott 2014, p. 222). Thus, when the international negotiations about a global climate regime, including the creation of the IPCC in 1988 and the UNFCCC in 1992, gained momentum in the late 1980s, climate change already was on the domestic agenda and the US was highly engaged in the beginnings of the global regime (Donner and Faltin 2007, p. 5; Harris 2000b, p. 16).

Building on this historical context, the next section takes a closer look at the discourses that shaped the US climate security debate between 1989 and 2015. The aim is to understand to what extent the particular discursive constellation has transformed the broader climate debate and thus has rendered climate change governable as a specific security issue.

The Sovereign Discourse: Climatising the Defence Sector

The most common and intensively debated framing in the US debate was the sovereign discourse that constructed climate change as a threat to US national security. In the late 1980s, sovereign articulations appeared specifically in Congressional debates. They were particularly strong in the 101st Congress (1989–1990) at a time when the international climate negotiations gained momentum and climate change in general became more important in US debates (US Senate 1989a, pp. S5252; US House of Representatives 1990, pp. H7684). The leading proponents of these early climate security debates were Gore, Joe Liebermann and John Kerry (US Senate 1989a, pp. S5252, 1992a, pp. S6275, 1992b, pp. S7511) who later would regularly reappear as sponsors of climate (security) bills. Especially Gore repeatedly highlighted the potential of climate change and ozone depletion to threaten US national security as well as the international order:

> However, the greenhouse effect and stratospheric ozone depletion fit the profile of national security issues of global significance. The phenomena certainly will in time produce effects big enough to threaten international order, even at the level of war and peace. (US Senate 1989a, pp. S5252ff)

In contrast to the later debate that focused on the defence sector (Interview 2014j), the principal aim of these early articulations of the sovereign discourse was to increase the attention for environmental issues and to reroute funding to the environmental policy, as this quote from Democratic Senator Claiborne Pell exemplifies:

> It is time that we recognize that the threat to our national security posed by environmental degradation is just as great in consequence, if not in immediacy, as global war. In the United States, we currently spend $296.5 billion for national defense and $5.6 billion at the Federal level, on environmental protection. While it is not realistic to expect these numbers to be reversed, they can be brought into better balance. We must understand that both are expenditures to preserve our national security. (US Senate 1990b, pp. S7410)

After this short peak period at the end of the 1980s, the sovereign discourse—as well as the environmental and climate security debate in general—lost in importance and their articulation decreased considerably in the debates of the 102nd (1991–1992) and 103rd (1993–1994) Congress.

It was not until the Clinton/Gore administration had taken over in 1993 and when climate change (Harris 2000c, pp. 38–43; Falkner 2005, p. 592; Vig 2013, p. 91) as well as environmental security resurfaced as key topics (Allenby 2000, p. 10; Matthew 2013, p. 346) that the sovereign discourse reappeared on a broader scale. Once again, the academic debate played an important role. Works such as Robert Kaplan's 'The Coming Anarchy' and Homer-Dixon's inquiries into environmental scarcity and conflict (Homer-Dixon 1994) were well received in the highest political circles (Sherbinin 1995, p. 31). In 1994, Gore even invited Homer-Dixon to Washington, DC, to brief him on environmental and climate security (Floyd 2010, p. 75). Subsequently, members of the Clinton administration linked the environment and climate change to national security at various occasions, mostly with the aim to improve environmental policy (Simmons 1995, p. 2; Sherbinin 1995, pp. 29–30).

At the same time, members of the 104th (1995–1996) and 105th (1997–1998) Congress again began to refer to the sovereign discourse more frequently. In many Congressional debates, the direct connection to defence planning played a more important role than in the articulations of the Clinton administration. For instance, climate change was discussed in relation to the melting poles and the effects on the US submarine strategy (US House of Representatives 1995, pp. H9946). The heated debates about the Kyoto Protocol in the late 1990s as well sparked more references to the sovereign discourse and a few Members of Congress presented signing the protocol as serving US national security:

> Indeed, the Kyoto Protocol will improve the national security of the United States by reducing the risk of catastrophic climate change, which would create upheaval and unrest throughout the world, including the potential for millions of environmental refugees. (US House of Representatives 1998a, pp. H3577)

Yet, in line with the decreasing general interest in climate issues at the end of the 1990s, the sovereign discourse increasingly lost in relevance and almost vanished during George W. Bush's first term in office (2001–2005).

The Think Tank Driven Sovereign Transformation of the Debate

The first precursor of the comeback of the sovereign discourse was the publication of a Department of Defense (DOD) report on the national security consequences of abrupt changes in 2003—later referred to as 'Pentagon-Report' (Schwartz and Randall 2003). Its core argument was that 'although uncertain and quite possibly small' the consequences of climate change 'should be elevated beyond a scientific debate to a U.S. national security concern' (Schwartz and Randall 2003, p. 3). The report thus exemplified the beginning of a far-reaching transformation of climate change from a mainly scientific and environmental concern—that might have some side effects on national security—, to a real and immediate defence issue that has to be taken seriously by the appropriate institu tions in this political sector.

The immediate resonance of the DOD report was moderate, among other things because its very drastic scenarios were seen as too speculative and implausible (Campbell et al. 2007, p. 19; Interview 2014i) but also because of its publication during the government of the climate sceptical Bush administration (Fincham 2014; Townsend and Harris 2004). Nonetheless, it prepared the ground for the success of this line of argumentation in the years to follow. It had reintroduced a specific sovereign argumentation, which enabled and empowered actors from the traditional security and defence sector—especially think tanks—to participate in the debate. The proponents of this discourse genuinely believed that climate change constituted a tangible national security problem and that US defence policy had to be adjusted accordingly (Interview 2014c, 2014e, 2014i). In addition, however, they also wanted to improve climate policy and used this angle in a tactical manner to put pressure on the Bush administration (Albright et al. 2006, p. 7) and to overcome the political standstill and polarisation between liberals and conservatives concerning climate change.

Particularly from 2007 onwards, the sovereign discourse clearly began to dominate the US debate. Under the lead of Bill Clintons former Deputy Under Secretary of Defense for Installations and Environment, Sherri Goodman, the semi-governmental think tank CNA had assembled a military advisory board (MAB) of high ranking retired military officials (Interview 2014i, 2014e). In 2007 they published a report on the national security implications of climate change (CNA 2007) and thereby initiated a whole series of think tank reports that increasingly dominated the climate security debate in the US. Articulations of the sovereign discourse appeared in all of the 28 analysed central think tank reports and are the dominant argumentation in most of them. Besides CNA, the most influential organisations were: Center for Strategic & International Studies (CSIS), the Center for a New American Security (CNAS), the Council on Foreign Relations (CFR), the American Security Project (ASP), the Center for American Progress (CAP), the Center for Climate Change Energy and Security (C2ES), and the Center for Climate and Security (CCS).

Besides articulating the same defence-oriented sovereign discourse and hence approximating a form of 'discourse coalition' (Hajer 1997), these think tanks were linked together through various personal connections. For instance, important individuals such as Kurt Campbell, Sherri Goodman and John Kerry—who later became Secretary of State under President Obama (Davenport 2014a)—founded or worked at several of these think tanks and often personally knew each other (Interview 2014d). Beyond that, from 2010 onwards, the Center for Climate Change and Security (CCS) established itself as an important convenor between all actors involved in the sovereign security focused climate debate and organised periodic meetings between all relevant actors (Interview 2014a). Finally, a few influential foundations, seeing a window of opportunity to increase the attention for climate change through this specific framing, played an important role in financing many of the think tank projects on climate security. Examples are the Energy Foundation, the Rockefeller Foundation and the Skoll Global Threats Fund (Interview 2014a, 2014d, 2014i).

Constructing Climate Change as a Threat to US National Security

These reports problematised climate change as a 'serious threat to the national security of the United States' (CNA 2007, p. 6; Vagg 2012, p. 1) and hence constructed it as a direct threat to the well-being and even survival of the sovereign. This often included a direct focus on possible military threats arising from climate change and the respective consequences to be taken in US defence strategies (CNA 2007, p. 37). Concerning the specific threat construction, the focus was mostly on the second-order socio-economic and political effects of climate change. Climate change would 'exacerbate regional and local tensions in 'hot-zones' around the world' (Foley and Holland 2012c, p. 11) and hence increase the demands for US civilian but also military interventions (Rogers and Gulledge 2010,

p. 16). A recurring element, originally coined by the 2007 CNA report, was the phrasing of climate change as a 'threat multiplier' (CNA 2007, p. 32) that destabilises states and regions and contributes to more fragile and failed states around the world (CNA 2007, pp. 13, 20; Rogers and Gulledge 2010, p. 16). This led to the conclusion that climate change would act as an 'accelerant for instability' (Foley and Holland 2012c, p. 12) and contribute to violent conflicts, resource wars—especially over water (Campbell et al. 2007, p. 16)—and terrorism around the globe (CNA 2007, p. 13). At the end of the day, even nuclear war was considered as an indirect consequence of climate change (Campbell et al. 2007, p. 78).

Within this sovereign representation, climate change progressively ceased to be an environmental problem and instead became a fairly traditional security issue with severe consequences for the security of nations and global peace and stability (Campbell et al. 2007, p. 5). Consequently, reports feared that it would facilitate confrontational geopolitics and eventually worsen the neorealist 'security dilemma'. In this vein, alliances might weaken and it would become increasingly difficult for states to 'look beyond their own salvation' (Campbell et al. 2007, p. 77). One report even compared a world hit by severe climate change with the Cold War and feared a new 'heat war' (Burke and Parthermore 2008, p. 5).

While the prevailing sovereign discourse constructed developing countries as being hit first and hardest by the security implications of climate change, it drew a direct connection to the national security of industrialised countries such as the US (CNA 2007, p. 32) and eventually predicted a destabilisation of the entire global security architecture. A central mechanism for the spread of the security effects of climate change across the globe was the prediction of millions of climate refugees potentially causing border conflicts or struggles over scarce resources: 'Some migrations cross international borders. Environmental degradation can fuel migrations in less developed countries, and these migrations can lead to international political conflict' (CNA 2007, p. 18). Increased global instability and the further weakening of instability hotspots around the world could eventually directly affect US national security interests (Werz and Manlove 2009, pp. 1–3; Campbell et al. 2007, p. 8). In this dire future, according to some reports resembling the post-apocalyptic world of the 'Mad Max' movies (Campbell et al. 2007, p. 9), the demands for the US military would increase considerably (Rogers and Gulledge 2010, p. 16). Due to its global reach, humanitarian interventions and disaster aid as well as robust stability and peacekeeping missions would become more common and more dangerous (McGrady et al. 2010, p. 3). Climate change would lead to an increase in 'complex emergencies' (CNA and Oxfam 2011, p. 1), meaning situations in which economic, political, humanitarian and security problems arise simultaneously, thus posing new challenges for US interventions. In this vein, many reports presented the violent conflicts in the Darfur region and in Somalia as contemporary examples of a world ravaged by unchecked climate change (C2ES 2009, p. 2; CNA 2007, p. 15; Campbell et al. 2007, p. 5).

Eventually, the US military could be forced to intervene more often to keep the problems from spiralling out of control and to secure domestic energy demand (Schwartz and Randall 2003, p. 14). Hence, this argumentation saw direct consequences for '(...) the way the Department of Defense (DOD) does business' (Carmen et al. 2010, p. 1) and strongly advised the appropriate US institutions to incorporate climate change into their defence planning: 'The U.S. military and other segments of the U.S. national security community have begun to recognize climate change as a threat multiplier that must be considered in long-term security planning' (C2ES 2009, p. 2). Beyond more missions abroad, the sovereign discourse also constructed the US military and its physical infrastructure as threatened by climate change (Foley and Holland 2012b, p. 47), for instance, through sea-level rise (Foley 2012, p. 1), extreme weather and heat waves (Carmen et al. 2010, p. 1; Foley 2012; McGrady et al. 2010).

Sovereign Recommendations: Climate Proofing the Defence Sector

To tackle the presented threats all reports called for decisive mitigation measures including laws for improved energy and fuel efficiency and a greening of the US economy and the US military (CNA 2007, p. 45; CNA Military Advisory Board 2014, p. 24; Carmen et al. 2010, p. 1). Moreover, they recommended multilateral cooperation, US leadership in the international negotiations and more ambitious and robust commitments to GHG cuts (CNA 2007, p. 23; Campbell et al. 2007, p. 19). However, in contrast to earlier articulations, the specific national security framing was not primarily supposed to increase attention and funding for an intrinsically environmental problem and for tackling its root causes. Instead, climate change was constructed as a defence issue that had to be tackled directly and immediately by appropriate (adaptation) measures in this sector:

> Because the links between climate change and national security are worthy of concern in their own right, and because some significant climate change is inevitable, strategies that go beyond long-run efforts to rein in greenhouse gas emissions are required. (Busby 2007, p. 2)

Thus, most reports demanded increased security and military planning and a preparation of the military for climate change (CNA 2007, p. 16). Consequently, they often directly addressed the DOD (Rogers and Gulledge 2010, p. 9), the intelligence sector (CNA 2007, p. 23) and the Armed Forces (Carmen et al. 2010). Most reports recommended to integrate climate change into the overarching planning schemes of the security and defence sector such as the National Security Strategies (NSS), the National Defense Strategies, the Military Strategies, the Quadrennial Defense Reviews (QDR), and the National Intelligence Estimates (CNA 2007, p. 46; Parthermore and Rogers 2010) and called for far-reaching institutional reforms (CNA 2007, p. 20). This included preparing military bases and equipment for climate change but also improving training exercises and scenario planning to prepare for more and different missions concerning natural disaster relief as well as stability operations (CNA 2007, p. 16; McGrady et al. 2010, p. 5; Campbell et al. 2007, p. 108; Werz and Conley 2012, p. 2). Beyond that, reports recommended to prepare critical infrastructure throughout the US for the impacts of climate change (CNA 2007, pp. 37, 48). Moreover, to tackle mass-migration, some reports suggested to strengthen border security in the US (Campbell et al. 2007, p. 59).

To facilitate these measures, a common demand within this discourse was the need for more detailed and 'actionable' (Parthermore and Rogers 2010, p. 9; Rogers and Gulledge 2010, p. 41) social-scientific data concerning the security implications of climate change and for a better cooperation between climate science and defence actors (Parthermore and Rogers 2010, p. 37; Carmen et al. 2010, p. 6). Thus, on the one hand, the securitisation of climate change facilitated what others have called a 'climatisation' (Oels 2012) of the security sector. On the other hand, this had also consequences for climate and social scientific knowledge production, which was expected to deliver appropriate security data. This contributed to the creation of new security concepts at the intersections of traditional military and new or civilian understandings of the term such as 'networked security' and 'sustainable security' (Rogers and Gulledge 2010, p. 8; Werz and Conley 2012, pp. 33–34). Ultimately, this exemplifies the bidirectional qualities of securitisation, which will be discussed in more detail in the concluding Chap. 5.

The Sovereign Discourse in Parliamentary Debates and High-Level Political Discussions

Notwithstanding the central role of think tanks in this phase, the US Congress soon began to pick up the defence-oriented sovereign argumentation as well and it became the dominant discourse between 2007 and 2012. The tenor of the debates resembled the think tank reports and represented climate change as a serious driver of conflicts around the world (US Senate 2010, pp. S10998), which would eventually directly affect the national security of the US (US Senate 2008a, pp. S4868), as this quote from Republican Senator John Warner exemplifies:

> These serious implications of climate change will have security consequences for the United States. For example, there will be an increased potential for failed nations and growth of global terrorism. Another serious implication of climate change is the mass migrations of people that are likely to occur. [...]. Climate change will add stress to our weapons system, threaten U.S. bases throughout the world, and have a direct effect on military readiness. (US Senate 2008a, pp. S4885)

At various occasions, the debates mentioned or even directly quoted the think tank reports or documents from the defence and intelligence sector. In particular, they regularly referred to the 2003 DOD study (US Senate 2007b, pp. S4990; US House of Representatives 2009b, pp. H8481), to the 2007 CNA report (US Senate 2008a, pp. S4868), to the 2007 CSIS/CNAS report (US Senate 2008b, pp. S4990) and to assessments of US intelligence organisations (US Senate 2010, pp. S10998). Consequently, many Members of Congress adopted the exact think tank phrasing. For instance, climate change as a 'threat multiplier' (US Senate 2008b, pp. S4989; CNA 2007, p. 1), the notion that 'we are now in an age of consequences regarding the foreign policy and national security implications of global climate change' (US Senate 2008b, pp. S4990; Campbell et al. 2007) and the connection of climate change to 'the spread of terrorism and failed states' (US Senate 2008a, pp. S4885; CNA 2007, p. 1). Moreover, they also often pointed to the Darfur crisis as an example of the security implications of climate change in the future (US Senate 2007b, pp. S13502).

Beyond the adoption of certain arguments, Members of Congress directly acknowledged the discursive transformation of climate change from an environmental towards a defence issue, as this quote from Republican Senator Elizabeth Dole exemplifies: 'I understand this bill is viewed by most as an environmental bill—which it is—but it is also essential to our national security. [...]' (US Senate 2008b, pp. S4989; US House of Representatives 2009b, pp. H8481). Accordingly, they also openly stressed that the connection between climate change and national security is not drawn from long-known environmentalists but from actors in the defence sector, which considerably increased the weight of the argument (Interview 2014c, 2014j), as this quote from Democratic Senator Barbara Boxer exemplifies:

> This is all from retired admirals and generals. This is not from BARBARA BOXER. This isn't from Al Gore. This isn't from MARIA CANTWELL. This isn't from Senator WARNER. It isn't from Senator LIEBERMAN. This is from our own retired admirals and generals [...]. (US Senate 2007b, pp. S13502)

Besides Congress, the leadership of the DOD as well as members of the Obama administration began to refer to climate change as a serious threat to US national security (Interview 2014k). One of the most influential articulations came in 2013 from Navy Admiral Samuel L. Locklear (chief of the US Pacific forces), who described climate change as the 'biggest long-term security threat in the Pacific region' (Bender 2013). Whereas Hillary Clinton as Secretary of State (2009 to 2013) only occasionally drew the connection, her successor John Kerry (2013–2017) (Davenport 2014a, 2014b) and both Secretaries of Defense, Leon Panetta (2011–2013) (Munoz 2012) and Chuck Hagel (2013–2015) (Davenport 2014c; Bendery 2014), were quite outspoken on the national security implications of climate change. Eventually, in May 2015, the White House compiled a report that summarised all mentions of climate change as national security issue in other federal reports (The White House 2015a).

The prevalence of the defence-oriented sovereign discourse also transformed the political landscape in terms of who was accepted as legitimate actor to speak climate security. While the above-mentioned think tanks have been a driving force themselves in advancing this discourse, they also increasingly were empowered by this specific framing. The focus on national security and on the military and defence sector revalued their expertise in security policy and at the same time almost closed down the debate for actors with a more environmentalist background such as Greenpeace, World Wildlife Fund (WWF), climate scientists or Gore (see also Grundmann and Scott 2014, p. 227). These actors would have liked to use climate security argumentations (and in the case of Gore did so for some time) but eventually realised that they could not compete with the security think tanks, at least not concerning the dominant sovereign discourse (Interview 2014g). On top of that, the liberal image of environmental groups and their linkages to climate science would have hurt the whole argumentation, which was supposed to overcome differences between conservatives and liberals concerning climate change (Interview 2014a, 2014c, 2014f, 2014g).

In general, Republicans were much less likely to believe in scientific argumentations concerning climate change (McCright and Dunlap 2011), which was reinforced by liberal bogey man Gore arguing extensively with climate science in the past (Gibbs and Lyall 2007). Thus, a further important role of the sovereign discourse and the discursive coalition of think tanks was to provide 'cover' for politicians that publicly spoke on climate change. Because of the progressively dominant representation of climate change as a threat to national security, it became much easier for politicians to talk about the issue without referring to traditional climate science and without being accused of as liberal environmentalists (Interview 2014i). The political weight of national security and the favourable reputation of actors from the security sector and the US military (Gallup 2014; US Senate 2007b, pp. S13502; Interview 2014j) contributed to move the climate debate, at least partly, away from the polarised partisan issue it used to be in the first half of the 2000s.

The Disciplinary Discourse: Recoding Human Security

Besides constructing climate change as a threat to US national security, the disciplinary discourse with its focus on the human security implications of climate change was the second-most important representation in the US. In the late 1980s, the disciplinary discourse above all appeared in Congressional debates and proponents often highlighted the serious threats of climate change for the human well-being to legitimise US leadership in the negotiations about the UNFCCC:

> Potential climate change presents such a serious threat to human well-being throughout the world, that the United States should undertake urgent action to support and encourage negotiations necessary to bring about a framework convention for international cooperation on limiting the emission of greenhouse gases [...]. (US Senate 1990c)

However, after this short-lived peak around 1990, the discourse lost in importance in the 103rd and 104th Congress between 1993 and 1996. While the Clinton/Gore administration occasionally pointed to it to legitimise their environmental agenda, it was not until 1997 that it became more influential again, especially in the wake of the debates about the Kyoto Protocol (US Senate 1997a, pp. S10872). While the human security of people around the world kept playing a role (US Senate 2001b, pp. S8895), Members of Congress now also began to focus on the direct implications for US citizens (US House of Representatives 1998b, pp. H6224): 'Global warming is not just a global challenge; it is also a very local one, impacting lives of Americans in critical and potentially disastrous ways' (US Senate 2003b, pp. S13485).

During the early 2000s, some proponents of the earlier environmental security debates reappeared and actively participated in the construction of climate change as a threat to human security, amongst others Lieberman (US Senate 2001b, pp. S8895, 2003b, pp. S13485) and Kerry (US Senate 2001a, pp. S2301). Beyond these two, particularly Gore kept articulating the disciplinary discourse and was able to reach a broad audience within and outside the US (Grundmann and Scott 2014, p. 229; Gore 2007; Guber and Bosso 2013, p. 54).

The Ambiguity of Human Security and Links to the Sovereign Discourse in Think Tank Reports

Besides these Congressional debates, think tank reports also contained numerous references to the disciplinary discourse. Most reports acknowledged that: 'Security cannot be defined in purely military terms. Instead, governments must prepare for threats to the security of people, not just states' (Foley and Holland 2012b, p. 2). In this vein, they constructed threats to the human security of people in developing countries (Werz and

Manlove 2009, p. 5) and especially in Africa (Stern and Antholis 2007, p. 176; Werz and Conley 2012, p. 3) arguing that they would be hit 'first and also most deeply' (Campbell et al. 2007, pp. 7, 56, 72) and therefore were in need of help from industrialised countries. Concerning the concrete threats, the think tank reports especially emphasised the physical effects of climate change on the daily lives of individuals. Examples were land degradation, water scarcity, the spread of diseases, an increase in bush fires (McGrady et al. 2010, p. 20; CNA and Oxfam 2011, p. 5) and also the increase in natural disasters, such as severe storms, floods or droughts, which would threaten millions of people around the world (CNA 2007, p. 6).

On the one hand, these articulations of the disciplinary discourse refocused the attention on the problems for individuals and away from solely militaristic or statist understandings and hence legitimised support for people directly affected by climate change (Campbell et al. 2007, p. 106). However, resting on the concept of 'normation' inherent to disciplinary power, this specific construction of climate threats also created a dichotomy between an ideal norm and (problematic and dangerous) deviations from it. Thus, while industrialised countries (and their inhabitants) such as the US were constructed as sufficiently prepared and 'able to cope with such change' (CNA 2007, p. 16), poor and underdeveloped nations and people were framed as 'greatly challenged' (CNA 2007, p. 16) and believed to have 'less stamina to deal with climate change' (Campbell et al. 2007, p. 7). In this argumentation, it is not climate change itself that determines the level of the threat to human security, but the insufficient coping capacity of affected populations:

> Poverty and underdevelopment are key factors in South Asia's vulnerability to climate change because they reduce people's capacity to cope with large-scale disasters as well as adapt for future disruptions. (Foley and Holland 2012b, p. 7)

Consequently, to reduce this deviation from the norm, reports first and foremost pointed to external interventions conducted by industrialised countries, thereby reinforcing existing dependencies of developing countries on external help (Campbell et al. 2007, p. 42). Closely connected, to identify people that deviate from the norm and to transform their behaviour, the disciplinary argumentation legitimised the increased surveillance of people in high-risk countries. Thus, many reports urged to establish 'early warning systems' (Werz and Conley 2012, p. 18) and 'monitoring activities' in affected countries (CNA 2007, p. 23).

This disciplinary argumentation also increased the pressure on developing countries to mitigate and adapt to climate change themselves. It hence added to existing 'Western' demands to behave in a specific manner such as democratisation or economic liberalisation. In line with the productive nature of disciplinary power, these demands were mostly connected to calls for increased support of industrialised countries (CNA and Oxfam 2011, p. 15; Werz and Conley 2012, p. 22). However, together with several remarks on the rapidly increasing GHG contributions of developing countries (Campbell et al. 2007, pp. 84, 97), it also deflected the attention away from the responsibility of industrialised countries to first of all mitigate themselves to tackle the root causes of climate change.

Especially in the US debate, pointing towards the human security implications of climate change in the Global South often directly intersected with the interests of industrialised countries. Accordingly, many reports connected the increased vulnerability of poor people towards the direct physical effects of climate change to threats for industrialised countries and their national security (CNA and Oxfam 2011, p. 18; Oels 2011), as this quote from the 2007 CNA report exemplifies:

> While the developed world will be far better equipped to deal with the effects of climate change, some of the poorest regions may be affected most. This gap can potentially provide an avenue for extremist ideologies and create the conditions for terrorism. (CNA 2007, p. 13)

Thus, while human security argumentations did not disappear with the rise of the sovereign discourse from 2007 onwards, they were increasingly recoded to eventually strengthen the dominant sovereign argumentation and to legitimise sovereign interventions by the military, which many saw as especially well equipped to deal with climate-induced crisis (McGrady et al. 2010, p. 9).

Ultimately, many articulations of the disciplinary discourse framed threats to human security as the first element of a chain reaction, which overburdened local governments (Rogers and Gulledge 2010, p. 15) and eventually led to threats to the US national security and required a response from the US military:

> A health emergency involving large numbers of casualties and deaths from disease can quickly expand into a major regional or global security challenge that may require military support, ranging from distribution of vaccines to full-scale stability operations. (CNA 2007, p. 15)

The reports often envisioned these interventions as humanitarian measures and as linked to developed cooperation (Busby 2007, p. 15; CNA and Oxfam 2011, p. 3). Yet, referring to the responsibility to protect, they also legitimised military interventions in states such as Sudan that were deemed unable to protect their citizens (see also McCormack 2010; Hartmann 2010, p. 241). The fear was that these countries could become the nucleus of large-scale instability and threats to the global order or the national security of industrialised countries (CNA 2007, p. 15).

Thus, although starting out from the other end of the continuum, the disciplinary discourse has also advanced concepts such as 'networked security' or 'sustainable security' (Werz and Manlove 2009, p. 5) and helped to facilitate a merging of climate, development and disaster response measures with military- and security-centred approaches (Werz and Conley 2012, pp. 5–34; CNA and Oxfam 2011, p. 3; McGrady et al. 2010, p. 3).

Combining National and Human Security and Focusing on US Citizens in Congressional Debates

Echoing the think tank reports, articulations of the disciplinary discourse in the US Congress often combined human and national security, especially after 2007. The physical effects of climate change were constructed as a threat to the human security of people around the world (US Senate 2008f, pp. S5341), which eventually could have geopolitical repercussions for US national security (US Senate 2008a, pp. S4885):

> A very large percentage of the poor people in the world live at or below sea level. The effect that rising tides will have in disrupting their lives, we should understand will have a very significant geopolitical implication. (US House of Representatives 2014, pp. H5346)

In addition, Congressional debates increasingly constructed the human security of US citizens as threatened by climate change (US House of Representatives 2014, pp. H5346). This refocusing towards the US populations became more common in the aftermath of hurricane Katrina in

2005 and particularly pronounced after 2010 (US House of Representatives 2011, pp. H2351, 2014, pp. H5346).

One important trigger for this change was that politicians and media reports had begun to link extreme weather in the US such as the 2012 'superstorm' Sandy to climate change (Brulle et al. 2012; Bloomberg 2013; US House of Representatives 2013, pp. H1895). Beyond that, the proponents of the climate security debate also saw a tactical advantage in highlighting the direct security implications of climate change for US citizens because the US American public was increasingly losing interest in problems that only affected distant peoples or futures (Leiserowitz 2005, p. 1437; Interview 2014j; CNA Military Advisory Board 2014; Melillo et al. 2014). As a consequence of this shift, demands for improved disaster management capabilities of the Department of Homeland Security (DHS) and the Federal Emergency Management Agency (FEMA) increased (Werz and Conley 2012, p. 9; CNA 2007, p. 7; Foley and Holland 2012b).

The Governmental Discourse: Introducing Risk Management and 'Black Swans'

While being least common as a standalone discourse, linking climate change to a range of risk conceptions always played a role in the US debate, often in direct connection with the other two discourses. In the 1990s, traces of the governmental discourse appeared on a regular basis, often in connection to the planet as a whole, to the environment and to human security. The key argumentation was that climate change would slowly but constantly increase the risks for adverse agricultural changes and humanitarian disasters around the world. One reason for the reliance on diffuse and long-term risk conceptions was that the magnitude and specific temporal impact of climate change were still a matter of debate in the political discussion (US Senate 1990a, pp. S4345). Thus, Members of Congress frequently constructed climate change as a long-term risk that in the future could bring the earth's ecosystem out of balance and increase the risk for various negative impacts, yet without linking the problem to definitive referent objects or threats. In connection to this, they regularly highlighted the benefits of investing in mitigation now to avoid higher costs in the future: 'The OTA estimates that a 35-percent reduction in emissions could save us $20 billion per year or could cost us up to $150 billion per year' (US Senate 1991a, pp. S1756).

While the focus of the debates in the early and mid-1990s was on the international level (US Senate 1991a, pp. S1756), it increasingly shifted towards a US-centred perspective at the end of the decade (US Senate 1999, pp. S18461). The governmental discourse also appeared quite frequently during the first Bush administration throughout the early and mid-2000s. In accordance with governmental power, Members of Congress often pointed to scientific debates and findings of the IPCC and the National Academy of Sciences to highlight the potential risks of climate change (US Senate 2001a, pp. S2301; US House of Representatives 2005, pp. H4291).

The Governmental Discourse in Think Tank Reports: Uncertainty and Military Contingency Planning

When the think tank driven sovereign discourse gained momentum from 2007 onwards, the governmental discourse lost in importance. Instead of relying on scientific risk assessments that often construct climate change as a serious, though long-term, risk without identifying specific threats and referent objects (see, e.g., IPCC 2007, p. 8), traditional security actors and knowledge became the main reference point of climate security argumentations. However, governmental argumentations did not disappear entirely. Thus, most think tank reports did contain references to risk conceptions and often connected them to the other two discourses to highlight the uncertainty of all climate predictions and the associated risks of socio-political consequences (Rogers and Gulledge 2010, p. 21).

Interestingly, think tank reports particularly mentioned probabilistic, future-oriented risk conceptions in connection to military planning because they believed that the security community and the military were most competent concerning long-term risk management and anticipating unknown and dangerous futures (McGrady et al. 2010, p. 23; CNA and Oxfam 2011, p. 13; CNA Military Advisory Board 2014, p. 1):

> The security community is also accustomed to long-term planning and preparing for a range of uncertain outcomes. These attributes are essential for managing the risks of climate change, but are lacking in most other policy communities. (C2ES 2009, p. 2)

In contrast, political actors were believed to be too caught up in short election cycles to really plan ahead for future climate risks (Interview

2014o). Thus, most think tank reports linked a risk-focused argumentation to the dominant sovereign discourse that privileges defence actors and solutions. Climate change was thus constructed as a 'black swan event' (CNA and Oxfam 2011, p. 13), meaning a risk with a 'low probability' to materialise, though with possibly 'high consequences' (CNA 2007, p. 41).

A further central diagnosis was that one should not tackle complex threats such as climate change by extrapolating from the past in a linear fashion:

> In a world where the past is no longer prologue, decision makers need new methods and analytical support to accommodate uncertainty about how climate changes could affect the future security environment. (Rogers and Gulledge 2010, p. 20)

Thus, the reasoning shifted from a focus on largely calculable or traditional risks towards new, even more uncertain and non-linear forms or risk. The aim became to be able to cope with upcoming climate effects and to anticipate these uncertain, yet dangerous future events by relying on probabilistic risk assessments and resilience schemes (CNA 2007, p. 46; Campbell et al. 2007, p. 6; CNA Military Advisory Board 2014, p. 9). Ultimately, many reports saw the need for a transformation of traditional security thinking to cope with these new and uncertain threats: 'The multi-faceted nature of 21st Century security threats requires a "fresh take on security," which allows the U.S. to be better prepared for contingencies related to climate change' (Foley and Holland 2012a, p. 7).

Consequently, the recommendations to tackle climate risks focused on concepts such as 'contingency planning' (McGrady et al. 2010, p. 77), 'risk-assessment' (Campbell et al. 2007, p. 36), 'vulnerability-assessments' (McGrady et al. 2010, pp. 38–39), 'scenario-planning' (Campbell et al. 2007, p. 36) and 'resilience building' (Rogers and Gulledge 2010, p. 3; Busby 2007, pp. 14–15; Foley and Holland 2012b, pp. 6–8). The increasing focus on climate change effects on the US homeland and parts of its population also led to calls to integrate climate risks into the National Infrastructure Protection Plan and the Strategic National Risk Assessment (CNA Military Advisory Board 2014, p. 21).

Based on the logic of normalisation and in the context of probabilistic and complex threat calculations, the referent objects and adequate behaviour in the face of climate-induced risks are not fixed a priori. Thus, many reports focused on statistically identifying certain risk groups—for example, poor people, women and children (CNA Military Advisory Board 2014, p. 27; CNA and Oxfam 2011, p. 7)—and areas—for example, developing countries, already fragile or failed states, regions near the sea, or arid-areas (Busby 2007, p. 8; Werz and Manlove 2009, p. 4; McGrady et al. 2010, p. 36)—that are particularly endangered by first- and second-order climatic effects in contrast to the general population (Busby 2007, p. 17; CNA Military Advisory Board 2014, p. 9).

Within this argumentation, climate change was constructed as a problem of circulation with the aim to ensure good circulation—for example, global trade, the acquisition of natural resources, business travel (CNA 2007, p. 20; CNA Military Advisory Board 2014, p. 4)—and prevent the bad one—for example, excessive GHG emissions, the spread of instability and climate refugees (Werz and Conley 2012, pp. 6–7; Werz and Manlove 2009; Busby 2007, p. 18). Reports also recommended to closely monitor and contain the risks before they get out of control, reach tipping points and self-reinforcing dynamics (CNA 2007, p. 15; Campbell et al. 2007, p. 6). The recurring connection of climate change to the spread of dangerous viruses and the 'risk of pandemic explosions of disease' exemplify this point (Campbell et al. 2007, p. 77; see also Elbe 2011).

In line with the theoretical rationale of governmental power, the aim was not to stop global warming altogether. Instead, the objective was bringing the average risk associated with climate change down to a tolerable level—at least for the developed world—by intervening as indirectly as possible in the natural dynamics of the population. In this vein, reports either suggested indirect and pre-emptive measures such as birth control, 'demographic management', or gradual changes in patterns of consumption to contain the risks of climate change (Campbell et al. 2007, p. 78) or they recommended to focus on the worst security implications (Foley and Holland 2012a, p. 8). The latter strategy was epitomised in 'no-regrets' policies that prioritise actions 'that it [the US] would not regret having pursued even if the consequences of climate change prove less severe than feared' (Busby 2007, p. 11). In most cases, these policies were not aimed at mitigation but focused on military preparedness, adaptation, resilience building and disaster aid measures (Busby 2007, p. 11).

Attempting to calculate the complex future threats and ways to anticipate them, many reports highlighted the need for social-scientific knowledge and closer cooperation between science and military actors (Rogers and Gulledge 2010, p. 9). The focus here was particularly on demands for fine-grained 'actionable data' concerning future climate risks (Rogers and

Gulledge 2010, p. 14; Parthermore and Rogers 2010, p. 9; Burke and Parthermore 2008, p. 7). Climate predictions and social scientific knowledge alone were not enough but had to be 'translated' (Rogers and Gulledge 2010, p. 8) to the security community and to actors in government to become useful—a task, for which many reports saw the think tanks best equipped (Rogers and Gulledge 2010).

Parliamentary Articulations: Broad Risks and Economic Considerations

In Congressional debates from 2007 onwards, linking the governmental discourse to defence actors was not as pronounced as in the think tank reports but did happen occasionally (US Senate 2008d, pp. S5191). In general, the focus was broader. Debates constructed climate change as a system-wide problem that would gradually increase the risk for all kinds of referent objects, ranging from humans all over the world and increasingly also in the US (US House of Representatives 2014, pp. H5346), to the environment (US Senate 2013, pp. S2825) and the entire planet (US Senate 2012a).

A further common argumentation pointed to the fact that even the 'hard calculating' insurance industry considers climate change as serious risk and incorporates it into their risk assessment schemes:

> But it is not just scientists around the world, not just government agencies in the United States; you have a business whose life and death, whose profit margin depends upon understanding this issue and that is the insurance industry. If the insurance industry ends up paying out a whole lot of money when there are disasters, they are going to lose money. (US Senate 2011a, pp. S8501)

In relation to this discussion, the economic risks of doing too little now to mitigate climate change played an increasingly important role in Congressional debates. Proponents of this argumentation feared exorbitant economic costs due to increased insurance premiums (US Senate 2014a, pp. S1450) and thus emphasised that climate risks would become one of the major economic risks for the US in the future (US House of Representatives 2013, pp. H1893).

Contextualising Climate Security Discourses in the US

The detailed analysis of the US climate security debate shows that all three discourses have played a role in constructing climate change as security issue. However, it also became clear that while risk and human security conceptions played an important role in the 1990s, at least from the mid-2000s onwards debates unambiguously focused on the national security implications of climate change. This raises the question why we find this discursive mixture in the US case. Thus, building on Didier Bigo's famous metaphor that speech acts might only be the 'tip of the iceberg' (Bigo 2009, p. 126) and on Foucault's ethos of problematising underlying societal preconditions, let us take a closer look at the cultural, political and institutional context (Ciuta 2009) in the US that has enabled this unique form of securitisation.

A key factor in understanding the dominance of the sovereign discourse is a specific political culture concerning questions of security, defence and foreign policy. Thus, questions of national security have been a central factor in the construction of the otherwise often diverse US identity (Bialasiewicz et al. 2007; Campbell 1992). Lacking a common ancestry, patriotism, 'rallying around the flag' and supporting the troops against a dangerous 'other' have been recurring elements in the US (Kutz 2011, p. 43; Campbell 1992, p. 3; Sperling 2010, p. 174). Thus, pointing out threats to US national security is an essential practice to constantly renew the identity as a nation (Campbell 1992, p. 13). Closely connected, there exists an influential discourse of exceptionalism, according to which the US is constructed as a 'shining city upon the hill', as the indispensable nation and 'chosen country' that ensures order and peace around the world (Kutz 2011; Koschut 2011, p. 31). Based on this narrative, talking about issues of national security, the self-interest of the US and the direct projection of its power to advance its righteous goals is a normal political position and does not spark much resistance (Sperling 2010, p. 174; Dalby 1990). Consequently, actors in the defence sector, such as specialised think tanks and the armed forces themselves, have a remarkably favourable reputation in the US today (Gallup 2014; Sperling 2010, p. 173). Many perceive the military as positive, pragmatic and neutral actor that stands above the political turf wars in Washington and only problematises the 'real threats'. Thus, the opinion of the military or retired military personnel was crucial in constructing climate change as a threat to US national security (Interview 2014d).

In addition, as largest military in the world (Sperling 2010, pp. 172, 197), the US armed forces play a more active role on the domestic and global level compared to other countries (Rid 2011, p. 107). They perceive themselves as a key force for global order and beyond that are often the first actors that are able to react in situations of (natural disaster induced) crisis (Interview 2014m; Sperling 2010, pp. 172, 197). Accordingly, the US military is well aware that they will have to deal with the effects of climate change and therefore have a pronounced self-interest to plan ahead for these situations (Interview 2014h, 2014k; US Navy 2009). Moreover, as Hurricane Katrina in 2005 exemplified, the Army Corps of Engineers (2015) and the National Guard (2015) play a vital role in domestic disaster response efforts (Giroux 2006). Given these circumstances, it is understandable that the armed forces play a much more active role in the US climate security debates than in other countries and that hence the sovereign discourse and especially the focus on security and defence policy found such a resonance.

Another important prerequisite for the success of the sovereign discourse is the US institutional setup. The presidential system gives the President as commander in chief a considerable degree of freedom in foreign and defence policy (Oldopp 2014, p. 59). On the one hand, this reinforces sovereign power tendencies and contributes to the political value of designating something as an issue of national security (Interview 2014h). On the other hand, lacking an automatic parliamentary majority, the President and his party have less direct influence when it comes to domestic politics and to ratifying binding international commitments. In the context of an increasingly divided political landscape concerning climate change (McCright and Dunlap 2011, p. 156; Grundmann and Scott 2014, p. 233), getting climate legislation through Congress can be a tiresome task (Oldopp 2014, p. 62; Braml 2011, p. 123). Besides relying on the unifying qualities of constructing climate change as a threat to national security, this can also give incentives to govern climate issues through executive orders or by integrating them into the planning processes of the governmental agencies over which one wields some control such as the DOD and the intelligence agencies.

Besides the largely stable political traditions discussed above, there are also contextual factors that have changed over the study period and help to understand the transformations concerning the dominant climate security discourses in the US. A first one concerns the US-American tradition to oscillate between phases of isolationism and interventionism concerning its foreign policy (Cox and Stokes 2012; Roskin 1974; Koschut 2011, p. 33), which also applies to its climate policies (Harris 2000a, 2001, 2009). While the US was highly involved during the founding days of the IPCC and UNFCCC, the concern for international solutions decreased considerably towards the end of the 1990s. This had repercussions for the climate security debates as well, which from the 2000s onwards increasingly focused on the US itself. Towards the end of the second Obama administration, this changed again to a certain extent and made possible several bilateral arrangements with China and the approval of the Paris Agreement in 2015. The election of Donald Trump in 2016 and the announcement to drop out of the Paris Agreement again point into the opposite direction.

Another important issue is the general role and value of scientific knowledge in political debates. During the 1990s, politicians regularly referred to scientific findings to justify their securitising articulations in Congressional debates. This contributed to the prevalence of more cautions risk-based discourses at that time. Some actors such as Gore kept pursuing this line of argumentation during the 2000s, however, they increasingly had problems to convince the relevant audiences. The reason was that lobbyists sceptical of climate change and several Republican politicians had successfully linked scientific findings to liberal, left-wing and environmentalist positions (Eshelman 2014). Thus, many conservatives perceived an argumentation based on scientific findings not as apolitical and neutral expert information, but as aligned with a specific political position. This anti-scientific sentiment grew even stronger in the 2000s with the formation of the tea party movement. Ultimately, it culminated in the 'post factual' or 'post truth' age of the 2016 Presidential election and the subsequent behaviour of the Trump administration (Pazzanese 2016). Together with the tradition of the US-American media for 'balanced' reporting—which often culminates in giving mainstream positions and fairly obscure climate sceptical theories an equal amount of coverage—this considerably devalued scientific findings in the political debate (Eshelman 2014).

In parts because of this development, think tanks specialised in security policy and engaging in the national security discourse began to fill the gap in the 2000s. They replaced the findings of climate science or at least got into the mediator position and 'translated' these findings into a more appealing political language. This development laid the ground for the success of the sovereign discourse from the mid-2000s onwards. It also helps to understand the absence of environmental organisations in the US climate security debates. For one thing, these organisations and their arguments were too close to climate science and actors such as Gore and hence increasingly perceived as partisan actors from the liberal, left-wing spectrum. In addition, lacking a background or credibility in security and defence policy, they were not able to provide the 'correct' kind of knowledge at that time. Consequently, they could not have participated in the climate security debate without hurting the argumentative strategy that tried to overcome the polarisation by securitising climate change as threat to the national security of the US (Interview 2014a, 2014g; Abrahams 2019, p. 338).

Finally, the success of the sovereign discourse in the 2000s was closely linked to the specific role of the mentioned think tanks in the US political system, which clearly distinguishes the US case from other countries (McGann 2007; Braml 2011, p. 117; Gellner 1995). Due to the high value of questions of national security, but also less strictly organised political parties, these security policy focused think tanks are very well connected to political circles and exert a considerable degree of influence (Abelson 2006, 2010; Interview 2014n). This applies especially to more or less non-partisan think tanks such as CNA, CSIS, the Council on Foreign Relations, CCS, the Wilson Centre and the Brookings Foundation, which are citable for government agencies. It even more so relates to those that have a security clearance, and thus carry out classified work for the government, for instance, CNA or RAND (Interview 2014n). Politicians currently not working in the government often stay a few years in one of these think tanks to stay connected and to later take on a position in government again after the next elections (Interview 2014b, 2014n; Braml 2011, p. 117). Moreover, these think tanks often carry out the development of certain policies for politicians who—as isolated political entrepreneurs—do not have the time to dig into a topic in this depth (Braml 2011, p. 123). At the same time, they can provide cover for politicians that come out with a new policy by organising media coverage and support for the proposal (Interview 2014a; Braml 2011, p. 123).

The Political Impact of Climate Security Discourses in the United States

While undoubtedly not being the sole driving force (McCright and Dunlap 2011; Harrison 2007), climate security discourses had a considerable impact on the US climate debate but also on other political sectors. Apart from rendering climate change governable in a novel way, this discursive constellation had a specific impact on the legitimisation and the content of concrete policies and governance practices. The following section discusses the concrete political consequences in key policy sectors, namely environment and climate, defence/security, development and disaster management.

Environmental and Climate Policy: Overcoming Partisan Divides?

Whereas not permanently taking the centre stage, abating climate change and legitimising international and domestic policies has always been a part of the climate security debate in the US. In 1989, mostly engaging in the disciplinary discourse, Republican Senator John Chafee (US Senate 1989b, pp. S13293) tried to gather support for his eventually failed Stratospheric Ozone and Climate Protection Act—its aim was to 'reduce the atmospheric concentration of these greenhouse gases' (US Senate 1989b, pp. S13293). Shortly after, the House of Representatives passed the House Concurrent Resolution 248 (US House of Representatives 1990), which entailed direct references to national and human security (US House of Representatives 1990, pp. H7684, H7684).

The first successful climate legislation was the establishment of the US Global Change Research Program in 1990 (US Government 1990). Although climate security argumentations were not the only driver, several Members of Congress pointed to the threats posed by climate change to argue in favour of the bill. For instance, Representative Christopher H. Smith highlighted that the 'devastating effects which a rise of only a few degrees of temperature could have on our planet, demand that we take steps now to prevent such manmade changes on our global environment' (US House of Representatives 1989, pp. E3575). As the name of the programme implies, the global dimension of climate change as well as finding global solutions was important at that time (US House of Representatives 1989, pp. E3575). Thus, the advocated solutions mostly focused on mitigation measures to tackle the root causes of climate change and on international cooperation (US House of Representatives 1989, pp. E3575). While not changing the debates altogether, these discursive interventions helped to put pressure on the Republican Presidents Ronald Reagan and George H. W. Bush to acknowledge climate change and to engage in the international debate, especially concerning the negotiations about the UNFCCC in 1992 (Harris 2001, pp. 5, 8, 2002, p. 150; Vig 2013, p. 91).

When Bill Clinton and his environmentalist Vice President Gore took office in 1993, environmental issues and climate security articulations received more attention. However, because the Republicans had gained a majority in Congress in 1994, this did not translate into immediate domestic action on climate change (McCright and Dunlap 2011, p. 158; Kraft 2013, p. 112). Nevertheless, the specific global focus of most securitising articulations at that time helped to sustain the dedication to the international negotiations. The US acknowledged the first findings of the IPCC, published a Climate Action Plan that foresaw significant, though voluntary, emission reductions in 1993 (Brunner and Klein 1999), and in 1994 the then US Permanent Representative to the United Nations Madeleine K. Albright emphasised that: 'We believe that America should be the world's environmental leader, not foot-dragger' (Albright 1995, p. 53). After ratifying the UNFCCC, Undersecretary for Global Affairs Timothy Wirth further pushed the United States' commitment to binding GHG regulations. Together with various securitising articulations—amongst others from Bill Clinton himself who made clear that 'Many previous threats could be met within our own borders, but global warming requires an international solution. [...]' (US Senate 1997b, pp. S11019)—this eventually paved the way for the initial acceptance of the Kyoto Protocol by the US in 1998.

Around the turn of the millennium, however, outright climate sceptical positions became more common again. Key proponents were, for instance, Senator James Inhofe (Harris 2001, p. 20; US Senate 2003b, pp. S13489), a well-organised non-governmental and business lobby (Falkner 2005, p. 590; Leiserowitz 2005, p. 1435; McCright and Dunlap 2011, p. 158), and since 2001 President George W. Bush (Grundmann and Scott 2014, p. 222; Barnett 2004). At the same time, climate security articulations, especially those linked to the sovereign discourse, declined. Key reasons were the terror attacks on September 11th 2001 and the following 'war on terror', which led to the emergence of another dominant discourse of danger (Floyd 2010, p. 122). Thus, particularly in Bush's first term in office, the problematic atmosphere for progressive climate legislation consolidated (Harris 2002, p. 153; Vig 2013, p. 94; Fletcher 2009, p. 806).

Despite this difficult political environment, attempts for concrete climate policies rooted in climate security discourses did not disappear entirely. In 2003 (and again in 2005 and 2007 as Climate Stewardship and Innovation Act) the Democrat Joe Lieberman and the Republican John McCain introduced the Climate Stewardship Act. It was supposed to strengthen the research on abrupt climatic changes—a threat that became particularly prominent in the wake of the 2003 Pentagon study—to reduce US emissions and to introduce a cap and trade system. In various debates on the Act, they constructed climate change as a threat to the 'environment', to the 'economy' and to 'public health' (US Senate 2003b, pp. S13485). However, eventually, none of these attempts was able to gain a majority in Congress.

Notwithstanding the absence of successful climate legislation on the federal level, several US states began to adopt fairly progressive laws concerning climate regulation (Donner and Faltin 2007, pp. 11–14; Mildner and Richert 2010, pp. 30–32; C2ES 2012). The climate security framing was not the only driver for this development as other factors such as more environmental-friendly governments at the state level (for instance Governor Arnold Schwarzenegger in California) also played an important role (Donner and Faltin 2007, pp. 14–15). However, mostly relying on the disciplinary discourse, Congressional debates had begun to highlight the domestic security implications of climate change from the early 2000s onwards (US Senate 2003a, pp. S3348). Especially the Democratic Senators Dianne Feinstein and Barbara Boxer from California at several occasions had emphasised the security implications of climate change (US Senate 2003c, pp. S13575, S13599). Subsequently, California became one of the most progressive states concerning climate regulation. In 2006, it adopted the Global Warming Solutions Act aimed at reducing GHG emissions through various regulations and market mechanisms, and in 2012 a cap and trade system (C2ES 2006, 2014).

On the federal level, Bush became slightly more open towards climate policies in his second term in office (Fletcher 2009, p. 805; Donner and Faltin 2007, p. 15; Mildner and Richert 2010, p. 31), inter alia due to a changed attitude in the business community (Guber and Bosso 2013, p. 55) and international pressure (Scherwitz 2014). Nevertheless, the domestic political debate on climate change was still heavily polarised between Democrats and Republicans (McCright and Dunlap 2011, pp. 158–159; Guber and Bosso 2013, pp. 62–64). Moreover, some aspects of the climate security discourses of the 1990s, especially representing climate change as a serious, though temporally and spatially distant risk, did not work anymore (Leiserowitz 2005, pp. 1438–1440). Thus, the disciplinary and governmental discourses were not able to effectively bridge the political divide. A key reason for this was that both discourses kept constructing climate change as an environmental or development issue and long-term risk, which did not resonate well in conservative circles. As a result, it was not until 2007 and the re-emergence of the sovereign discourse that the climate security argumentation became more important again and began to contribute to overcoming the political polarisation as well as to the legitimisation of several concrete policies.

Together with the Democrats gaining the Congressional majority in 2006, the common denominator 'national security' raised attention and enabled a whole range of debates and attempts to adopt concrete policies on climate change in Congress. Thus, the number of climate legislation attempts rose from 106 in 2005/2006 (109th Congress) to an all-time high of 235 in the 110th Congress (2007–2008) (C2ES 2008). While not all of these attempts can be linked to climate security argumentations, many of the key pieces of legislation and proponents of these bills (mostly introduced by Democrats or moderate Republicans) drew a connection to security threats, often articulating the sovereign discourse (US Senate 2008b, pp. S4990).

In 2007 Democrat Bernard Sanders introduced the Global Warming Pollution Reduction Act pointing out that 'global warming poses a significant threat to the national security and economy of the United States, public health and welfare, and the global environment' (US Senate 2007a, pp. S309IS). Shortly afterwards, Joseph Lieberman and John Warner introduced the Lieberman-Warner Climate Security Act (2008) that was supposed to install a cap and trade system. The corresponding Congressional debates highlighted the national security consequences of climate change and referred to the 2003 Pentagon report (US Senate 2008c, pp. S5017, 2008e, pp. S5197). Only a year later, the American Clean Energy and Security Act (also referred to as Waxman-Markey Bill) highlighted the benefits of clean energy sources for mitigating climate change and argued for energy independence, which eventually would enhance US national security (US House of Representatives 2009b, pp. H8477). All these bills, which were often of a bipartisan nature, demonstrate how the construction of climate change as national security issue was able to again bring the issue onto the agenda (see Richert 2009, p. 7; Brzoska 2012, p. 172) and how it at least partially bridged the divide between Democrats and Republicans (see Fletcher 2009, p. 807). However, despite this development and the much more favourable political environment after the election of Barack Obama in 2009 (US Department of State 2009; Mildner and Richert 2010), none of these attempts was able to generate sufficient political support to pass both chambers of Congress.

Key reasons were the still strong opposition in both parties against any measures that could negatively affect the US economy (Interview 2014l; Mildner and Richert 2010, p. 26; Guber and Bosso 2013, p. 69), the rise of the Tea Party and the 2008 economic crisis. Finally, the failed 2009 climate summit in Copenhagen made it even more difficult to gather support for progressive national climate policies. After the Democrats had lost their Congressional majority in the midterm elections of 2010 (Mildner et al. 2012, p. 3; Leggett and Lattanzio 2009) and the US saw a renewed oil and gas boom (IEA 2012, p. 2), Congressional action on climate change stagnated and the interest in the public declined (Pew Research Center 2012, p. 1). As a consequence, Obama increasingly began to govern through technical regulations issued by the Environmental Protection Agency (EPA) and the Department of Transportation (DOT) (Mildner and Richert 2010, p. 21)—a thoroughly sovereign for of governance—and in 2015 announced his Clean Power Plan that aimed to reduce US emissions by 32 per cent from 2005 levels by 2030 (The White House 2015c). Beyond that, the US struck a bilateral agreement with China in 2014 (Taylor and Branigan 2014) and helped to negotiate the 2015 Paris Agreement.

Altogether, even though invigorating the debate, climate security articulations were not able to substantively contribute to concrete climate policies on the federal level or to significantly alter the international position of the US. The most important inhibiting factor was that most attempts for federal climate legislation aimed at the environmental sector and tried to install measures that would mitigate climate change. This did not fit well with the specific sovereign discourse at that time that highlighted the immediate and direct consequences of climate change for US national security and aimed at legitimising institutional responses in the defence sector. Thus, many people perceived the existing policy proposals as political tricks that only used the national security label to hide their environmental agenda (Interview 2014c, 2014i; US Senate 2009, pp. S10148; US House of Representatives 2009a, pp. H5555). Consequently, the main impact of the securitisation from 2007 onwards was not on climate policy but on security and defence policy.

Defence and Security Policy: Climatisation and the Bidirectional Effects of Securitisation

While becoming particularly influential from 2007 onwards, the first impacts of the sovereign discourse on the defence sector materialised in the 1980s. After Reagan had already acknowledged environmental degradation as security problem for the US (Harris 2002, p. 150), his successor Bush for the first time included climate change in the National Security Strategy (NSS) in 1991 (The White House 1991, p. 2). Under the Clinton-Gore administration, climate change became an even more important issue in the NSS, which now recognised it as an 'environmental risk[s] serious enough to jeopardize international stability' (The White House 1994, p. 15; Harris 2002, p. 151; Below 2007, p. 709). Beyond that, the administration increased the funding for research into the socio-economic dimensions of climate change, which included the national security implications (Floyd 2010, pp. 78–79), and created a Global Environmental Affairs Directorate at the National Security Council (Simmons 1995, p. 2). Additionally, they appointed several people to important positions in the government who were sympathetic to the idea that environmental problems constituted important security threats. Most importantly, Timothy Wirth at the State Department and Sherri Goodman as Deputy Undersecretary for the DODs newly created Office for Environmental Security (ODUSD-ES) (Harris 2001, p. 151; Interview 2014j; Matthew 2013, p. 353). The main tasks of this office were to oversee the DOD's environmental impact and to facilitate the cooperation with the EPA (EPA 1994; US Senate 1991b).

Environmental and climate security argumentations also led to the establishment of several programmes at the intersections between the environment and security (Floyd 2010, p. 102). In line with the most common variant of the sovereign discourse at that time, the focus was not primarily on preparing the actors in the defence sector for the effects of climate change, but rather on using assets from this sector for environmental monitoring purposes. One example is the Strategic Environmental Research and Development Program (SERDP), which allowed the use of military and intelligence capabilities, for instance satellite imagery, to monitor the environment (US Senate 1990d, pp. S12406; The White House 1994; DOD 2015a). Beyond that, constructing climate change as national security issue did have some immediate implications for the planning of the defence sector. Thus, the US Central Command (CENTCOM) under General Anthony Zinni, who later would become the lead author of an influential think tank report on climate security, began to consider questions of environmental security and to cooperate with the ODUSD-ES. Additionally, the administration established a Task Force on State Failure to assess whether environmental changes posed a threat to instable states (Matthew 2013, pp. 355–356). Towards the end of the 1990s and especially under newly elected President Bush, the sovereign discourse and hence the impact on defence and security policy declined. Bush even defunded or dismantled several measures of the Clinton administration concerning the linkages between the environment and security (Harris 2002, p. 153; Matthew 2013, p. 356) and did not mention direct connections between climate change and security in the National Security Strategies of 2002 and 2006 (The White House 2002, 2006).

This picture again changed entirely with the re-emergence of the defence-oriented sovereign discourse particularly from 2007 onwards. Instead of primarily legitimising climate policy, the focus on threats to US national security first and foremost emphasised short-term solutions to defend against these threats and thus led to a heightened impact on defence policy. Another reason for the failure to influence climate policy but success in inducing changes in the defence sector was the only partial bridging of the political divide on climate change. A number of former conservative sceptics now accepted that the climate was changing and that this could have severe implications for the national security of the US (Interview 2014i). However, many still doubted that anthropogenic emissions played an important part. They rather believed—or chose to believe—in natural factors, and hence kept opposing mitigation measures. Instead, they favoured adaptation or the already mentioned 'no regrets' strategies, such as integrating climate change into military planning, which also were the logical outcome of the dominant defence-oriented sovereign discourse (Interview 2014i; Busby 2007, p. 11). This also was in line with the original motives of many of the think tanks that first of all aimed at transforming security policy and only in second place wanted to influence climate policy (Interview 2014j). They were very sceptical about merely using a national security or military framing for getting climate legislation through Congress as this could easily be dismissed as 'political branding' with little substance (Interview 2014c, 2014j). As one interview partner put it, 'standing beside a soldier does not make an environmental bill security relevant' (Interview 2014i).

Against this background, Congress quickly began to adopt the national security framing and to discuss respective legislation for the defence sector (US Senate 2008a, pp. S4885, 2009; US House of Representatives 2012, pp. H2451). For instance, Republican Senator John Warner stated that '[…] there are also direct impacts on U.S. military systems, infrastructure and operations. Climate change will add stress to our weapons system, threaten U.S. bases throughout the world, and have a direct effect on military readiness' (US Senate 2008a, pp. S4885). The primary aim was not anymore to address the root causes of climate change by mitigation measures, but to directly address the security implications of climate change, as this quote from Democratic Senator Whitehouse summarises:

> I rise to discuss the importance of assessing and planning for and mitigating the national security effects of climate change. Our changing climate is not simply a green issue invented by environmentalists and conservationists; climate change threatens our strategic interests, our military readiness, and our domestic security in many ways. (US Senate 2012b, pp. S7017)

Although the first attempt for legislation, the 2007 Global Climate Change Security Oversight Act, died in the Senate (Govtrack 2007), it set the stage for the adoption of the National Defense Authorization Act of 2008. This obligated the DOD to integrate climate change into the next NSS and National Defense Strategy as well as into the Quadrennial Defense Review, the most important and influential publicly available planning document of the DOD and the Armed Forces (US Congress 2008, p. 122).

Accordingly, the National Defense and Military Strategy of 2008 and 2011 included climate change (DOD 2008, p. 5) and considered the possibility that it could weaken the ability of already instable regions to withstand natural disasters (DOD 2011a, p. 2). The QDR 2010 contained a whole paragraph on climate change and, using formulations put forward by many of the think tank reports, constructed it as an 'accelerant of instability or conflict' and as a serious geopolitical threat that will profoundly change 'the operating environment, roles, and missions that we undertake' (DOD 2010, p. 84). As a consequence, the report advised the DOD to take the effects of climate change into account concerning its 'operating environment, missions, and facilities' (DOD 2010, p. xv), thus promoting the 'climatisation' (Oels 2012) of the defence sector. The QDR 2014 as well included climate change and predicted that:

> The impacts of climate change may increase the frequency, scale, and complexity of future missions, including defense support to civil authorities, while at the same time undermining the capacity of our domestic installations to support training activities. (DOD 2014b, p. 8)

Furthermore, the NSS 2010 and 2015 again mentioned climate change as 'key global challenge' (The White House 2010, p. 47) and 'urgent and growing threat to our national security' (The White House 2015b, p. 12).

Climate change also began to appear in various more fine-grained planning schemes of the DOD. The DOD's Defense Science Board Task Force discussed climate change in a report of 2011 and emphasised the central role of the DOD concerning the issue (DOD 2011b, p. xv). In 2012 (DOD 2012) and 2014, the DOD developed even more detailed Climate Adaptation Roadmaps (in 2010 and 2014), which depicted the issue as a 'threat multiplier' (a phrase coined by the 2007 CNA report) and 'immediate risk to national security' (DOD 2014a, p. 1). Finally, the DOD integrated climate change into its Science, Infrastructure, Research, Development & Acquisition Plan as well as into its Strategic Sustainability Performance Plan 2014 (DOD 2014c).

In 2014, the United States' Central Command (USCENTCOM) too incorporated climate change into one of its regional risk assessments (DOD 2014d). Responding to a Congressional request (US Senate 2014b), in 2015 the DOD published a report on the National Security Implications of Climate-Related Risks and a Changing Climate, which underlines the far-reaching transformation of the climate security debate by stating that 'The Department of Defense sees climate change as a present security threat, not strictly a long-term risk' (DOD 2015b, p. 14). Subsequently, the US combatant commands adopted a range of concrete measures to deal with the security threats posed by climate change. Examples are training activities, the improvement of infrastructure, vulnerability assessments, the planning of humanitarian disaster relief operations and the close cooperation with partners (civilian and military) all around the globe (DOD 2015b, pp. 6–7).

Following the recommendations of many think tank reports (Werz and Manlove 2009, pp. 4–5) and as a result of the integration of human security conceptions into the sovereign discourse, the DOD increasingly began to cooperate with civilian and development actors such as the DHS, United States Agency for International Development (USAid) but also with academics (Hartmann 2010; DOD 2010, p. 87). The securitisation of climate change hence transformed the appropriate practices in this field to a considerable extent. In fact, under catchphrases such as 'Great Green Fleet' (US Navy 2010, p. 6, 2016), the DOD and the armed forces have already implemented some of these measures and until 2010 had already spent $2.7 billion on energy efficiency (Gilbert 2012, pp. 6–7). Beyond this, the increased cooperation with civilian actors also applied to disaster response and humanitarian missions. According to several Congressional debates, the 'Department of Defense recognizes that it has a role to play in supporting humanitarian assistance and disaster relief missions' (US Senate 2014a, pp. S1469). Thus, it promoted 'whole-of-government' or 'whole-of-community' approaches that included very different types of actors to tackle today's complex security threats and in particularly climate change (DOD 2010, p. 70, 2012, p. 6, 2014b, p. 22; Interview 2014k). One concrete example is the close cooperation of the African Command (USAFRICOM) with development organisations such as USAid (DOD 2015b, p. 9; see also Hartmann 2010).

Looking at the armed forces in more detail, the Navy was the most active branch when it comes to climate change (Interview 2014h). Key reasons for this were its global areas of operations and already important contribution to disaster relief operations, but also the realisation that its coastal bases—particularly Norfolk, Virginia—would soon seriously be affected by sea level rise (US House of Representatives 2014, pp. H5348). Furthermore, the Navy itself had contributed to the research on climate change, for example, when its submarines helped to measure the thickness of the arctic ice shelves (US Navy 2009; Interview 2014h, 2014l; US Navy 2010, p. 5). Interestingly, David Titley, who later became a member of the second Military Advisory Board (MAB) of the CNA and an outspoken proponent of the sovereign discourse, had been involved in the Navy's research on climate change (Interview 2014h). Consequently, the Navy was one of the first branches of the armed forces that established a Task Force on Climate Change in 2009 (US Navy 2009), and in 2010 published a Climate Roadmap, which acknowledged that 'Climate change is a national security challenge with strategic implications for the Navy' (US Navy 2010, p. 3). An important side-effect—that also applies to the other branches of the armed forces—is that in order to spend money on climate change, the Navy had to classify it as a threat to national security (Interview 2014h). Thus, the success of the sovereign discourse was also partly owed to these institutional and financial constraints.

Finally, the construction of climate change as national security issue also had repercussions for the intelligence sector. While the Intelligence Authorization Act for the Fiscal Year 2008 failed, it nevertheless opened the doors for important changes in how the intelligence sector understood climate change. Thus, in the same year, the House Permanent Select Committee on Intelligence and the House Select Committee on Energy Independence and Global Warming published an influential National Intelligence Assessment on the National Security Implications of Climate Change (Fingar 2008). Additionally, climate change was mentioned in the 2009 Annual Threat Assessments for the Senate Select Committee on Intelligence (Blair 2009). Moreover, in 2009 the CIA founded a Center for Climate Change and National Security (CIA 2009). Even though it discontinued to exist as independent office from 2012 onwards, the CIA continued to monitor the security implications of climate change in several regional departments (Broder 2012; Interview 2014j).

Development Policy: The Rise of 'Networked' and 'Sustainable' Security

Due to the incorporation of the disciplinary discourse into the overarching sovereign argumentation and the construction of endangered Southern populations as major threat, the climate security debates since 2007 also affected the development sector. Thus, USAid has acknowledged the security dimension of climate change in several of its reports. The central USAid Climate Change and Development Strategy 2012–2016, for instance, mentioned climate change in relation to food security (USAid 2012, p. vi), but also directly linked the disciplinary to the sovereign discourse when it adopted the think tank phrasing of climate change as 'threat multiplier' (USAid 2012, p. 4) and stated that:

> If not well managed, climate change may exacerbate water scarcity and increase conflicts among water users; in some cases it could trigger displacement and contribute to national and regional resource governance tensions, threatening U.S. national security objectives in key regions of the world. (USAid 2012, p. 3)

The proposed solutions ranged from supporting developing countries with mitigation and adaptation measures, to measures enhancing their resilience, or plans to stabilise regions particularly affected by climate effects in the future. It is difficult to assess whether this has generated additional funding. Yet, the close interlinkages between the disciplinary and sovereign discourse have certainly legitimised a closer cooperation between civilian agencies and the military sector (see also Hartmann 2010; Gilbert 2012) epitomised in concepts such as 'new security threats', 'networked security' and 'sustainable security' (Rogers and Gulledge 2010, p. 8; Werz and Conley 2012, pp. 33–34).

The USAid Climate Change Strategy 2012–2016 mentioned key documents and findings of the defence sector such as the NSS, the QDR and the National Intelligence Estimates to argue for the need to tackle climate change. Referring to the Executive Order 13514 and the Interagency Task Force on Adaptation (The White House 2013), it furthermore proposed a close cooperation of all US branches of government to approach climate change (USAid 2012, pp. 7–8). Directly referring to the implications of climate change for national security, it also has advised its Office of Conflict Management and Mitigation to research 'how specific climate factors contribute to the risk of conflict and affect the resilience of social structures and institutions' (USAid 2012, p. 16). It furthermore recognised the close interlinkages between US defence and development policy in relation to climate change:

> The current U.S. National Security Strategy recognizes that development is a core pillar, together with defense and diplomacy, of national security policy. It is therefore imperative to ensure a strong development voice in the debates and dialogues that shape that policy. (USAid 2012, p. 20)

A 2008 published report of CSIS on Integrating 21st Century Development and Security Assistance summarised these developments and found a 'growing involvement of the U.S. Department of Defense as a direct provider of "nontraditional" security assistance' and that the 'Pentagon's role as a direct provider of foreign assistance has surged' (CSIS 2008, pp. v, vi). Finally, another concrete example of this kind of cooperation is the Denton Program, which is jointly administered by USAid, the Department of State, the Defense Security Cooperation Agency (DSCA) and the DOD. The programme allows certain aid organisations to use the DODs transport capabilities and is mentioned in the 2015 DOD report and by USAid as an important cornerstone in the plans to cope with the security threats of climate change (DOD 2015b, p. 6; USAid 2016).

Disaster Management and Insurance Solutions

Disaster management became a major concern from 2010 onwards when key reports and Congressional debates began to link climate change to threats to the human security of US citizens and urged the respective US agencies (e.g. FEMA, the Army Corps of Engineers and the DHS) to act (US Senate 2011b, pp. S8711; US House of Representatives 2013, H1895). Subsequently, FEMA included climate change into various disaster management plans (FEMA 2012) and the DHS mentioned climate issues in its National Infrastructure Protection Plans (DHS 2013; see also Brzoska 2012, pp. 174–175). Moreover, articulations pointing to climate risk conceptions increasingly led to a focus on the insurance sector (US Senate 2014a, pp. S1422). In line with the concept of normalisation and the aim of governmental power to stabilise risks at a tolerable level, risk management and insurance solutions increasingly played a more central role in coping with climate change. In parts, this approach shifted the attention away from tackling the root causes of climate change through mitigation measures. Instead, it led to an acceptance of climate change while insuring against the worst impact and increasing resilience to future shocks.

While many actors in the US hence seemed to accept a high level of risk, some also cautioned that eventually the insurance system could be overwhelmed by too extreme climatic changes (US Senate 2014a, pp. S1472). Hence, some feared that climate change could spiral out of control, exceeding the bearable level of risk for the US and its economy.

Key Developments Between 2015 and 2020: A Climatised Defence and Intelligence Sector Trying to Withstand President Trump

The election of Donald Trump as President in 2016 has led to a considerable decrease in the importance of climate change on the federal level (Selby 2019; Abrahams 2019, p. 324). Trump immediately began trying to dismantle domestic climate regulation (e.g. the Climate Action Plan, Clean Power Plan), appointed climate change sceptics to influential climate-related positions (Selby 2019, pp. 471–472; Greshko et al. 2019), and in 2017 announced to withdraw the US from the Paris Agreement (Trump 2017). Moreover, he has increasingly challenged the assessments of the defence and intelligence sector on climate change as an important threat and has tried to discourage them from further working on this connection (Revkin 2019). Most fittingly in this respect, the 2017 NSS does not directly mention climate change anymore but instead discusses the threats of global climate policy for US energy security (The White House 2017, p. 27).

Yet, while Trump's actions have undoubtedly damaged the international reputation of the US in climate matters and have slowed down climate action on the federal level, the manifold checks and balances of the US political system and more progressive climate action at the state level have prevented a total rollback so far (Selby 2019, p. 475; Green 2019). Moreover, especially the US defence and intelligence sector has continued to discuss the security implications of climate change (Abrahams 2019, pp. 324, 328, 334). The focus has remained on a mix of the disciplinary with the sovereign discourse. Thus, reports have often cited the global human security implications of climate change, its potential as a 'threat multiplier' (Abrahams 2019, p. 325), or 'catalyst for conflict' (Fleishman and Goodman 2018, p. 3) and then highlighted the relevance for US national security and the need for adaptation as well as risk management in the defence and intelligence sector (Abrahams 2019, p. 329).

Concerning the defence sector, the DOD has issued several reports considering the direct implications of climate change for its bases and underlined the necessity to improve adaptation and resilience as well a climate risk planning (DOD 2016, 2018). It has also repeatedly considered the relevance for its operations abroad and particularly emphasised the far-reaching implications for the Artic region (DOD 2016). Moreover, key figures such as Secretary of Defense James Mattis have stressed the continued relevance of climate change to US national security and the work of the DOD: 'Climate change can be a driver of instability and the Department of Defense must pay attention to potential adverse impacts generated by this phenomenon' (Revkin 2017). At a briefing on 'The National Security Implications of Climate Change' in June 2017, John Conger, Former Principal Deputy Undersecretary of Defense, retired Air Force General Ron Keys, and Sherri Goodman reaffirmed the sovereign discourse and the climatisation of the defence sector by calling climate change a threat multiplier and by emphasising its importance for the work of the DOD (CCS et al. 2017). Lastly, the 'National Defense Authorization

Act for the Fiscal Year 2020' repeatedly mentions climate change and asks about 'how climate impacts affected the ability of the Department of Defense to accomplish its mission, and the short- and long-term actions the Department can take to ensure military installation resilience' (US House of Representatives 2019, p. 256).

In the intelligence sector, a 2016 report by the NIC on the 'Implications for US National Security of Anticipated Climate Change' states that climate change will have considerable negative effects especially on vulnerable populations, which in turn can destabilise countries, heighten social and political tensions and directly stress US military operations and bases (National Intelligence Council 2016, pp. 6–7, 10). Beyond that a 2017 Global Trends Report discusses climate security implications (National Intelligence Council 2017) and two Worldwide Threat Assessments in 2018 (Director of National Intelligence 2018) and 2019 mention climate change and urge to consider its global human security implications (Director of National Intelligence 2019, p. 21).

Finally, in 2019, 58 senior military and national security leaders wrote a letter to President Trump urging him to stop disputing and undermining 'military and intelligence judgments on the threat posed by climate change' and to once again take climate change as its security implications seriously (American Security Project and CCS 2019). Thus, similar to the period under George W. Bush in the 2000s, the US defence and intelligence sector is trying to keep climate security on the political agenda against the will of the current President. The main motivation undoubtedly remains a real concern for national and international security as well as for US operations and bases. Yet, linking climate change to national security, especially coming from the influential defence and intelligence sector, undoubtedly also can once again help to keep climate change as such on the federal political agenda (Abrahams 2019, pp. 329, 338).

Conclusion

Compared to other world regions, securitising the environment and climate change started early in the US. Thus, security already played a role at the very beginning of the political debate on climate change in the late 1980s and early 1990s and contributed to put climate change on the political agenda in the first place. This first phase was influential in numerous ways: it raised the attention for environmental destruction and climate change in general and helped to establish these issues as key themes in the political debate. It transformed the political meaning of climate change away from an environmental concern debated by small groups of experts towards a pressing political issue. At the same time, the environmental and climate security debate also had an impact on how people understood security in general. It helped to open up traditional state-centred conceptions of security, introduced new referent objects, practices and new modes of response, hence exemplifying what others have labelled as the 'climatisation of the security field' (Oels 2012) and which I would call the 'bidirectional qualities' of securitisation.

Securitising climate change also had concrete implications for policy-making. Due to the global focus of the climate security debate at that time, it helped to keep the US involved in the international negotiations on climate change. Moreover, it facilitated a number of research programmes on climate change, helped to adopt policies at the intersections between security and environmental policy, and led to the inclusion of climate threats into US National Security Strategies. However, despite this initial success, it eventually fell short of legitimising far-reaching policies and increasingly lost its appeal towards the end of the 1990s.

While the late 1990s and early 2000s hence saw fewer securitising articulations, the debate reappeared in the mid-2000s. This second phase of securitisation largely constructed climate change within the sovereign discourse and as main threat to US national security. While delegitimising actors without experience in national security policy such as scientists and environmental NGOs, this representation established a range of security policy focused think tanks and military institutions as the main proponents of the climate security debate. It also raised the attention for climate change, especially as an issue of 'high or hard politics', reached more conservative audiences and helped to bridge the political divide between Democrats and Republicans. Eventually, this led to increasing attempts for often bipartisan climate legislation, which, however, only had a limited impact on federal climate policy, mainly because of the narrow understanding of national security. Instead, the dominant construction of climate change as immediate national security threat helped to further 'climatise' the security and defence sector by influencing a range of key security and defence policies and associated best practices.

The US case hence demonstrates the power securitisations anchored in the sovereign discourse and traditional conceptions of security still can have when it comes to raising attention, setting the agenda and transforming political practices. Moreover, it provides insights concerning the bidirectional qualities of securitisation, in this case illustrated by the extensive climatisation of the security and defence sector as well as by the growing demand for 'actionable' climate security data from climate and social science. At the same time, it exemplifies the limits of the sovereign discourse and narrow national security conceptions when it comes to influencing climate policy. Beyond that, the analysis of the securitisation in the US underscores Foucault's claim concerning the close interconnections between different forms of power and particularly elucidates how a dominant discourse can incorporate and transform other representations.

References

Abelson, D. E. (2006). *A Capitol Idea: Think Tanks and US Foreign Policy.* Montreal: McGill-Queen's University Press.

Abelson, D. E. (2010). Do Think Tanks Matter?: Opportunities, Constraints and Incentives for Think Tanks in Canada and the United States. *Global Society, 14*, 213–236. https://doi.org/10.1080/13600820050008458.

Abrahams, D. (2019). From Discourse to Policy: US Policy Communities' Perceptions of and Approaches to Climate Change and Security. *Conflict, Security & Development, 19*, 323–345. https://doi.org/10.1080/14678802.2019.1637080.

Albright, M. K. (1995). Ambassador Albright's Keynote Address to the 1994 Symposium for the Environmental Defense Fund on the Global Environment: International Issues and Institutions. In Wilson Center (Ed.), *Environmental Change and Security Project Report 1* (p. 53). Washington, DC.

Albright, M. K., Berger, S. R., Browner, C., & Danvers, W. (2006). *Energy Security Energy and Security in the 21st Century. A New National Strategy.* Report of the National Security Task Force on Energy. Center for American Progress. Retrieved March 12, 2014, from http://www.americanprogress.org/kf/energy_security_report.pdf.

Allenby, B. R. (2000). Environmental Security: Concept and Implementation. *International Political Science Review, 21*, 5–21. https://doi.org/10.1177/0192512100211001.

American Security Project, & CCS. (2019). *Letter to the President of the United States: 58 Senior Military and National Security Leaders.* Washington, DC. Retrieved February 16, 2020, from https://climateandsecurity.files.wordpress.com/2019/03/letter-to-the-president_senior-military-and-national-security-leaders-denounce-nsc-climate-panel_2019_3_05-1.pdf.

Barnett, A. (2004). Bush Attacks Environment "Scare Stories": Secret Email Gives Advice on Denying Climate Change. Retrieved March 8, 2013, from http://www.guardian.co.uk/environment/2004/apr/04/usnews.theobserver.

Below, A. (2007). The Missing Link: Regionalism as a First Step Toward Globalizing U.S. Environmental Security Policy. *Politics & Policy, 35*(4), 702–715.

Bender, B. (2013, March 9). Chief of US Pacific Forces Calls Climate Biggest Worry. Retrieved August 25, 2014, from http://www.bostonglobe.com/news/nation/2013/03/09/admiral-samuel-locklear-commander-pacific-forces-warns-that-climate-change-top-threat/BHdPVCLrWEMxRe9IXJZcHL/story.html.

Bendery, J. (2014, October 29). Chuck Hagel: We Should Worry About Climate Change Like We Worry About ISIS. *Huffington Post.* Retrieved June 22, 2015, from http://www.huffingtonpost.com/2014/10/29/chuck-hagel-climate-change_n_6068922.html.

Bialasiewicz, L., Campbell, D., Elden, S., Graham, S., Jeffrey, A., & Williams, A. J. (2007). Performing Security: The Imaginative Geographies of Current US Strategy. *Political Geography, 26*, 405–422. https://doi.org/10.1016/j.polgeo.2006.12.002.

Bigo, D. (2009). International Political Sociology. In P. Williams (Ed.), *Security Studies: An Introduction* (pp. 116–129). London, New York: Routledge.

Blair, D. C. (2009). *Annual Threat Assessment of the Intelligence Community for the Senate Select Committee on Intelligence.* Washington, DC.

Bloomberg, M. (2013). Bloomberg: Why Sandy Forced Cities to Take Lead on Climate Change. *CNN.* Retrieved February 24, 2016, from http://edition.cnn.com/2013/08/21/world/europe/bloomberg-why-sandy-force-cities/.

Braml, J. (2011). Interessengruppen und Think Tanks. In S. Koschut (Ed.), *Außenpolitik USA: Theorie – Prozess – Politikfelder – Regionen* (pp. 117–132, UTB, 8472: Politik). Stuttgart: UTB.

Broder, J. M. (2012, November 20). C.I.A. Closes Its Climate Change Office. *New York Times.* Retrieved June 23, 2015, from http://green.blogs.nytimes.com/2012/11/20/c-i-a-closes-its-climate-change-office/?_r=0.

Brown, L. R. (1977). *Redefining National Security* (Worldwatch Paper) (Vol. 14). Washington, DC: Worldwatch Institute.

Brulle, R. J., Carmichael, J., & Jenkins, J. C. (2012). Shifting Public Opinion on Climate Change: An Empirical Assessment of Factors Influencing Concern over Climate Change in the U.S., 2002–2010. *Climatic Change, 114*, 169–188. https://doi.org/10.1007/s10584-012-0403-y.

Brunner, R. D., & Klein, R. (1999). Harvesting Experience: A Reappraisal of the U.S. Climate Change Action Plan. *Policy Sciences, 32*, 133–161.

Bryner, G. (2000). Congress and the Politics of Climate Change. In P. G. Harris (Ed.), *Climate Change and American Foreign Policy* (pp. 111–130). New York, NY: St. Martin's Press.

Brzoska, M. (2012). Climate Change as a Driver of Security Policy. In J. Scheffran, M. Brzoska, H.-G. Brauch, M. P. Link, & J. P. Schilling (Eds.), *Climate*

Change, Human Security and Violent Conflict. Challenges for Societal Stability (Hexagon Series) (Vol. 8, pp. 165–179). Berlin: Springer.

Burke, S., & Parthermore, C. (2008). *Climate Change War Game: Major Findings and Background*. Washington, DC.

Busby, J. W. (2007). *Climate Change and National Security: An Agenda for Action* (Council Special Report No. 32). Washington, DC. Retrieved February 22, 2013, from http://i.cfr.org/content/publications/attachments/ClimateChange_CSR32.pdf.

C2ES. (2006). California Global Warming Solutions Act (AB 32). Retrieved March 3, 2016, from http://www.c2es.org/us-states-regions/action/california/ab32.

C2ES. (2008). Legislation in the 110th Congress Related to Global Climate Change. *Center for Climate and Energy Solutions*. Retrieved March 8, 2013, from http://www.c2es.org/federal/congress/110.

C2ES. (2009). *The National Security Implications of Climate Change*. Arlington.

C2ES. (2012). State Legislation from Around the Country. Center for Climate and Energy Solutions. Retrieved March 8, 2013, from http://www.c2es.org/us-states-regions/key-legislation.

C2ES. (2014). California Cap and Trade. Retrieved March 3, 2016, from http://www.c2es.org/docUploads/calif-cap-trade-01-14.pdf.

Campbell, D. (1992). *Writing Security: United States Foreign Policy and the Politics of Identity*. Manchester: Manchester University Press.

Campbell, K. M., Gulledge, J., McNeill, J. R., Podesta, J., Odgen, P., Fueth, L., et al. (2007). *The Age of Consequences: The Foreign Policy and National Security Implications of Global Climate Change*. Washington, DC.

Carmen, H. E., Parthermore, C., & Rogers, W. (2010). Broadening Horizons: Climate Change and the U.S. Armed Forces. *Center for a New American Security (CNAS)*. Retrieved March 14, 2014, from http://www.cnas.org/files/documents/publications/CNAS%20Publication_Climate%20Change%20and%20the%20US%20Armed%20Forces_April%2020.pdf.

CCS, EESI, & Jackson Foundation. (2017). The National Security Implications of Climate Change. Retrieved February 16, 2020, from https://www.eesi.org/briefings/view/060517security.

CIA. (2009). CIA Opens Center on Climate Change and National Security. Retrieved June 22, 2015, from https://www.cia.gov/news-information/press-releases-statements/center-on-climate-change-and-national-security.html.

Ciuta, F. (2009). Security and the Problem of Context: A Hermeneutical Critique of Securitisation Theory. *Review of International Studies, 35*, 301. https://doi.org/10.1017/S0260210509008535.

CNA. (2007). *National Security and the Threat of Climate Change*. Alexandria. Retrieved October 17, 2010, from http://securityandclimate.cna.org/.

CNA Military Advisory Board. (2014). *National Security and the Accelerating Risks of Climate Change*. Alexandria, VA.

CNA, & Oxfam. (2011). *An Ounce of Prevention: Preparing for the Impact of a Changing Climate on US Humanitarian and Disaster Response*. Alexandria, VA, Boston, MA. Retrieved March 12, 2014, from http://www.oxfamamerica.org/static/oa3/files/an-ounce-of-prevention.pdf.

Cox, M., & Stokes, D. (Eds.). (2012). *US Foreign Policy* (2nd ed.). Oxford: Oxford University Press.

CSIS. (2008). *Integrating 21st Century Development and Security Assistance: Final Report of the Task Force on Nontraditional Security Assistance*. Washington, DC. Retrieved October 17, 2010, from http://africacenter.org/wp-content/uploads/2008/07/Integrating-21st-Century-Development-Assistance.pdf.

Dabelko, G. D., Dabelko, D. D., Matthew, R., Sherbinin, A. d., Claussen, E., & Levy, M. A. (1995). *Environmental Change and Security Program (ECSP) Report 1*. Washington, DC.

Dalby, S. (1990). American Security Discourse: The Persistence of Geopolitics. *Political Geography Quarterly, 9*, 171–188. https://doi.org/10.1016/0260-9827(90)90017-5.

Dalby, S. (2002). *Environmental Security* (Borderlines) (Vol. 20). Minneapolis, MN: University of Minnesota Press.

Dalby, S. (2009). *Security and Environmental Change*. Cambridge: Polity.

Davenport, C. (2014a, January 2). Kerry Quietly Makes Priority of Climate Pact. *New York Times*. Retrieved June 22, 2015, from http://www.nytimes.com/2014/01/03/world/asia/kerry-shifts-state-department-focus-to-environment.html?_r=0.

Davenport, C. (2014b, May 13). Climate Change Deemed Growing Security Threat by Military Researchers. *New York Times*. Retrieved June 22, 2015, from http://www.nytimes.com/2014/05/14/us/politics/climate-change-deemed-growing-security-threat-by-military-researchers.html.

Davenport, C. (2014c, October 13). Pentagon Signals Security Risks of Climate Change. *New York Times*. Retrieved June 15, 2015, from http://www.nytimes.com/2014/10/14/us/pentagon-says-global-warming-presents-immediate-security-threat.html?_r=0.

DHS. (2013). *NIPP 2013 Partnering for Critical Infrastructure Security and Resilience*. Washington, DC.

Director of National Intelligence. (2018). *Statement for the Record: Worldwide Threat Assessment of the US Intelligence Community*. Washington, DC.

Director of National Intelligence. (2019). *Statement for the Record: Worldwide Threat Assessment of the US Intelligence Community*. Washington, DC.

DOD. (2008). *National Defense Strategy*. Washington, DC.

DOD. (2010). *Quadrennial Defense Review Report 2010*. Washington, DC. Retrieved March 8, 2013, from http://www.defense.gov/qdr/images/QDR_as_of_12Feb10_1000.pdf.

DOD. (2011a). *The National Military Strategy of the United States of America: Redefining America's Military Leadership*. Washington, DC.

DOD. (2011b). *Trends and Implications of Climate Change for National and International Security: Report of the Defense Science Board Task Force*. Washington, DC.

DOD. (2012). *Climate Change Adaptation Roadmap (CCAR)*. Washington, DC. Retrieved September 8, 2014, from http://www.acq.osd.mil/ie/download/green_energy/dod_sustainability/2012/Appendix%20A%20-%20DoD%20Climate%20Change%20Adaption%20Roadmap_20120918.pdf.

DOD. (2014a). *2014 Climate Change Adaptation Roadmap*. Washington, DC. Retrieved May 28, 2015, from http://de.scribd.com/doc/242845848/Read-DoD-report-2014-Climate-Change-Adaptation-Roadmap.

DOD. (2014b). *Quadrennial Defense Report 2014*. Washington, DC.

DOD. (2014c). *Strategic Sustainability Performance Plan FY 2014*. Washington, DC. Retrieved May 28, 2014, from http://denix.osd.mil/sustainability/upload/DoD-SSPP-FY14-FINAL-w_CCAR.pdf.

DOD. (2014d). *USCENTCOM Climate Change Assessment*. Washington, DC. Retrieved May 28, 2015, from http://web.ornl.gov/sci/knowledgediscovery/QDR/docs/USCENTCOM_Climate_Change_Assessment.pdf.

DOD. (2015a). About SERDP. Retrieved June 10, 2015, from https://www.serdp-estcp.org/About-SERDP-and-ESTCP/About-SERDP.

DOD. (2015b). *National Security Implications of Climate-Related Risks and a Changing Climate*. Washington, DC.

DOD. (2016). *DOD Directive 4715.21: Climate Change Adaptation and Resilience*. Washington, DC.

DOD. (2018). *Climate-Related Risk to DoD Infrastructure: Initial Vulnerability Assessment Survey. (SLVAS) Report*. Washington, DC.

Donner, S., & Faltin, F. (2007). *Klimapolitische Entwicklungen in den USA: Initiativen auf bundesstaatlicher und regionaler Ebene – Info Brief – Deutscher Bundestag Wissenschaftliche Dienste*. Berlin. Retrieved from http://www.bundestag.de/dokumente/analysen/2008/Klimapolitische_Entwicklung_USA.pdf.

Dröge, S. (2016). The Paris Agreement 2015: Turning Point for the International Climate Regime. *SWP Research Paper* (RP4).

Ehrlich, P. (1968). *The Population Bomb*. New York, NY: Ballantine.

Elbe, S. (2011). Pandemics on the Radar Screen: Health Security, Infectious Disease and the Medicalisation of Insecurity. *Political Studies, 59*, 848–866. https://doi.org/10.1111/j.1467-9248.2011.00921.x.

EPA. (1994). Memorandum of Understanding Between the U.S. Environmental Protection Agency and the U.S. Department of Defense. Retrieved June 10, 2015, from http://www2.epa.gov/fedfac/memorandum-understanding-between-us-environmental-protection-agency-and-us-department-defense.

Eshelman, R. S. (2014, May 1). The Danger of Fair and Balanced. *Columbia Journalism Review*. Retrieved October 20, 2016, from http://www.cjr.org/essay/the_danger_of_fair_and_balance.php.

Falkner, R. (2005). American Hegemony and the Global Environment. *International Studies Review, 7*, 585–599.

FEMA. (2012). *Crisis Response and Disaster Resilience 2030: Forging Strategic Action in an Age of Uncertainty*. Washington, DC.

Fincham, M. W. (2014). The Day Before Yesterday: When Abrupt Climate Change Came to the Chesapeake Bay. Retrieved August 18, 2015, from https://www.climate.gov/news-features/features/day-yesterday-when-abrupt-climate-change-came-chesapeake-bay.

Fingar, T. (2008). *National Intelligence Assessment on the National Security Implications of Global Climate Change to 2030*. Washington, DC.

Fleishman, R., & Goodman, S. W. (2018). Climate Change and the U.S. Military. *Defense Dossier, 23*, 3–7.

Fletcher, A. L. (2009). Clearing the Air: The Contribution of Frame Analysis to Understanding Climate Policy in the United States. *Environmental Politics, 18*, 800–816. https://doi.org/10.1080/09644010903157123.

Floyd, R. (2010). *Security and the Environment: Securitisation Theory and US Environmental Security Policy*. Cambridge: Cambridge University Press.

Foley, C. (2012). *Military Basing and Climate Change*. Washington, DC. Retrieved July 4, 2013, from http://americansecurityproject.org/wp-content/uploads/2012/11/Military-Basing-and-Climate-Change.pdf.

Foley, C., & Holland, A. (2012a). *Climate Security Report: Part One Climate Change and Security*. Washington, DC: Retrieved from http://americansecurityproject.org/wp-content/uploads/2012/11/CSR-INTRODUCTION-email.pdf.

Foley, C., & Holland, A. (2012b). *Climate Security Report: Part Three Climate Change & the Homeland*. Washington, DC: Retrieved from http://americansecurityproject.org/wp-content/uploads/2012/11/CSR-INTRODUCTION-email.pdf.

Foley, C., & Holland, A. (2012c). *Climate Security Report: Part Two Climate Change & Global Security*. Washington, DC: Retrieved from http://americansecurityproject.org/wp-content/uploads/2012/11/CSR-INTRODUCTION-email.pdf.

Gallup. (2014). Confidence in Institutions. Retrieved August 25, 2014, from http://www.gallup.com/poll/1597/confidence-institutions.aspx.

Ge, M., Friedrich, J., & Damassa, T. (2014). 6 Graphs Explain the World's Top 10 Emitters. *World Resources Institute (WRI)*. Retrieved January 19, 2016, from http://www.wri.org/blog/2014/11/6-graphs-explain-world's-top-10-emitters.

Gellner, W. (1995). *Ideenagenturen für Politik und Öffentlichkeit: Think Tanks in den USA und in Deutschland* (Studien zur Sozialwissenschaft). Wiesbaden: VS Verlag für Sozialwissenschaften.

Gibbs, W., & Lyall, S. (2007, October 13). Gore Shares Peace Prize for Climate Change Work. *New York Times*. Retrieved August 28, 2015, from http://www.nytimes.com/2007/10/13/world/13nobel.html?pagewanted.

Gilbert, E. (2012). The Militarization of Climate Change. *ACME: An International E-Journal for Critical Geographies, 11*(1), 1–14.

Giroux, H. A. (2006). Reading Hurricane Katrina: Race, Class, and the Biopolitics of Disposability. *College Literature, 33*(3), 171–196.

Gore, A. A. (2007). *Nobel Peace Prize Lecture*. Oslo. Retrieved October 17, 2010, from http://nobelprize.org/nobel_prizes/peace/laureates/2007/gore-lecture_en.html.

Govtrack. (2007). Summaries for the Global Climate Change Security Oversight Act. Retrieved March 4, 2016, from https://www.govtrack.us/congress/bills/110/hr1961/summary.

Green, C. (2019). 7 Ways U.S. States Are Leading Climate Action. Retrieved February 16, 2020, from https://unfoundation.org/blog/post/7-ways-u-s-states-are-leading-climate-action/.

Greshko, M., Parker, L., Brian Clark, H., Stone, D., Borunda, A., & Gibbens, S. (2019, May 3). A Running List of How President Trump Is Changing Environmental Policy. *National Geographic*. Retrieved February 16, 2020, from https://www.nationalgeographic.com/news/2017/03/how-trump-is-changing-science-environment/.

Grundmann, R., & Scott, M. (2014). Disputed Climate Science in the Media: Do Countries Matter? *Public Understanding of Science, 23*(2).

Guber, D. L., & Bosso, C. J. (2013). "High Hopes and Bitter Disappointment": Public Discourse and the Limits of the Environmental Movement in Climate Change Politics. In N. J. Vig & M. E. Kraft (Eds.), *Environmental Policy: New Directions for the Twenty-First Century* (8th ed., pp. 54–82). Thousand Oaks, CA: CQ Press.

Hajer, M. A. (1997). *The Politics of Environmental Discourse*. Oxford: Oxford University Press.

Harris, P. G. (Ed.). (2000a). *Climate Change and American Foreign Policy*. New York, NY: St. Martin's Press.

Harris, P. G. (2000b). Climate Change and Foreign Policy: An Introduction. In P. G. Harris (Ed.), *Climate Change and American Foreign Policy* (pp. 3–25). New York, NY: St. Martin's Press.

Harris, P. G. (2000c). Climate Change: Is the United States Sharing the Burden? In P. G. Harris (Ed.), *Climate Change and American Foreign Policy* (pp. 29–49). New York, NY: St. Martin's Press.

Harris, P. G. (2001). International Environmental Affairs and U.S. Foreign Policy. In P. G. Harris (Ed.), *The Environment, International Relations, and U.S. Foreign Policy* (pp. 3–41). Washington, DC: Georgetown University Press.

Harris, P. G. (2002). Environmental Security: Will Bush Follow Clinton's Lead? *Pacifica Review: Peace, Security & Global Change, 14*, 149–157. https://doi.org/10.1080/13239100220141191.

Harris, P. G. (Ed.). (2009). *Climate Change and Foreign Policy: Case Studies from East to West* (Routledge Advances in International Relations and Global Politics) (Vol. 71). London, New York: Routledge.

Harrison, K. (2007). The Road not Taken: Climate Change Policy in Canada and the United States. *Global Environmental Politics, 7*, 92–117. https://doi.org/10.1162/glep.2007.7.4.92.

Hartmann, B. (2010). Rethinking Climate Refugees and Climate Conflict: Rhetoric, Reality and the Politics of Policy Discourse. *Journal of International Development, 22*(2), 233–246.

Homer-Dixon, T. (1994). Environmental Scarcities and Violent Conflict: Evidence from Cases. *International Security, 19*(1), 5–40.

IEA. (2012). *World Energy Outlook 2012: Zusammenfassung. German Translation*. Paris.

Interview. (2014a). *Interview on Climate Change and Security with Think Tank Experts in Washington D.C., 06.05.2014*. Washington, DC.

Interview. (2014b). *Interview on Climate Change and Security with Think Tank Experts in Washington D.C., Center for Climate and Security, Frank Femia, 06.05.2014*. Washington, DC.

Interview. (2014c). *Interview on Climate Change and Security with Think Tank Experts in Washington D.C., 08.05.2014*. Washington, DC.

Interview. (2014d). *Interview on Climate Change and Security with Think Tank Experts in Washington D.C., 13.05.2014*. Washington, DC.

Interview. (2014e). *Interview with an NGO Expert on Climate Change and Security, 13.05.2014*. Berlin.

Interview. (2014f). *Interview on Climate Change and Security with Think Tank Experts in Washington D.C., 14.05.2014*. Washington, DC.

Interview. (2014g). *Interview on Climate Change and Security with NGO Experts in Washington D.C., 16.05.2014*. Washington, DC.

Interview. (2014h). *Interview on Climate Change and Security with Government Officials in Washington D.C., 27.05.2014*. Washington, DC.

Interview. (2014i). *Interview on Climate Change and Security with Government Officials in Washington D.C., 28.05.2014*. Washington, DC.

Interview. (2014j). *Interview on Climate Change and Security with Think Tank Experts in Washington D.C., 28.05.2014.* Washington, DC.
Interview. (2014k). *Interview on Climate Change and Security with Government Officials in Washington D.C., 30.05.2014.* Washington, DC.
Interview. (2014l). *Interview on Climate Change and Security with Think Tank Experts in Washington D.C., 03.06.2014.* Washington, DC.
Interview. (2014m). *Interview on Climate Change and Security with Think Tank Experts in Washington D.C., Center for American Progress, Michael Werz and Max Hoffmann, 03.06.2014.* Washington, DC.
Interview. (2014n). *Interview on Climate Change and Security with Think Tank Experts in Washington D.C., 09.06.2014.* Washington, DC.
Interview. (2014o). *Interview on Climate Change and Security with Experts for US Intelligence Policy, 10.06.2014.* Washington, DC.
IPCC. (2007). *A Report of Working Group I of the Intergovernmental Panel on Climate Change: Summary for Policymakers.* New York.
Koschut, S. (2011). Exzeptionalismus, Interventionismus und Isolationismus. In S. Koschut (Ed.), *Außenpolitik USA: Theorie – Prozess – Politikfelder – Regionen* (pp. 31–38, UTB, 8472: Politik). Stuttgart: UTB.
Kraft, M. E. (2013). Environmental Policy in Congress. In N. J. Vig & M. E. Kraft (Eds.), *Environmental Policy: New Directions for the Twenty-First Century* (8th ed., pp. 109–134). Thousand Oaks, CA: CQ Press.
Kutz, M.-S. (2011). Zivilreligion und politische Kultur. In S. Koschut (Ed.), *Außenpolitik USA: Theorie – Prozess – Politikfelder – Regionen* (pp. 39–46, UTB, 8472: Politik). Stuttgart: UTB.
Leggett, J., & Lattanzio, R. (2009). *Status of the Copenhagen Climate Change Negotiations* (CRS Report for Congress). Retrieved February 14, 2013, from http://www.fas.org/sgp/crs/misc/R40910.pdf.
Leiserowitz, A. (2005). American Risk Perceptions: Is Climate Change Dangerous? *Risk Analysis, 25*(6), 1433–1442.
Matthew, R. A. (2013). Environmental Security. In N. J. Vig & M. E. Kraft (Eds.), *Environmental Policy: New Directions for the Twenty-First Century* (8th ed., pp. 344–368). Thousand Oaks, CA: CQ Press.
McCormack, T. (2010). *Critique, Security and Power: The Political Limits to Empanicipatory Approaches.* London: Routledge.
McCright, A., & Dunlap, R. E. (2011). The Politicization of Climate Change and Polarization in the American Public's View of Global Warming, 2001–2010. *The Sociological Quarterly, 52*, 155–194.
McGann, J. G. (2007). *Think Tanks and Policy Advice in the United States: Academics, Advisors and Advocates* (Routledge Research in American Politics) (Vol. 1). Abingdon [England], New York: Routledge.

McGrady, E., Kingsley, M., & Steward, J. (2010). *Climate Change: Potential Effects on Demands for US Military Humanitarian Assistance and Disaster Response*. Alexandria, VA.

Meadows, D. L., & Meadows, D. H. (1972). *Die Grenzen des Wachstums: Bericht des Club of Rome zur Lage der Menschheit*. Stuttgart: Deutsche Verlags-Anstalt.

Melillo, J. M., Richmond, T. C., & Yohe, G. W. (2014). *Climate Change Impacts in the United States: The Third National Climate Assessment*. Washington, DC.

Mildner, S.-A., & Richert, J. (2010). Obamas neue Klimapolitik: Möglichkeiten und Grenzen eines klimapolitischen Wandels in den USA. *SWP Studie, Februar 2010*.

Mildner, S.-A., Petersen, H., & Wodni, W. (2012). Klimaschutz bleibt Nebensache für die USA. *SWP Aktuell, 69*(November 2012).

Munoz, C. (2012, May 4). Panetta Warns Climate Change Having 'Dramatic Impact' on National Security. *The Hill*. Retrieved June 22, 2015, from http://thehill.com/policy/energy-environment/225475-panetta-climate-change-hits-national-security.

National Academy of Sciences. (1975). *Understanding Climatic Change: A Program for Action*. Washington, DC: National Academy of Sciences.

National Academy of Sciences. (1977). *Energy and Climate: Studies in Geophysics*. Washington, DC: National Academy of Sciences.

National Intelligence Council. (2016). *Implications for US National Security of Anticipated Climate Change*. Washington, DC.

National Intelligence Council. (2017). *Global Trends: Paradox of Progress*. Washington, DC: National Intelligence Council.

Oels, A. (2011). Rendering Climate Change Governable by Risk: From Probability to Contingency. *Geoforum, 45*, 1–13. https://doi.org/10.1016/j.geoforum.2011.09.007.

Oels, A. (2012). From 'Securitization' of Climate Change to 'Climatization' of the Security Field: Comparing Three Theoretical Perspectives. In J. Scheffran, M. Brzoska, H.-G. Brauch, M. P. Link, & J. P. Schilling (Eds.), *Climate Change, Human Security and Violent Conflict. Challenges for Societal Stability* (Hexagon Series) (Vol. 8, pp. 185–205). Berlin: Springer.

Oldopp, B. (2014). *Das politische System der USA: Eine Einführung* (2nd ed., EBL-Schweitzer). Dordrecht: Springer.

Park, J. (2000). Governing Climate Change Policy. In P. G. Harris (Ed.), *Climate Change and American Foreign Policy* (pp. 73–87). New York, NY: St. Martin's Press.

Parthermore, C., & Rogers, W. (2010). Promoting the Dialogue: Climate Change and the Quadrennial Defense Review. *Center for a New American Security (CNAS)*. Retrieved March 14, 2014, from http://www.cnas.org/files/documents/publications/QDR&ClimateChange_Parthemore_Rogers_Jan2010_code406_workingpaper_2.pdf.

Pazzanese, C. (2016, July 14). Politics in a 'Post-Truth' Age: In This Topsy-Turvy Presidential Campaign, the Old Laws May No Longer Apply. *Harvardgazette*. Retrieved October 20, 2016, from http://news.harvard.edu/gazette/story/2016/07/politics-in-a-post-truth-age/.

Pew Research Center. (2012). Tough Stance on Iran Endorsed. Public Priorities: Deficit Rising, Terrorism Slipping. Retrieved March 27, 2013, from http://www.people-press.org/files/legacy-pdf/1-23-12%20Priorities%20Release.pdf.

Pielke, R. A., Jr. (2000a). Policy History of the US Global Change Research Program: Part I. Administrative Development. *Global Environmental Change, 10*, 9–25.

Pielke, R. A., Jr. (2000b). Policy History of the US Global Change Research Program: Part II. Legislative Process. *Global Environmental Change, 10*, 133–144.

Revkin, A. (2017, March 14). Trump's Defense Secretary Cites Climate Change as National Security Challenge. *ProPublica*. Retrieved February 16, 2020, from https://www.propublica.org/article/trumps-defense-secretary-cites-climate-change-national-security-challenge?utm_campaign=bt_twitter&utm_source=twitter&utm_medium=social.

Revkin, A. (2019, February 22). Does the U.S. Need a 'Presidential Climate Security Committee'? *National Geographic*. Retrieved February 16, 2020, from https://www.nationalgeographic.com/environment/2019/02/trump-presidential-climate-security-committee/.

Richert, J. (2009). *Klimawandel und Sicherheit in der Amerikanischen Politik* (Diskussionspapier). Berlin. Retrieved August 14, 2012, from http://www.swp-berlin.org/fileadmin/contents/products/arbeitspapiere/2009_DisP_Richert05_ks.pdf.

Rid, T. (2011). Militär. In S. Koschut (Ed.), *Außenpolitik USA: Theorie – Prozess – Politikfelder – Regionen* (pp. 107–116, UTB, 8472: Politik). Stuttgart: UTB.

Rogers, W., & Gulledge, J. (2010). Lost in Translation: Closing the Gap Between Climate Science and National Security Policy. *Center for a New American Security (CNAS)*. Retrieved March 12, 2014, from http://www.cnas.org/files/documents/publications/Lost%20in%20Translation_Code406_Web_0.pdf.

Roskin, M. (1974). From Pearl Harbor to Vietnam: Shifting Generational Paradigms and Foreign Policy. *Political Science Quarterly, 89*, 563. https://doi.org/10.2307/2148454.

Scherwitz, E. (2014). *Welche Wege führen nach Washington?: Eine Analyse des EU-Einflusses auf die US-Politik* (Internationale Beziehungen) (Vol. 20, 1st ed.). Baden-Baden: Nomos.

Schwartz, P., & Randall, D. (2003). *An Abrupt Climate Change Scenario and Its Implications for United States National Security*. Washington, DC. Retrieved

February 28, 2013, from http://stephenschneider.stanford.edu/Publications/PDF_Papers/SchwartzRandall2004.pdf.
Selby, J. (2019). The Trump Presidency, Climate Change, and the Prospect of a Disorderly Energy Transition. *Review of International Studies, 45*, 471–490. https://doi.org/10.1017/S0260210518000165.
Sherbinin, A. D. (1995). World Population Growth and U.S. National Security. In Wilson Center (Ed.), *Environmental Change and Security Project Report 1* (pp. 24–39). Washington, DC.
Simmons, P. J. (1995). Introduction. In Wilson Center (Ed.), *Environmental Change and Security Project Report 1* (p. 2). Washington, DC.
Sperling, J. (2010). United States: A Full Spectrum Contributor to Governance? In E. J. Kirchner & J. Sperling (Eds.), *National Security Cultures: Patterns of Global Governance* (pp. 172–210). London, New York: Routledge.
Stern, T., & Antholis, W. (2007). A Changing Climate: The Road Ahead for the United States. *The Washington Quaterly, 31*(1), 175–188.
Taylor, L., & Branigan, T. (2014, November 12). US and China Strike Deal on Carbon Cuts in Push for Global Climate Change Pact. *The Guardian*. Retrieved May 28, 2015, from http://www.theguardian.com/environment/2014/nov/12/china-and-us-make-carbon-pledge.
The White House. (1991). *National Security Strategy of the United States*. Washington, DC.
The White House. (1994). *A National Security Strategy of Engagement and Enlargement*. Washington, DC.
The White House. (2002). *The National Security Strategy of the United States of America, 2002*. Washington, DC.
The White House. (2006). *The National Security Strategy of the United States of America, 2006*. Washington, DC.
The White House. (2010). *National Security Strategy of the United States 2010*. Washington, DC.
The White House. (2013). Climate Change Resilience. Retrieved October 11, 2017, from https://obamawhitehouse.archives.gov/administration/eop/ceq/initiatives/resilience.
The White House. (2015a). *Findings from Select Federal Reports: The National Security Implications of Climate Change*. Washington, DC: Retrieved May 28, 2015, from https://www.whitehouse.gov/sites/default/files/docs/National_Security_Implications_of_Changing_Climate_Final_051915.pdf.
The White House. (2015b). *National Security Strategy of the United States 2014*. Washington, DC.
The White House. (2015c). *The Clean Power Plan*. Washington, DC: Retrieved September 14, 2015, from https://www.whitehouse.gov/climate-change.
The White House. (2017). *National Security Strategy 2017*. Washington, DC.

Townsend, M., & Harris, P. (2004, February 22). Now the Pentagon Tells Bush: Climate Change Will Destroy Us. *The Guardian/The Observer.* Retrieved August 18, 2015, from http://www.theguardian.com/environment/2004/feb/22/usnews.theobserver.

Trump, D. J. (2017). Statement by President Trump on the Paris Climate Accord. Retrieved February 16, 2020, from https://www.whitehouse.gov/briefings-statements/statement-president-trump-paris-climate-accord/.

US Army Corps of Engineers. (2015). About Us. Retrieved June 9, 2015, from http://www.usace.army.mil/About.aspx.

US Congress. (1988). *Global Climate Protection Act of 1987.* Washington, DC: Retrieved from http://www.gpo.gov/fdsys/pkg/STATUTE-101/pdf/STATUTE-101-Pg1331.pdf.

US Congress. (2008). National Defense Authorization Act for Fiscal Year 2008. 110th US Congress. Retrieved March 4, 2016, from https://www.gpo.gov/fdsys/pkg/PLAW-110publ181/html/PLAW-110publ181.htm.

US Department of State. (2009). Appointment of Special Envoy on Climate Change Todd Stern. Retrieved March 8, 2013, from http://www.state.gov/secretary/rm/2009a/01/115409.htm.

US Government. (1990). *National Global Change Research Act of 1990: Public Law 101-606 101st Congress.* Retrieved August 21, 2015, from http://www.gpo.gov/fdsys/pkg/STATUTE-104/pdf/STATUTE-104-Pg3096.pdf.

US House of Representatives. (1989). *The Global Change Research Act of 1989 Hon. Christopher H. Smith (Extension of Remarks October 26, 1989): Congressional Record 101st Congress (1989–1990).* Washington, DC.

US House of Representatives. (1990). *Sense of Congress Regarding Linkage Between the Environment and National Security (House of Representatives – September 17, 1990): Congressional Record – 101st Congress (1989-1990).* Washington, DC.

US House of Representatives. (1995). *Omnibus Civilian Science Authorization Act of 1995.* Washington, DC.

US House of Representatives. (1998a). *Personal Explanation: Congressional Record – House May 20, 1998 (H3577).* Washington, DC.

US House of Representatives. (1998b). *Departments of Veterans Affairs and Housing and Urban Development, and Independent Agencies Appropriations Act: Congressional Record – House, 105th Congress, 2nd Session.* Washington, DC.

US House of Representatives. (2005). *Announcing the Introduction of the New Apollo Energy Project: Congressional Record.* Washington, DC.

US House of Representatives. (2009a). *Cap and Trade: Congressional Record—House May 13, 2009 (H5555).* Washington, DC.

US House of Representatives. (2009b). *Energy Independence Is a Matter of National Security: Congressional Record.* Washington, DC.

US House of Representatives. (2011). *Energy Tax Prevention Act: Congressional Record 112th Congress (2011–2012)*. Washington, DC.
US House of Representatives. (2012). *Climate Change and National Security: Congressional Record 112th Congress (2011–2012), 09.05.2012.* Washington, DC.
US House of Representatives. (2013). *The Safe Climate Caucus: Congressional Record 113th Congress (2013–2014)*. Washington, DC.
US House of Representatives. (2014). *Climate Change Denial: Congressional Record 113th Congress (2013–2014)*. Washington, DC.
US House of Representatives. (2019). *National Defense Authorization Act for the Fiscal Year 2020*. Washington, DC.
US National Guard. (2015). About the Guard. Retrieved June 9, 2015, from http://www.nationalguard.mil/AbouttheGuard.aspx.
US Navy. (2009). Task Force on Climate Change. Retrieved September 8, 2014, from http://greenfleet.dodlive.mil/climate-change/.
US Navy. (2010). *Climate Change Roadmap*. Washington, DC: Retrieved September 8, 2014, from http://www.navy.mil/navydata/documents/CCR.pdf.
US Navy. (2016). The Great Green Fleet (GGF). Retrieved March 17, 2016, from http://greenfleet.dodlive.mil/energy/great-green-fleet/.
US Senate. (1989a). *The Global Environment, Senate Debate, May 12, 1989: Congressional Record – 101st Congress (1989–1990)*. Washington, DC.
US Senate. (1989b). *The Stratosperic Ozone and Climate Protection Act of 1989 (Senate – October 12, 1989): Congressional Record – 101st Congress (1989–1990)*. Washington, DC.
US Senate. (1990a). *Global Warming: What Can the World's Legislators Do About It?: Congressional Record – 101st Congress (1989–1990)*. Washington, DC.
US Senate. (1990b). *World Environment Day: Congressional Record – 101st Congress (1989–1990)*. Washington, DC.
US Senate. (1990c). *S.RES.316 – Concerning the Interparliamentary Conference on the Global Environment. (Introduced in Senate – IS): Congressional Record – 101st Congress (1989–1990)*. Washington, DC.
US Senate. (1990d). *National Defense Authorization Act for Fiscal 1990*. Washington, DC.
US Senate. (1991a). *Global Warming – Senate Resolution 53: Congressional Record 102nd Congress (1991–1992)*. Washington, DC.
US Senate. (1991b). *Global Climate System: Congressional Record 102nd Congress (1991–1992)*. Washington, DC.
US Senate. (1992a). *Earth Summit: Congressional Record 102nd Congress (1991–1992)*. Washington, DC.
US Senate. (1992b). *The Earth Summit (Senate – June 04, 1992): Congressional Record 102nd Congress (1991–1992)*. Washington, DC.

US Senate. (1997a). *Global Warming: Congressional Record – 105th Congress (1997–1998)*. Washington, DC.
US Senate. (1997b). *The Climate Change Treaty (Senate – October 23, 1997): Congressional Record – 105th Congress (1997–1998)*. Washington, DC.
US Senate. (1999). *Statements on Introduced Bills and Joint Resolutions: Congressional Record*. Washington, DC: Retrieved from https://www.gpo.gov/fdsys/pkg/CRECB-1999-pt13/html/CRECB-1999-pt13-Pg18460-4.htm.
US Senate. (2001a). *Clean Air and Global Warming: Congressional Record, 107th Congress*. Washington, DC.
US Senate. (2001b). *Cap and Trade Approach to Climate Change*. Washington, DC.
US Senate. (2003a). *Air Pollution and Global Warming: Congressional Record – Senate, 03.03.2003*. Washington, DC.
US Senate. (2003b). *Climate Stewardship Act: Congressional Record – Senate, 29.10.2003*. Washington, DC.
US Senate. (2003c). *Climate Stewardship Act: Congressional Record – Senate, 30.10.2003*. Washington, DC.
US Senate. (2007a). *Global Warming Pollution Reduction Act of 2007*. Washington, DC.
US Senate. (2007b). *Global Warming: Congressional Record – Senate October 29, 2008 (S13498)*. Washington, DC.
US Senate. (2008a). *Climate Security Act of 2008 – Motion to Proceed: Congressional Record – Senate*. Washington, DC.
US Senate. (2008b). *Congressional Record – Senate: S3036 – the Lieberman-Warner Climate Security Act of 2008*. Washington, DC.
US Senate. (2008c). *S5014 Lieberman-Warner Climate Security Act of 2008*. Washington, DC.
US Senate. (2008d). *Climate Security Act: Congressional Record – Senate June 5, 2008 (S5188)*. Washington, DC.
US Senate. (2008e). *Debates on Climate Security Act: Congressional Record – Senate June 5, 2008 (S5194)*. Washington, DC.
US Senate. (2008f). *Lieberman Warner Climate Security Act: Congressional Record – Senate June 6, 2008*. Washington, DC.
US Senate. (2009). *Department of Defense Appropriations Act: Congressional Record – Senate, October 6, 2009 (S10143)*. Washington, DC.
US Senate. (2010). *Debates on Energy Reform and Climate Change: Congressional Record – Senate, December 22, 2010*. Washington, DC.
US Senate. (2011a). *Expiry of Section 1603 Treasury Grant Program: Congressional Record – Senate, 13.12.2011*. Washington, DC.
US Senate. (2011b). *Disasters in 2011: Congressional Record*. Washington, DC.
US Senate. (2012a). *Flood Insurance Reform and Modernization Act – Motion to Proceed: Congressional Record*. Washington, DC.

US Senate. (2012b). *National Defense Authorization Act: Congressional Record – Senate, 28.11.2012*. Washington, DC.
US Senate. (2013). *Climate Change: Congressional Record*. Washington, DC.
US Senate. (2014a). *All Night Session on Climate Change: Congressional Record – Senate, 10.03.2014*. Washington, DC. Retrieved from https://www.congress.gov/crec/2014/03/10/modified/CREC-2014-03-10-pt2-PgS1421-2.htm.
US Senate. (2014b). *Senate Report 113-211, Accompanying H.R. 4870, the Department of Defense Appropriations Bill, 2015*. Washington, DC.
USAid. (2012). *Climate Change and Development Strategy 2012–2016: Clean Resilient Growth*. Washington, DC.
USAid. (2016). Denton Program for Private Donations. *USAid*. Retrieved March 14, 2016, from https://www.usaid.gov/work-usaid/partnership-opportunities/humanitarian-responders/denton-program.
USGCRP. (2015). Legal Mandate of the U.S. Global Change Research Program (USGCRP). Retrieved August 25, 2015, from http://www.globalchange.gov/about/legal-mandate.
Vagg, X. (2012). *American Security: The Impact of Climate Change*. Washington, DC. Retrieved from http://americansecurityproject.org/wp-content/uploads/2012/11/Ref-0093-American-Security-The-Impact-of-Climate-Change.pdf.
Vig, N. J. (2013). Presidential Powers and Environmental Policy. In N. J. Vig & M. E. Kraft (Eds.), *Environmental Policy: New Directions for the Twenty-First Century* (8th ed., pp. 84–108). Thousand Oaks, CA: CQ Press.
Werz, M., & Conley, L. (2012). Climate Change, Migration, and Conflict. Addressing Complex Crisis Scenarios in the 21st Century. *Center for American Progress; Heinrich Böll Stiftung*. Retrieved March 12, 2014, from http://www.americanprogress.org/wp-content/uploads/issues/2012/01/pdf/climate_migration.pdf.
Werz, M., & Manlove, K. (2009). *Climate Change on the Move Climate Migration Will Affect the World's Security*. Washington, DC: Retrieved March 12, 2014, from http://www.americanprogress.org/wp-content/uploads/issues/2009/12/pdf/climatechangeonthemove.pdf.
World Bank. (2013). CO_2 Emissions (kt). Retrieved April 2, 2013, from http://data.worldbank.org/indicator/EN.ATM.CO2E.KT/countries.

CHAPTER 3

Germany: Climate Change, Human Security and Southern Populations

Introduction

> *Rising global temperatures will jeopardize the bases of many people's livelihoods, especially in the developing regions, increase vulnerability to poverty and social deprivation, and thus put human security at risk. (WBGU* 2008*, p. 1)*

The above quote stands at the beginning of one of the most influential reports on climate change and security published by the German Advisory Council on Global Change (WBGU) at the peak of the global climate security debates in the year 2007. The quote pointedly summarises the dominant climate security discourse in Germany, which has primarily constructed climate change as a threat to the human security of poor individuals in developing countries. Beyond that, while risk conceptions linked to the governmental discourse were also common, in stark contrast to the US case, traditional and state-focused national security conceptions in line with the sovereign discourse were much less influential in Germany. Consequently, climate security discourses mainly led to a focus on measures to mitigate GHG emissions and to support developing countries to adapt to the security implications of climate change.

Why include Germany as a case? Despite its rather small geographical expanse, due to its large population of over 80 million and its highly industrialised economy (OECD 2012, p. 8), Germany is one of the largest contributors to global GHG emissions (Global Carbon Atlas 2016) and

F. von Lucke, *The Securitisation of Climate Change and the Governmentalisation of Security*, New Security Challenges,
https://doi.org/10.1007/978-3-030-50906-4_3

also has well above average per capita emissions (World Bank 2016; Clark 2011; Jänicke 2011, p. 130). However, Germany has also been a vanguard for progressive domestic and international climate policies since the 1980s (Weidner and Mez 2008, pp. 356–357; Jänicke 2011, p. 129; Schreurs 2002, pp. 1–2, 10). It committed itself to ambitious emissions reductions from early on (Weidner 2008, p. 7) and was one of the driving forces behind the establishment of the UNFCCC and later the Kyoto Protocol (Böckem 2000, p. 7; Weidner 2008, p. 7). Consequently, Germany has ranked well in several comparative climate rankings and, for instance, always stayed in the top ten of the Climate Performance Index (CPI) between 2008 and 2013. Even though some of its GHG reductions in the past are due to the deindustrialisation of former East Germany (DIW/SPRU 2001, p. 9; Weidner and Mez 2008, p. 357) and the country fell in the climate rankings after 2015, it has still outperformed many other industrialised countries and especially the US (Germanwatch and CAN 2015).

The remainder of this chapter explores the role of climate security discourses in this development. Concerning the empirical material, it rests on a discourse analysis of over 60 parliamentary debates on climate change and 35 key publications from research institutions, NGOs and think tanks that have drawn connections between climate change and security. Moreover, it includes findings from 25 semi-structured interviews with government officials, scientists, and experts from NGOs and think tanks.

The Origins of the Climate and Security Debate in Germany

In contrast to its progressive role today, Germany was a latecomer when it comes to environmental policies (Weidner and Mez 2008, p. 358; OECD 1993) and only began to take these issues more seriously in the 1970s (Weidner and Mez 2008, p. 358; Mederake and Duwe 2014, pp. 5–6; Schreurs 2002, p. 17). At this time, the country initiated several environmental protection programmes, established the Federal Environment Agency (*Umweltbundesamt*, UBA) (Deutscher Bundestag 1971; Mederake and Duwe 2014, p. 5) and began to entrench concepts such as 'sustainable development' and later 'ecological modernization' into the political debate (Arnold et al. 2012, p. 194; Eastin et al. 2011, p. 17). Ultimately, a key precursor of the climate (and security) debate in Germany was the detection of the destruction of the ozone layer and the subsequent international debates in the 1980s, which already contained several securitising articulations.

These debates led to the realisation that it was impossible to handle environmental issues by national policies alone (Mederake and Duwe 2014, pp. 7, 9) and helped to popularise concerns for dangerous anthropogenic interference with the earth's atmosphere in Germany (Weidner and Mez 2008, p. 360; Ulbert 1997, pp. 21, 33; Schreurs 2002, p. 154). This growing awareness was then reinforced by the rise of influential social movements that campaigned against nuclear energy and transnational environmental problems in the 1970s and 1980s (Weidner and Mez 2008, p. 368; Schreurs 2002, pp. 76–77). Eventually, environmental concerns found entrance into the political landscape with the founding of the Green Party in 1980 (Ulbert 1997, p. 26; Raschke 1993) and its first electoral success in 1983 (Krück et al. 1999, p. 2; Schreurs 2002, p. 16).

At the same time, climate scientists played a central role in sensitising political actors for the dangers of climate change (Weber 2008, pp. 59–95). They were amongst the first to explicitly link climate change to threat conceptions and in the mid-1980s coined the notion of the 'climate catastrophe' (*Klimakatastrophe*). In 1985, the Working Group on Energy of the German Physical Society (DPG) issued a highly noticed appeal with the title 'A Warning of the Coming Climate Catastrophe' (DPG 1986). A year later this was followed by a joined appeal of the DPG and the German Meteorological Society (DMG), titled a 'Warning of Global Anthropogenic Climate Change' (DPG and DMG 1987). Originally, the intention of the DPG had been to push nuclear energy as GHG neutral energy source. However, this plan was largely diminished by the Chernobyl catastrophe in 1986, and instead, the debate shifted more towards renewable energy to prevent the climate catastrophe (Weidner and Mez 2008, pp. 368–369), which would later become a recurring theme of the German debate.

Because of the growing awareness for climate change and its possibly catastrophic consequences and further pushed by the Chernobyl disaster (Mederake and Duwe 2014, p. 7), in 1986 Chancellor Kohl founded the Federal Ministry for the Environment, Nature Conservation and Nuclear Safety (BMU) and in 1987 called climate change the single most important environmental concern (Weidner 2008, pp. 6–7; Weidner and Mez 2008, p. 362). Subsequently, the heightened attention facilitated various concrete political developments, which paved the way for climate change becoming an important and distinct policy field in Germany. A key development was the establishment of an Enquete Commission concerned with taking Precautions for the Protection of the Atmosphere (*Vorsorge zum Schutz der Erdatmosphäre*) by the German Parliament (Bundestag) in 1987 (Krueck et al. 1999, p. 1; Weidner 2008, p. 7). While it initially was a response to the ozone layer problematic, the commission played an increasingly important role for the development of Germany's climate policies (Mederake and Duwe 2014, pp. 8–9) and, as the next sections will show, also participated in the construction of climate change as security issue.

Starting out from these historical precursors, the next section takes a closer look at the period between 1990 and 2015 and discusses how the three climate security discourses shaped the German debate and helped to render climate change governable in a specific manner.

The Disciplinary Discourse: Mitigating at Home and Optimising the World's Poor

The disciplinary discourse with its focus on human security and the well-being of people, especially in developing countries, was the most prevalent discourse in Germany. A key intention of most articulations of the disciplinary discourse was to put emphasis on and support the most disadvantaged populations. Yet, despite these laudable aims, the discourse also fuelled a paternalistic 'normation' process. At its core was the juxtaposition of 'helpless and powerless victims' in the Global South with seemingly 'more advanced and climate resilient people' in the Global North, a portrayal, which was reinforced by Germany's self-proclaimed role as a climate vanguard, leader in green technology and 'good' global citizen.

Protecting Humanity, Planet Earth and Southern Victims

By the early 1990s, the disciplinary discourse had firmly arrived in the political debate in Germany. The Enquete Commission on the Protection of the Atmosphere had a significant role in this respect and was frequently cited in parliamentary debates (Deutscher Bundestag 1995a). Thus, already in 1989, the Bundestag Committee on the Environment recognised the Commissions' recommendations and echoed the disciplinary discourse, at this time with a decisively global focus: 'The threat to the Earth's atmosphere endangers the life on Earth if the present trend is not fully and comprehensively halted' (Deutscher Bundestag 1989, p. 1). And while climate security articulations in general faded somewhat from the spotlight amid the German reunification process in the early 1990s, the disciplinary discourse remained the dominant representation.

On the one hand, parliamentary articulations kept the global focus and portrayed climate change as an important threat to humanity and all human societies:

> The imminent climate catastrophe simply necessitates a decision between a reasonably secure and humane survival of human society as we know it, or the crash into chaos with unmanageable consequences. (Deutscher Bundestag 1993a, p. 2646)

On the other hand, they introduced a distinction between the developed and developing world that would become a recurrent argumentation in the German climate security debate. While acknowledging the heightened responsibility of industrialised countries, Members of Parliament frequently stressed that developing countries and their inhabitants, especially in Africa (Deutscher Bundestag 1997, p. 18328), would be hit much harder by climate change (Deutscher Bundestag 1991, p. 3786, 1996, p. 8942). Often using Biblical metaphors (see also Doyle and Chaturvedi 2010) several articulation also focused on the inhabitants of the small island states and depicted them as the first victims of climate change whose only hope for survival was to wait for 'Noah's Ark', if industrialised countries would not act soon to abate climate change (Deutscher Bundestag 1995b, p. 2345).

Concerning the concrete threats, almost all of the parliamentary debates at that time focused on the physical effects of climate change (Deutscher Bundestag 1993a, p. 12646) and hence cautioned against floods (Deutscher Bundestag 1995b, p. 2345), extreme weather events (Deutscher Bundestag 1995c, p. 2529), water scarcity, agricultural losses (Deutscher Bundestag 1998a, p. 19406) and widespread famines in developing countries and especially in Africa (Deutscher Bundestag 1997, p. 18328). To counter these problems, they mostly suggested mitigation measures in industrialised countries (Deutscher Bundestag 1993b, p. 14254, 1998a). These included changing consumption patterns (Deutscher Bundestag 1991), reducing energy use, investing in renewables and, especially towards the mid-1990s, an ecological tax reform in Germany (Deutscher Bundestag 1995b, p. 2345).

Linking Human Security to Climate Policy and the Rise of Adaptation

When the Red-Green (*Sozialdemokratische Partei Deutschlands*, SPD, and *Bündnis 90/Die Grünen*, Greens) government coalition under Chancellor Gerhard Schröder took office in 1998, the frequency and intensity of securitising articulations in the parliament increased. The disciplinary discourse still dominated the debates and most articulations kept pointing to the physical dangers of climate change such as extreme weather events including tornados, hurricanes and cyclones (Deutscher Bundestag 1999b, p. 5985, 2004c, p. 10247), the spread of tropical diseases and water scarcity (Deutscher Bundestag 1999b, p. 6000, 2003b, p. 6439). Members of Parliament feared that all of these events would have considerable effects on human health (Document 3, Müller, 6000. 98-02), would very soon displace millions of people in the developing world and hence would increase the number of environmental refugees (Deutscher Bundestag 1999b, p. 6000).

Even though they kept juxtaposing industrialised and developing countries, from 2002 onwards, similar to the US case, the debates also began to stress the adverse effects of climate change for people in industrialised countries. Key articulations referred to the flooding of the Elbe river in 2002 (Deutscher Bundestag 2004c, p. 10247), to the likeliness of droughts and heatwaves in the US and Southern Europe (Deutscher Bundestag 2004c, p. 10247, 2004d, p. 13417) or to the possibility of abrupt climate events that could lead to considerable sea level rise in Europe (Deutscher Bundestag 2004b, p. 8379). Besides improved scientific knowledge (IPCC 2001), the 2003 Pentagon study on abrupt changes in the climate system (Schwartz and Randall 2003), which was partly based on research by the German Potsdam Institute for Climate Impact Research (PIK) (Deutscher Bundestag 2004b, p. 8374), played a crucial role in introducing this argumentation. Another driver was the Hollywood movie *The Day After Tomorrow*, which depicted one of the scenarios of the Pentagon study and which was frequently mentioned in parliamentary debates in Germany (Deutscher Bundestag 2004c, p. 10244, 2005a, p. 14659).

The proposed countermeasures within the disciplinary discourse initially remained on mitigation. Examples were CO_2 reductions in airplane traffic, increasing energy efficiency, the ecological tax reform and the quick ratification of the Kyoto Protocol (Deutscher Bundestag 1999b, p. 6001,

1999b, p. 5985). With a view on the 'flexible mechanisms' of Kyoto, and in accordance with the disciplinary concept of normation, Members of Parliament also urged to include developing countries in the efforts to mitigate climate change (Deutscher Bundestag 1999b, p. 5985, 2005d, p. 193). Yet, not least due to new scientific evidence concerning the advanced stage of climate change (IPCC 2001) and the increased focus on the security implications, adaptation measures gradually became more central. Instead of preventing climate change by tackling its root causes, Members of Parliament increasingly suggested that the focus had to shift to alleviating its worst consequences:

> Ten years ago, we still discussed whether we could prevent climate change. Today, we merely talk about how we can mitigate the worst effects of climate change with our utmost effort. (Deutscher Bundestag 2004b, p. 8370)

Non-parliamentary Articulations at the Peak of the Climate Security Debate

Apart from the parliament, several other domestic actors played an important role in mainstreaming climate security argumentations in Germany. Beginning at the end of the 1990s, a growing number of reports by German governmental advisory bodies, research-institutions and NGOs began to articulate climate security discourses and hence contributed to the peak of the climate security debate in Germany between 2007 and 2011.

Although the debate was most influential between 2007 and 2011, some organisations, such as the environmental consulting firms adelphi and Ecologic or the Hans Seidel Foundation but also academics such as Hans Günter Brauch, had already begun to articulate climate security discourses around the turn of the millennium (Interview 2014b; Carius and Lietzmann 1998; Rotte 2001; Eberwein and Chojnacki 2001). These early reports were mostly part of the broader academic discussions about environmental security and hence often discussed other environmental stress factors besides climate change as well. The central argument was that 'scarce resources and environmental degradation' (Eberwein and Chojnacki 2001, p. 3) together with the problematic socio-economic situation and population growth in developing countries could have serious security implications (Eberwein and Chojnacki 2001, p. 13; Carius and Lietzmann 1998). Similar to earlier parliamentary debates, these reports had a genuine concern for the security of people in developing countries and stressed that climate change could become 'one of the most important challenges for human security' (Brauch 2002, p. 15). However, some of the reports included paternalistic argumentations that granted the industrialised countries a 'special role' that allowed them to intervene in the developing countries for the sake of climate policy (Rotte 2001, p. 63).

While these early securitisation moves had limited influence, they laid the ground for the ensuing climate security discussions from the mid-2000s onwards. Especially after 2007, these discussions were able to reach beyond academic and expert audiences and began to firmly influence the general discussion on climate change in Germany. Particularly in the wake of the publication of the 2007 WBGU report 'World in Transition. Climate Change as a Security Risk' (WBGU 2007b), climate security articulations in reports and parliamentary debates multiplied. Similar to the earlier parliamentary articulations, many reports stressed that because of problematic socio-economic and political preconditions the poorest—particularly in Africa (WBGU 2007b, pp. 3–4; adelphi 2013, p. 42)—will be hit hardest:

> Rising global temperatures will jeopardize the bases of many people's livelihoods, especially in the developing regions, increase vulnerability to poverty and social deprivation, and thus put human security at risk. (WBGU 2008, p. 1; see also GTZ 2008b, p. 8)

The influential annually published Climate Risk Index of the NGO Germanwatch further reinforced the recurring dichotomisation between people in developing and developed countries by stating that the ten countries that were most affected by climate change in terms of economic costs and human suffering between 1990 and 2010 were all developing countries (Germanwatch 2011).

Concerning specific threats, the focus continued to rest on the physical effects of climate change such as water stress (WBGU 2008, p. 2; Schünemann 2014, p. 4; Seiler 2011, p. 25), food insecurity (WBGU 2008, p. 2; Franke 2011, pp. 14–15; GTZ 2008b, p. 27), health problems (Jakobeit and Methmann 2007, p. 7; GTZ 2008b, p. 30) and sea level rise (Schünemann 2014, p. 6; GTZ 2008b, p. 29). As a study of Greenpeace summarised in 2013, taken together all the adverse effects of climate change are 'already taking 5 million lives a year', a number that could rise to 100 million in 2030 (Greenpeace 2013, p. 6). Moreover, climate change would 'undermine development and undo progress made in reducing poverty, attaining Millennium Development Goals, and improving the socio-economic well-being of broad sectors of the population' (adelphi 2013, p. 42). Finally, the deteriorating conditions in developing countries could induce 'historically unsurpassed' migration movements (Germanwatch 2013, pp. 3–4).

Based on this threat assessment—but also linked to a global justice argumentation that played some role in the German debate (Jakobeit and Methmann 2007, p. 2)—most reports continued to recommend measures that would benefit the mitigation of climate change in order to prevent the worst threats to human security (Seiler 2011, 18, 24–25). These included the promotion of renewables, energy efficiency and the abandonment of coal power plants (Greenpeace 2013, pp. 7, 27), which in combination would lead to a 'transformation of the energy systems in Europe' (WBGU 2008, p. 199). Further emphasis was on demands for increased international cooperation (Bayer 2011, p. 141) and political efforts of Germany and the EU to bring about effective global mitigation treaties to halt global warming below 2 °C (WBGU 2008, pp. 189, 193, 198).

Beyond that, reports increasingly suggested adaptation measures (adelphi and UBA 2013, p. 10). In line with the disciplinary concept of normation, the goal was to alleviate the problematic deviation from the norm, which was primarily vulnerable people in developing countries. Thus, a central argumentation was to point to the responsibility of industrialised countries to protect and support weak developing countries and their inhabitants from the worst effects of climate change (WBGU 2008, p. 5). More specifically, this entailed an increased monitoring of the most vulnerable people and regions (WBGU 2008, pp. 179, 185–186) and efforts to make them more resilient towards the hostile changes in their environment (WBGU 2007a, p. 116). Furthermore, reports urged to expand the technical and educational preparation of people in developing countries for disasters (GTZ 2008a, p. 55; WBGU 2008, pp. 10, 115).

Due to the prevalent focus on the Global South, there was a widespread agreement on 'establishing climate protection as a crosscutting theme in development cooperation' (WBGU 2008, p. 199). The goal was to abolish underdevelopment, to improve the coping capacities of people in developing countries (adelphi 2013, p. 15; GTZ 2008b, pp. 9–10; WBGU 2008, p. 9) and to foster the role of development institutions in international policy (GTZ 2008a, p. 49; WBGU 2008, p. 197). In this context, several reports also recommended to include climate security into

Germany's foreign policy (WBGU 2008, p. 193) and to establish concepts such as 'climate diplomacy' (adelphi 2012, 2013). The aim was to develop a holistic approach that linked foreign, development and disaster prevention policies (adelphi 2013, p. 15).

Focusing on Threatened Populations in the Global South

Especially the 2007 WBGU report (Deutscher Bundestag 2010, p. 6855, 2007b, p. 10403) and reports by the PIK (Bulling-Schröter 2009, p. 599) received plenty of attention in parliamentary debates. Consequently, discussions in parliament largely adopted the main discursive threads and focused on the physical effects of climate change (Deutscher Bundestag 2009a, p. 24255, 2008d, p. 21069). Members of Parliament also problematised possible consequences for Europe and Germany, for instance, the spread of tropical diseases (Deutscher Bundestag 2008d, p. 21069) or the increased frequency and intensity of flooding (Deutscher Bundestag 2013b, p. 31287). However, there was widespread agreement that 'the first victims of climate change are the people in the countries of the Global South' (Deutscher Bundestag 2007d, p. 13389, 2012, p. 25760).

Connecting to earlier narratives, discussions in parliament continued to build a stark distinction between the coping capacities of industrialised and developing countries (Deutscher Bundestag 2013b, p. 31304), as the following quote fittingly exemplifies:

> If we see the consequences of such disasters in an industrialised country with existing infrastructure and a certain level of prosperity, then, of course, is it all the more apparent what that means for Africa, parts of Asia or Latin America […]. There, climate protection is a matter of life and death (Deutscher Bundestag 2013b, p. 31291).

A specific focus was once more on poor populations in Africa (Deutscher Bundestag 2008b, p. 17247) and partly Asia (Deutscher Bundestag 2009a, p. 24255), which were constructed as 'especially vulnerable' (Deutscher Bundestag 2007c, p. 10951) and 'immediately threatened' (Deutscher Bundestag 2007c, p. 10950). Ultimately, parliamentary debates at that time argued that climate change threatened 'millions' of (poor) people around the world (Deutscher Bundestag 2006, p. 6191, 2005d, p. 192) and for some, such as the inhabitants of small island states, it even constituted 'a question of life and death' (Deutscher Bundestag 2011b, p. 17991).

Mitigate at Home, Adapt and Educate Abroad to Strengthen Human Security

In response to the prevalent threat construction in the parliament, mitigation measures continued to be important (Deutscher Bundestag 2009a, p. 24255). On the one hand, many Members of Parliament wanted to uphold the pioneering role of Germany and the EU concerning mitigation (Deutscher Bundestag 2006, 6192, 10922), not only to prevent threats to human security around the world but also to accomplish intergenerational climate justice (Deutscher Bundestag 2007a, p. 8144). Consequently, they frequently called for the adoption of a national climate law (Deutscher Bundestag 2013a, p. 31291). On the other hand, Members of Parliament also increasingly emphasised the need to include emerging economies and developing countries in the efforts to combat climate change (Deutscher Bundestag 2008a, p. 14278, 2013b, p. 31299) and welcomed that emerging economies and developing countries had accepted more responsibility at COP-13 in Bali (Deutscher Bundestag 2008a).

Besides mitigation, the ever-increasing debates on the human security implications of climate change for people in developing countries together with the realisation that 'climate change is a reality now' (Deutscher Bundestag 2006, p. 6192) further accelerated the discursive shift in Germany towards adaptation. Becoming particularly pronounced in the second half of the 2000s (Böcher and Töller 2012, pp. 64–65, 69), Members of Parliament increasingly presented adaptation measures as a crucial part of tackling climate change and its security implications (Deutscher Bundestag 2008c, p. 21069). Concrete examples were calls to double Germany's pledges for the UN Adaptation Fund (Deutscher Bundestag 2006, p. 6192), to help developing countries to adapt to climate change (Deutscher Bundestag 2007d, p. 10950), and to scale up climate and environmental education to raise 'ecological awareness' (Deutscher Bundestag 2007c, p. 10951).

The Governmental Discourse: Risk Management and Market Mechanisms

While the governmental discourse was an important part of the securitisation of climate change in Germany, it only became more widespread from the mid-2000s onwards. Overall, the governmental discourse was closely intertwined with the prevailing disciplinary discourse because most of the underlying risk assessments located the high-risk groups and areas in the Global South. However, it has also added new concepts and practices such as probabilistic assessments, prevention strategies, future-oriented scenario planning, a focus on risk groups and, in the spirit of 'laissez-faire', less direct or invasive countermeasures such as market instruments and insurance schemes.

Governmental Narratives in Parliament: Constructing Risk Groups and Strengthening Resilience

In the early 2000s, parliamentary articulations began to regularly refer to climate change as an overarching global hazard that would gradually increase the risk of extreme weather (Deutscher Bundestag 2007c, p. 10950), health problems (Deutscher Bundestag 2008d, p. 21071), flooding, droughts, and the extinction of flora and fauna (Deutscher Bundestag 2009a, p. 24255). Beyond that, they singled out specific high-risk groups and areas, which they primarily located in the Global South:

> In different regions of the world water availability decreases. This reduces agricultural yields. Also in this area one can predict that in parts of the world such as Africa, the Middle East and India, the consequences will be particularly pronounced. (Deutscher Bundestag 1999b, p. 6000)

Contrasting the disciplinary discourse, the focus was not on immediate and concrete threats to human security. Instead, the emphasis was on an aggregate risk analysis that while singling out various high-risk groups or outlier cases, always looks at the broader picture, i.e. at the average risk to the 'Foucauldian population'. Thus, the early governmental articulations in Germany did not exclusively look at developing countries but constructed specific areas such as low-lying coastal regions (Deutscher Bundestag 2004b) as particularly at risk from the adverse effects of climate change. Apart from developing countries, Members of Parliament also identified such high-risk areas in Germany, for instance, in the southwest, along the coasts and in the very east of the country (Deutscher Bundestag 2008d, p. 21071).

Resembling the emphasis of governmental power on cost-efficiency, parliamentary debates also stressed the economic risks of climate change and pointed to the insurance sector that already had started to integrate climate hazards into their risk assessments (Deutscher Bundestag 1999a, p. 3343, 2003b, p. 6439). Especially after the publication of the 2006 Stern Report, economic elements became more important and Members of Parliament feared that the financial risks of future adaptation could spiral out of control if current mitigation measures were neglected (Deutscher Bundestag 2006, p. 6184, 2008a, p. 14268).

To handle these multifaceted risks, articulations in the parliament initially almost exclusively focused on mitigation measures with the aim to lower the overall risk to a tolerable level. They recommended incentives for energy efficiency, renewable energy and the export of these technologies (Deutscher Bundestag 1999a, p. 3343, 2003b, p. 6439). Beyond that, parliamentary debates increasingly concentrated on less direct measures and market instruments such as emission trading schemes (Deutscher Bundestag 2005d, p. 197, 2007c, p. 10906). Especially for members of the market friendly liberal party (*Freie Demokratische Partei*, FDP) and the conservative party (*Christlich Demokratische Union Deutschlands*, CDU), the goal was to tackle climate change in an 'economically sound manner' (Deutscher Bundestag 2009b, p. 25879). This was not to be accomplished solely by top-down state interventions but had to include cost-efficient economic solutions and the business sector itself (Deutscher Bundestag 2009b, p. 25878). In this, vein, many debates revolved around fixing and improving the existing EU emission trading system, which had attracted considerable criticism due to the collapse in the prices of emission certificates (Böcher and Töller 2012, pp. 62–63; Deutscher Bundestag 2009b, p. 25878, 2013b, p. 31285).

To enhance the general resilience towards the increasingly complex risks of climate change, adaptation measures in risk areas became an important part of the debates, especially from the peak of the climate security debates in 2007 (Deutscher Bundestag 2009a, p. 24253). The discussions on adaptation entailed recommendations that focused directly on Germany, for instance, a national adaptation strategy (Deutscher Bundestag 2008d, p. 21071), but also calls to better support high-risk developing countries in their adaptation efforts (Deutscher Bundestag 2007c,

p. 10950). Furthermore, Members of Parliament began to discuss weather and climate insurance schemes to control and alleviate the—seemingly non-avoidable—risks of climate change (Deutscher Bundestag 2008d, p. 21070).

Non-parliamentary Articulations: Reinforcing International Risk Management

The first scattered articulations of the governmental discourse in reports of NGOs, advisory bodies, environmental consulting firms and academics appeared around the turn of the millennium. The 1998 WBGU report on Strategies to Cope with Environmental Risks dedicated a whole chapter to climate change. The report constructed it as a phenomenon that would gradually increase the risk in certain areas and for specific groups and also highlighted the considerable degree of uncertainty concerning the exact timing and magnitude (WBGU 1998, pp. 134–135). To cope with these problems, it not only emphasised adaptation measures but also highlighted the importance of the Kyoto Protocol as a first step of an 'international risk management' system (WBGU 1998, p. 142). Other early articulations discussed different concepts of risk and their applicability to environmental problems (Brauch 2002, p. 44) and identified several environmental risk factors (Brauch 2002, p. 57; Carius and Lietzmann 1998, 25, 37). They also singled out specific risk areas and groups, for instance, low-lying coastal dwellings, large areas in Africa and highly populated regions (Rotte 2001, p. 13). In this vein, the 'insufficient insurance level' in developing countries (Rotte 2001, p. 14) as well as insurance solutions to alleviate climate risks (Rotte 2001, p. 41) already began to play a role in the debates.

Yet, it was not until 2007 that climate risk argumentations became more common and influential. Especially the WBGU, adelphi and the PIK—which had compiled a series of reports under the heading 'Turn Down the Heat' together with the World Bank and the NGO Climate Analytics (World Bank et al. 2012, 2013, 2014)—engaged in the governmental discourse. They feared that climate change would 'greatly increase the risk of natural disasters' (WBGU 2008, p. 3) and similar to parliamentary articulations singled out specific risk groups and areas (adelphi 2012, p. 31; World Bank et al. 2013, pp. xviii, xx). In this respect, they relied on existing scientific findings, which

> [...] indicate a growing risk for low-latitude regions at quite low levels of temperature increase and a growing risk for systemic global problems above a warming of a few degrees Celsius. (World Bank et al. 2012, p. 43)

Reports also identified high-risk areas in low-lying coastal dwellings, often located in higher latitudes (Jakobeit and Methmann 2007, pp. 2–3; World Bank et al. 2012, p. xvii) but also in mid-latitude regions that already experienced water stress and would stand a much higher risk of droughts in the future (Jakobeit and Methmann 2007, p. 5; World Bank et al. 2012, pp. xvii, 48). In principle, these risk assessments applied to specific regions or groups in developed countries as well. However, mirroring the parliamentary debates, the focus was especially on developing countries and poor populations (World Bank et al. 2012, pp. 2, 64; Germanwatch et al. 2013, p. 8). Thus, many articulations constructed climate change as a complex long-term risk, which together with existing problems could gradually overstretch the coping capacity of societies in the Global South (WBGU 2008, p. 3).

Despite this Southern focus, most reports emphasised that the risks could escalate and reach developed countries. Thus, the argumentation was closely intertwined with the disciplinary and also partly with the sovereign discourse and entailed the fear that the 'vulnerable could become dangerous' (Methmann and Oels 2015; Oels 2009), for instance, by migrating towards industrialised countries or by radicalising themselves. Reports also feared that economic risks could further exacerbate these problems, undo previous development processes, increase poverty, and hence make regions and people even more vulnerable towards the effects of climate change (World Bank et al. 2012, pp. xiv, 17). To plan for such 'contingencies' in 'high-risk areas', to keep the risk of migration and instability at a tolerable level, and to prevent it from spiralling out of control, reports suggested measures such as 'national resettlement programmes' that would ensure that such movements would proceed in an 'orderly fashion' (WBGU 2008, p. 127).

Besides these fairly traditional risk assessments, many discussions also revolved around non-linear risks and new concepts to cope with the unknown and non-predictable (World Bank et al. 2012, pp. 59–60):

> And most importantly, a 4°C world is so different from the current one that it comes with high uncertainty and new risks that threaten our ability to

> anticipate and plan for future adaptation needs. (World Bank et al. 2012, p. ix)

In this context, reports cautioned against 'nonlinear tipping elements in the Earth system', which, even though unlikely, would have severe consequences and were difficult to prepare for (World Bank et al. 2012, pp. xvii, 61, 2012). In line with the governmental notion of controlling the risks and keeping them at a tolerable level, reports especially cautioned against 'cascading' and 'escalating' risks (World Bank et al. 2012, pp. 60, 44) that could spiral out of control if not closely monitored.

Governmental Recommendations: Control Risks Through Resilience and Insurance Schemes

To handle these multifaceted traditional and non-linear risks, the focus within the governmental discourse was primarily on what reports called 'effective risk management' (World Bank et al. 2012, p. x). This included not only the identification and the monitoring of risks but also keeping them under control without interfering too much in the normal functioning of the global circulation. Thus, as a *first step* reports suggested, 'the identification of geographical-climatic risks by means of efficient monitoring systems and technology-supported climatic forecasts' (WBGU 2008, p. 186). Beyond that they urged to 'identify climate-hot-spots' (GTZ 2008b, p. 56) and high-risk countries in an 'anticipatory manner' (GTZ 2008a, p. 10) by carrying out 'systematic risk analysis' (GTZ 2008b, p. 56) and by conducting scenario planning exercises in order to anticipate and control the future (WBGU 2008, pp. 77–78). Consequently, the *next step* was to control the risks and to prevent their escalation by taking precautionary measures and by increasing the resiliency and coping capacity of identified risk groups. This included mitigation measures to prevent the worst risks from materialising in the future (WBGU 2008, p. 6). Thus, 'avoiding dangerous climate change' (WBGU 2008, p. 208)—at least for developed countries—remained at the centre of all recommendations. Yet, in line with governmental power, top-down state regulations were not the weapon of choice to reach this goal. Thus, reports increasingly recommended less direct and invasive measures such as the clean development mechanism (CDM) or the introduction of 'internationally tradable quotas for renewable energies' (WBGU 2008, p. 208) to cost-efficiently control the crucial variables 'behind the scenes' (Elbe 2009, p. 76). This strategy ensured that the general processes of production, which have led to climate change in the first place, would not have to be disturbed too heavily (see Methmann 2010).

The focus on the seemingly overwhelming and already pressing risks of 'unavoidable climate change' (WBGU 2008, p. 209) and the perceived incalculability of these risks also led organisations to conclude that 'the case for resilience has never been stronger' (World Bank et al. 2013, p. xii) and contributed to emphasising adaptation measures (World Bank et al. 2012, p. 46). From a governmental power perspective, this strategy can make sense, especially if the high-risks groups or outliers are predominately identified at the periphery and if the overall risk for the total population can be kept at a tolerable level through these measures (see Dalby 2014, p. 7). In more detail, the suggested adaptation measures included early warning systems for disasters and climate-induced conflicts (GTZ 2008b, pp. 11–12; WBGU 2007b, pp. 11–13, 197; Germanwatch et al. 2013, p. 42) as well as the implementation of surveillance and educational programmes to prevent future disasters (World Bank et al. 2012, p. xvii; WBGU 2008, p. 10; Germanwatch et al. 2013, p. 42). Moreover, reports recommended financial and technological support as well as knowledge transfer for high-risk countries by increasing development aid (GTZ 2008b, pp. 53–54; WBGU 2007b, p. 12) and by mainstreaming early warning systems for disasters into development programmes (WBGU 2008, p. 10) to increase the resilience of the affected populations (World Bank et al. 2013, p. 162). That way reports hoped to considerably decrease the risk for far-reaching catastrophes and thus to bring the outlier cases down to the average risk of the total population (Germanwatch et al. 2013, p. 53).

A further instrument to keep climate risks at bay and to cope with their incalculable nature was the promotion of climate or weather insurance schemes. Reports recommended to 'strengthen microsinsurance' (WBGU 2008, p. 12), 'public insurance pools', 'catastrophe bonds' (adelphi and UBA 2013, p. 97) or 'insurance against crop failures' (World Bank et al. 2013, pp. 46, 2014, pp. 26, 225, 228) and to make these instruments part of development cooperation. This would contribute to lowering poverty and vulnerability and thus the risk of certain adverse impacts of climate change (WBGU 2008, pp. 12, 211). While primarily pertaining to high-risk groups in developing countries, these insurance solutions would eventually also benefit developed countries because they had the potential to prevent large-scale migration movements (WBGU 2008, pp. 12, 102,

127). Finally, reports also recommended a further integration of climate risk management schemes into German foreign policy. The goal was to 'move from early warning' to 'early action in order to prevent future crises and increase security' (adelphi 2012, p. 4). Especially adelphi frequently cooperated with the German Federal Foreign Office to develop a comprehensive 'climate diplomacy' (adelphi 2012, 2013), to reduce the risks of climate change and to strengthen the capacity to anticipate potential climate threats in high-risk areas (adelphi 2012, p. 24).

The Sovereign Discourse: Developing Climate Foreign Policy

Although never becoming as dominant as in the US, the sovereign discourse played a role in Germany. However, it was not until the end of the 1990s that it gradually became more common. In addition, German actors understood national security in a more multilateral way and did not link it to the defence sector as readily as in the US.

Non-parliamentary Articulations: Distributive Conflicts in the Global South

The first distinct articulations of the sovereign discourse mainly appeared outside the parliament at the end of the 1990s, for instance, in reports by the Hans Seidel Foundation, adelphi and the Institute for Development and Peace. These early sovereign articulations were closely entangled with academic debates about environmental and ecological security (Eberwein and Chojnacki 2001, p. 1; Carius and Lietzmann 1998, pp. ix, x). They emphasised that they operated with a 'modern' or 'non-traditional' concept of security that went beyond state-centric and military understandings and questioned a simplistic correlation between environmental problems and conflict (Rotte 2001, pp. 7, 62; Eberwein and Chojnacki 2001, pp. 3, 12, 16; Brauch 2002, p. 25). Yet, despite these theoretical reflections, many reports clearly articulated the sovereign discourse by underlining that climate change would be a central threat to the security of Europe and Germany in the twenty-first century and would become a recurring theme in discussions about security and defence policy (Rotte 2001, p. 7). The main line of argumentation was a combination of the disciplinary with the sovereign discourse. Accordingly, climate change would negatively affect human security, mostly in developing countries, which in turn could lead to instability and violent conflicts over scarce resources (Rotte 2001, pp. 13–15, 20; Eberwein and Chojnacki 2001, p. 7; Brauch 2002, p. 22). In the end, this could lead to 'domino effects', escalation dynamics and migration, which eventually would directly threaten Europe's and Germany's interests as well as international stability and security (Rotte 2001, pp. 19, 20, 54–55; Carius and Lietzmann 1998, p. vii; Brauch 2002, p. 23).

To prevent this from materialising, the reports, on the one hand, called for mitigation measures (Rotte 2001, pp. 24–26, 28–29) and emphasised the importance of early warning mechanisms, hence partly linking the sovereign to the governmental discourse (Brauch 2002, pp. 25–26). On the other hand, they highlighted the importance of backup plans in the case of a failure of mitigation and prevention. To cope with violent conflicts (Rotte 2001, pp. 50, 52), they highlighted that '[...] peacekeeping and conflict management in the third world' are 'a vested interest of Germany [...]' (Rotte 2001, p. 53). Thus, the security implications of climate change had to become a central pillar of German foreign policy (Rotte 2001, pp. 53–54) and should be included in the planning of the defence sector (Rotte 2001, pp. 57–58, 61).

The influence of these initial articulations of the sovereign discourse on the broader political debate was measured, and it was not until 2007 that, triggered by the fourth IPCC report, by the UNSC debates and by the ensuing discussion in the US (GTZ 2008a, p. 7), sovereign articulations became more widespread and influential in Germany. Here, the WBGU and adelphi were the most active proponents, but also several other organisations such as the GTZ and the Hans Seidel Foundation occasionally engaged in the discourse. Most organisations were certain that climate-induced security threats would go beyond human security (Tänzler 2011, p. 34) and eventually would have 'major implications for peace and stability' (adelphi 2012, p. 15). In this context, climate-induced migration would play an important role because mass migration to neighbouring states could escalate human security threats towards national security issues (Bayer 2011, p. 141). Moreover, many reports once again focused on the Global South where climate change could lead to 'distributive conflicts' and could exacerbate the problem of 'failing states', which eventually would trigger violent tendencies in many societies (WBGU 2007b, p. 2).

Often reports linked the sovereign argumentation to the disciplinary and governmental discourses by, for example, stating that climate change '[...] has the potential to cause significant and highly uncertain impacts on societies, undermining human security and increasing the risks of conflict and instability' (adelphi 2012, p. 4; see also WBGU 2008, p. 3). Yet, especially in relation to the small island states, climate change also became a traditional national security threat:

> Climate change threatens lives, livelihoods, sovereign territory and, in some cases, the existence of states. The situation we face is as dire and serious as civil war, terrorism and nuclear weapons (adelphi 2013, p. 21, quote by Tony de Brum, Minister in Assistance to the President of the Marshall Islands).

Even though the focus of most articulations was on the developing world, several reports emphasised that threats could soon spill over to industrialised countries. Thus, they picked up the earlier argumentation about the danger of escalation dynamics, according to which the effects of climate change could 'act as a multiplier' of existing problems (adelphi 2012, p. 8) and overstretch the 'adaptive capacities' of developing countries. At the end of the day, this could lead to 'domino' effects concerning the destabilisation of fragile states (Tänzler 2011, p. 34) and contribute to 'jeopardizing national and international security' (WBGU 2008, p. 1; adelphi and UBA 2013, p. 13; GTZ 2008b, p. 56; adelphi 2012, p. 8). Reports saw a real danger that the failure of 'preventive and adaptive strategies' could 'overstretch the global governance system', which eventually could lead to 'far-reaching (global) structural policy or even geostrategic consequences' (GTZ 2008a, p. 8).

However, even though many reports articulated the sovereign discourse around that time, none of them predominately considered Germany's national security as threatened or primarily focused on a role for the armed forces. Instead, the reports acknowledged the debates about new or non-traditional concepts of security (adelphi 2012, p. 15) and did not uncritically adopt a straightforward link between climate change and national security or conflict. The tenor of the sovereign discourse in Germany was that although climate change could have implications for the national security of certain developing countries and could (indirectly) affect international security, the solutions were to be found beyond traditional security and defence policy (WBGU 2008, pp. 21, 23). One adelphi report directly quoted the then Foreign Minister Guido Westerwelle (FDP), who described climate change as a 'non-traditional security threat' that required 'climate diplomacy' as well as 'resource efficiency, energy transformation and technology cooperation' (adelphi 2013, p. 21).

Mainstreaming Climate Change in Development and Foreign Policy

Concerning policy recommendations, most reports included mitigation measures to prevent threats to national security from materialising. Thus, they urged to adopt 'an ambitious global climate policy' in the 'next 10–15 years' (WBGU 2008, pp. 1, 6) or advised that 'efforts to promote sustainable energy systems aimed at ending energy poverty and reducing resource competition need to move ahead' (GTZ 2008a, p. 10). However, even more so than in connection to the other two discourses, the focus on the traditional security implications of climate change also led to an increased focus on adaptation measures to tackle the immediate security concerns. For instance, reports suggested 'conflict sensitive adaptation' strategies (adelphi and UBA 2013, p. 2; GTZ 2008a, p. 9) that could help to contain problematic situations in affected countries in the Global South. In this context, many reports underlined the importance of development policy, which could support local populations and increase their resilience to prevent the escalation of human security threats to conflicts. Some even saw the discussion about the security implications of climate change as 'a great window of opportunity' to increase the importance of the German Federal Ministry for Economic Cooperation and Development (BMZ) (GTZ 2008a, p. 8) and to 'upgrade climate-related development cooperation in terms of staffing and funding' (GTZ 2008a, p. 10). Moreover, reports linked the sovereign discourse to governmental articulations by emphasising the importance of 'prevention', which would 'reduce fragility, and lower the risk of future crisis and violent conflict' (WBGU 2008, p. 44).

Finally, while reports largely avoided defence policy recommendations, they nevertheless considered climate change as one of the 'key foreign policy challenges of the 21st century' (adelphi 2013, p. 5). In this vein, they frequently stressed that it was necessary to 'strengthen the profile of climate change within the foreign policy agenda' (adelphi 2012, p. 12; see also GTZ 2008a, p. 8). Thus, reports recommended a close cooperation of all OECD countries with the emerging economies and developing countries to prevent global instability and conflicts (WBGU 2007b, p. 5). Especially, the rise of new global powers such as India and China and related conflicts about climate policy necessitated a stronger integration of climate change into German and European foreign policy agendas (WBGU 2008, p. 193). Furthermore, particularly adelphi was very keen to foster a dialogue between climate security scholars and political practitioners especially in the German Federal Foreign Office and in this context actively promoted the concept of 'climate diplomacy' (adelphi 2012, 2013).

Parliamentary Articulations: Supporting Endangered Developing Countries

In the 1990s, the sovereign discourse only occasionally appeared in the German Bundestag. The reports of the Enquete Commission, for instance, mentioned the possibility of 'distributive conflicts' (Deutscher Bundestag 1990, p. 89) and 'threats to world peace' (Deutscher Bundestag 1992, p. 15). Paralleling non-governmental articulations, the discourse gradually became more common in the early to mid-2000s. One important trigger was the publication of the 2003 Pentagon study, which was directly discussed in the German Bundestag to emphasise the traditional security dimension of climate change (Deutscher Bundestag 2004a, pp. 8374, 8379). However, the discourse remained a minority position and it was not until the mid-2000s and especially in the peak of the climate security debate between 2007 and 2011 that sovereign argumentations became more important in the parliament. Similar to the above-described reports, Members of Parliament most of the time did not discuss traditional security implications of climate change in isolation but always emphasised the close interconnectedness to human security (Deutscher Bundestag 2008a, p. 14269).

Throughout most debates, the emphasis was on threats in already instable developing countries. Thus, Members of Parliament highlighted how climate change could lead to new conflicts (Deutscher Bundestag 2009c, p. 604) or at least could exacerbate existing ones, for example, '[…] fierce wars and conflicts over water, food and the access to raw materials […]' (Deutscher Bundestag 2007c, p. 10953, 2011a, p. 17375). Once again, a particular focus was on Africa where, according to some Members of Parliament, climate change already had fuelled violent conflicts, for instance, in Darfur:

> There are already wars and civil wars as a result of climate change. For example, the deserts in Sudan have spread by 100 kilometres in the past 40 years. Part of the war and civil war happening there is a conflict over the land mass [...]. (Deutscher Bundestag 2008d, p. 21069)

In contrast to the US debate, the discussion only seldom revolved around direct threats to Germany. Thus, Members of Parliament occasionally argued that conflicts in the Global South could eventually pertain Europe's and Germany's security as well (Deutscher Bundestag 2008a, p. 14278). More specifically, they feared that climate change could lead to 'global tensions', 'conflicts', 'migration', 'extremism' and 'terrorism', which all could become direct threats to Germany if left unattended (Deutscher Bundestag 2009c, p. 604, 2011b, p. 17991).

However, direct threats to Germany's national security or to international peace and security were mostly portrayed as an unlikely scenario and never became the centre of the debate. Moreover, parliamentary articulations always made clear that instead of traditional military responses, the focus ought to be on preventive and holistic strategies (Deutscher Bundestag 2012, p. 25936). Thus, mitigation measures and helping developing countries were central pillars even within the sovereign discourse. Members of Parliament called for more CDM projects (Deutscher Bundestag 2008b, p. 17247), for technology transfer (Deutscher Bundestag 2007d, p. 13377, 2011b, p. 17991) and also for reducing CO_2 on a global scale through emission trading systems (Deutscher Bundestag 2007c, p. 10953).

Besides mitigation, the focus on climate change as a driver of traditional security threats increasingly contributed to a discursive shift towards adaptation measures during the end of the 2000s. Thus, Members of Parliament emphasised that '[...] climate and adaptation policies should be understood as an element of preventive security policy' and called for 'effective instruments to finance the high costs of adaptation' (Deutscher Bundestag 2007d, p. 13377). Once again, some articulations clearly displayed an acceptance of a considerable degree of global warming and merely focused on measures to cushion its worst blows, for instance, through an 'international fund for compensation' (Deutscher Bundestag 2008a, p. 14269) and also with a focus on Germany and its 'national adaptation strategy' (Deutscher Bundestag 2008d, p. 21069).

Contextualising Climate Security Discourses in Germany

In contrast to the US case, debates in Germany primarily articulated the disciplinary and governmental discourses and constructed climate change as a threat or risk to the human security of poor populations in the Global South. At the same time, the sovereign discourse and traditional national security conceptions were less prevalent and understood in a more multilateral fashion, which also led to a much smaller role of the defence sector in climate security debates. A look at the broader contextual features in Germany can help to understand these differences.

A central contextual characteristic is the specific political culture in Germany when it comes to questions of national security. Due to the country's problematic experiences with nationalism and the unilateral use of military force in the past and the subsequent management of the country by the allied forces, 'national security' has been largely discredited and thus does not play an important role in political discussions (Hellmann et al. 2006, p. 187). Instead of focusing on Germany's national security and the development of its own military capabilities or sovereign solutions, the emphasis is on ideas such as shared sovereignty and collective security that primarily strive to ensure international stability (Böckenförde 2014; Böckenförde and Gareis 2014). In this context, international law, the UN and particularly the European Union play a central role (Hellmann et al. 2006, p. 188). Germany tries to act as a civilian or normative power that primarily relies on soft power and leadership by example to facilitate multilateral solutions to global problems such as climate change (Interview 2014c; Maull 1993; Hellmann et al. 2006, p. 192). This is further reinforced by a quest to make up for its atrocities in the past, to become a good and respected global citizen and to help disadvantaged populations and states, which is supposed to increase human security around the world (Harnisch and Wolf 2010, pp. 45–46).

This specific political culture has had several concrete consequences for the climate security debate. Together with the influential peace movement and widespread 'anti-militarism' in Germany (Gareis 2005, p. 46), it has led to a less significant role of the armed forces, which has even further deteriorated since the end of the Cold War. Besides a comparatively small size and budget—Germany spends 1.2 per cent of GDP on defence, compared to 3.3 per cent in the US (Statistika 2017)—the German armed forces (*Bundeswehr*) or military solutions as such do not have a particularly favourable reputation in the public or amongst political actors. Thus, the

Bundeswehr usually does not participate in political debates. Especially concerning climate change, the involvement of the armed forces would not have helped the debate but would most likely have been received as dangerous 'militarisation' (Wagner 2008) of the issue. Consequently, even the more traditional security implications of climate change were mostly linked to foreign and development policy, which resonated much better with the general role of Germany in international society and its strive for collective solutions and preventive diplomacy (Harnisch and Wolf 2010, p. 49). Moreover, although this has begun to change since the 1999 Kosovo war, the German constitution has defined the role of the Bundeswehr rather narrowly as a 'defensive army' (Gareis 2005, pp. 166, 175; Deutscher Bundestag 1949, Art. 87a). This has for a long time prevented it from a more active role around the world and thus to directly operate in areas constructed as threatened by climate change. Accordingly, in contrast to the much more globally active US army, the German armed forces have not picked up climate change as a key security threat.

The absence of a strong national security culture has also led to a closer integration of Germany into international society, most importantly epitomised in its membership and role in the EU (Jänicke 2011, p. 142). The key role of multilateralism in its foreign policy (Hellmann et al. 2006, p. 217; Gareis 2005, p. 56) has also had a considerable effect on which climate security discourses and respective solutions were seen as appropriate. Thus, discursive interventions from the EU Commission (Solana and EU Commission 2008) and the UN (UNGA 2008, 2009; UNSC 2007a, 2011a) were highly influential and helped to forge a much more multilateral debate on climate security than, for instance, in the US. Together with the efforts to make up for past wrongdoings and global justice considerations, this has facilitated the focus on the threats of climate change for poor populations around the world and the focus on development policy.

Beyond the limited role of the sovereign discourse, climate security discourses in Germany are more directly linked to climate policy recommendations and, as the next section will show, have been more successful in legitimising legislation in this sector than it is the case in the US. One important explanation for this striking difference is that although climate security articulations in both countries have pointed to the necessity of mitigation measures, in Germany these demands met a much more favourable discursive context. Partly due to the higher density of the population and hence the seemingly more precious value of nature as such (Ulbert 1997, p. 22) as well as due to the fear of nuclear annihilation, beginning in the 1970s Germany saw the growth of an influential anti-nuclear and environmental movement (Schreurs 2002, pp. 5–7). Subsequently, a green consciousness and post-materialistic worldview (Inglehart 1977) have quickly expanded and, with the rise of the Green Party, also became part of the political establishment (Ulbert 1997, p. 26; Schreurs 2002, p. 6; Kuckartz et al. 2006).

Consequently, environmental and climate protection soon became mainstream positions in Germany. Over 9 per cent of the population are members of environmental organisations (Jänicke 2011, p. 131), and a huge majority feels directly threatened by climate change (Kuckartz et al. 2006, p. 20) and favours progressive policies in this sector (Jänicke 2011, pp. 141, 144; Geden and Tils 2013, p. 24). In this context, all parties—with a possible exception of the far-right and extremist parties such as the *Alternative für Deutschland*—have adopted decisive climate protection measures as important goal (Jänicke 2011, pp. 131, 141; Geden and Tils 2013, p. 24). From early on, the German climate research landscape became one of the most advanced and well financed worldwide (Krück et al. 1999; DWD 2013). Its findings were broadly disseminated (Schmidt 2012, p. 69) and today several German climate scientists such as Joachim Schellnhuber or Mojib Latif play key roles in global climate debates (Dehmer 2015). Thus, scientific organisations and knowledge have been a major driver of German climate security debates and helped to facilitate human security and risk-focused climate threat constructions in Germany (Jänicke 2011, p. 141). In this context, scientific 'expert knowledge' is usually not questioned or politicised—at least not to the same extent as in the US—and often has been directly integrated into Germany's climate policies. Together with the strong role of environmental NGOs and the absence of influential climate sceptics, this has facilitated the close linkages of climate security discourses to mitigation policy.

The Political Impact of Climate Security Discourses in Germany

Resting on the dominance of the disciplinary and governmental discourses as well as on the above-described political context, the concrete political impact of the securitisation of climate change in Germany differs considerably from the US. Instead of climatising the defence sector, it primarily helped to legitimise climate policy and gave new inputs for foreign and development policy. Yet, while often playing a decisive role, climate security discourses were never the only driving force behind these developments. Instead, climate justice argumentations and influential environmental movements also played an important part in the German climate debate. Beyond that, particularly since the mid-1990s, climate protection measures were increasingly presented as an economic opportunity (Beck et al. 2009, p. 19) and framed in the context of ecological modernisation and environmental industrial policy (Jänicke 2011, p. 134). Nevertheless, these concepts did not stand in opposition to climate security discourses but often helped to reinforce the focus on progressive and multilateral climate and development policies.

Environmental and Climate Policy: Legitimising Ambitious Policies and Market Solutions

Abating climate change with decisive mitigation measures has always been one of the key demands in the German climate security debate. A first tangible political consequence of the mostly globally focused securitising articulations in the late 1980s was the establishment of the influential Enquete Commission on the Preservation of the Earth's Atmosphere. In its first reports, it adopted the disciplinary discourse and the emphasis on the global dimension by constructing climate change as 'a danger of almost unimaginable proportions for all mankind' (Deutscher Bundestag 1990, p. 88), as a 'global threat' (Deutscher Bundestag 1992, p. 3), and by highlighting the increased frequency of climate-induced catastrophes (Deutscher Bundestag 1992, p. 10). Therefore, the commission favoured a global solution for the problem and decisive restrictions for harmful substances, similar to the Montreal protocol.

Accordingly, in 1991, Germany announced an ambitious CO_2 reduction target of -25 per cent between 1987—later 1990—and 2005 and began to convince other industrialised countries to do the same (Böcher and Töller 2012, p. 54; Weidner 2008, p. 7). Subsequently, as one of the first signatories, Germany became a dedicated promoter of the UNFCCC (Böckem 2000, p. 7; Weidner 2008, p. 7). These efforts cumulated in the first COP in Berlin and the adoption of the Berlin Mandate (UNFCCC 1995), which paved the way for the establishment of the Kyoto Protocol in 1997. Motivated by the recurring climate security articulations, Germany was one of the few industrialised countries that advocated for binding emission cuts (Jänicke 2011, p. 133) and pushed for the establishment of the precautionary principle in relation to climate change (Mederake and Duwe 2014, p. 11; Schreurs 2002, p. 4). Consequently, the country was also the driving force behind the EU's Kyoto pledge and agreed to bear the largest share of reduction within the EU (Weidner and Mez 2008, p. 363; Mederake and Duwe 2014, p. 14; Geden and Tils 2013, p. 24).

The early articulations of the disciplinary discourse also contributed to Germany's domestic climate policies. In 1990, the BMU established the Inter-Ministerial Working Group 'CO_2-Reduction' (IMA) (BMU 2006, pp. 4–5). The third IMA report published in 1994, called climate change a 'global threat', emphasised the importance of multilateral solutions and, in accordance with the dominant disciplinary discourse, highlighted the special responsibility of industrialised countries to reduce their emissions (Deutscher Bundestag 1994, pp. 5, 57, 160). Beyond that, throughout the 1990s, several Members of Parliament had advocated for measures that were supposed to combine a prevention of the security implications of climate change with (green) economic opportunities (Deutscher Bundestag 1995b, p. 2345, 1997, p. 18328, 1999b, p. 5985). This eventually led to the adoption of the ecological tax reform in 1999 by the newly elected SPD/Greens coalition government (Böcher and Töller 2012, p. 58), which aimed at increasing the tax for fossil fuels and to promote energy efficiency as well as renewables (Deutscher Bundestag 1998b, p. 1). One year later, the Renewable Energy Law (*Erneuerbare Energien Gesetz*, EEG) followed (Böcher and Töller 2012, p. 59; Jänicke 2011, p. 138). The draft bill explicitly articulated the disciplinary discourse by emphasising that the increasing frequency of climate-induced natural disasters necessitates immediate legislative action to mitigate GHG emissions, but also highlighted the economic opportunities of renewables (Deutscher Bundestag 2000a, pp. 18, 1).

Later that year, the government coalition established the German National Climate Protection Programme, which underlined Germany's mitigation target of -25 per cent between 1990 and 2005 (Deutscher Bundestag 2000b) and also emphasised the human security-related dangers of climate change (Deutscher Bundestag 2000b, pp. 5, 137). In 2005, the new conservative led coalition government (CDU/SPD) prolonged the programme and set a new target of -21 per cent between 2008 and 2012 relative to 1990 (Deutscher Bundestag 2005c, p. 4). The initial report about the programme again articulated the disciplinary discourse by emphasising that 'Extreme events cause billions in damage and threaten the health and livelihoods of the population in many regions of the world' (Deutscher Bundestag 2005c, p. 4). To handle these problems, the report highlighted the importance of decisive mitigation measures and underlined Germany's ambition to remain a climate forerunner (Deutscher Bundestag 2005c, p. 4).

Whereas Germany had been hesitant to accept market solutions for the climate problem for a long time, this began to change with the rise of the governmental discourse in the early to mid-2000s. Parliamentary debates and policy documents began to mention emission trading programmes as important tools to fight climate change in a cost-efficient manner (Deutscher Bundestag 2003a, p. 6249) and emphasised the economic opportunities that CDM as well as Joint Implementation Projects would bring for German businesses (Deutscher Bundestag 2005b, p. 16616). Subsequently, this also contributed to a more favourable attitude in Germany towards the European emission trading system (EU ETS), which entered into effect in 2005 (European Commission 2013, p. 1). In 2007, Germany announced the Integrated Energy and Climate Programme (*Integriertes Energie und Klimaprogramm*), which aimed at a 40 per cent GHG reduction target by 2020 (Jänicke 2011, p. 135; Böcher and Töller 2012, p. 63) and pledged to cut emissions by 80–94 per cent until 2050 (BMU 2014). In line with the governmental discourse, the programme intended to control the risks of climate change as cost-efficient as possible by increasing energy efficiency and promoting renewables, and at the same time ensuring economic growth (BMU 2007a, p. 1).

Between 2007 and 2011, the climate security debate intensified and diversified, which further raised the political profile of climate issues and led to several tangible consequences. In 2007, the BMU published an influential report about the Relevance of Renewable Energies for Security Policy written by adelphi and the Wuppertal Institute (BMU et al. 2007), which frequently articulated the sovereign discourse. In contrast to the US, however, the emphasis was not on the defence sector but on preventing resource conflicts and geopolitical tensions by promoting renewable energy sources (BMU et al. 2007, pp. xi, xiii). In the run up to the COPs in Bali (2007) and Copenhagen (2009) (Dröge 2014, p. 199), during its EU and G8 Presidency, as well as at the G8 summit in Heiligendamm (Jänicke 2011, p. 135), Germany reassumed its international leadership role. Consequently, Chancellor Angela Merkel called climate change the 'biggest challenge of the twenty-first century' (TAZ 2007). Moreover, German politicians at several occasions articulated climate security discourses to emphasise the central importance of decisive mitigation measures. A prominent example is the then Environmental Minister Sigmar Gabriel, who called climate change a 'threat to all of humanity' and urged to integrate climate protection into the EU constitution (FAZ 2007).

While the climate security debate had become less pronounced after the failed summit in Copenhagen in 2009 and in the wake of the financial crisis (Geden and Tils 2013, p. 24), it never disappeared entirely and helped to sustain a fairly progressive climate path under the new conservative-liberal CDU/CSU-FDP government from 2009 onwards (Jänicke 2011, p. 135). A key political development at that time was the German Energy Turn, which was initiated in 2011. Whereas the Fukushima incident in March 2011 clearly was the final trigger for the phase out of nuclear energy (Dröge 2016, p. 17), the far-reaching transformation of the German energy production system towards renewable energies had long been legitimised by the security implications of climate change (WBGU 2007a, pp. 1–9; Richert 2012). Thus, parliamentary debates (Deutscher Bundestag 1999b, p. 5987, 1999c, p. 7252), the Green Party (Bündnis 90/Die Grünen Bundestagsfraktion 2008, p. 3) and several reports had frequently called for a transition of German and European energy production to avoid dangerous climate change and its security implications (WBGU 2003, 2008, pp. 193, 199; BMU et al. 2007, p. 1). However, while Germany continued its fairly progressive climate policies from 2011 onwards, it did not come up with new substantive ideas or ambitious reduction pledges at the COPs from Durban (2011) to Paris (2015) (Geden 2013, pp. 1–2; Dröge 2016, pp. 6, 19, 21). While the frustration after the failed COP in Copenhagen was one important reason, the gradual decline of the climate security debate at that time also contributed to the fading enthusiasm for climate issues.

Apart from mitigation policy, especially since the mid-2000s, adaptation measures became more important (Böcher and Töller 2012, p. 69). While this was also due to new research about the unavoidable effects of climate change, climate security discourses played a significant legitimising role. They contributed to shifting the attention to the immediate security implications or 'symptoms' of climate change that had to be kept under control by adaptation and resilience measures (see also Oels and von Lucke 2015, pp. 45, 61). Thus, in 2006, the German Environment Agency UBA founded the Competence Centre on Climate Impacts and Adaptation (*Kompetenzzentrum Klimafolgen und Anpassung*, KomPass) that was supposed to prepare Germany and Europe for the unavoidable effects of climate change (UBA 2016). It also compiled the German Adaptation Strategy (*Deutsche Anpassungsstrategie,* DAS), which was initially published in 2008. Articulating the by then fairly common governmental discourse, the DAS mainly revolved around an 'integrated approach' to identify the relevant risks of climate change for Germany and to lower the resulting vulnerability as well to develop 'strategies to cope with uncertainties' (Bundesregierung Deutschland 2008, pp. 4–5). In this context, it also highlighted the importance of the insurance industry to keep climate risks at a tolerable level and for instance referred to the influential Munich Climate Insurance Initiative (MCII) (Bundesregierung Deutschland 2008, p. 37).

In 2011, the DAS was developed further by the publication of the Action Plan Adaptation (*Aktionsplan Anpassung*) (Bundesregierung Deutschland 2011b) and the founding of an Inter-Ministerial Working Group on Adaptation. The 2011 Action Plan again adopted the governmental discourse and aimed at 'prioritising different climate risks' to develop an 'integrated vulnerability assessment' in order to lower those risks and vulnerabilities in a cost-efficient manner (Bundesregierung Deutschland 2011b, p. 12). Beyond that, it emphasised the principles of subsidiarity and underlined that 'the responsibility for adaptation was mainly among citizens and companies themselves' (Bundesregierung Deutschland 2011b, p. 9), thereby again relying on a core principle of governmental power.

Development Policy: Saving Poor People in the Global South and Stabilising the World

One of the defining traits of the German climate security debate was the emphasis on the threats of climate change for poor people in developing countries, which led to a profound impact on German development policy. Climate change and its security implications became more influential towards the end of the 1990s and particularly in the course of the 2000s in German development policy. Apart from a general reorientation of development policy towards security aspects (Ziai 2010, p. 140; Pospisil 2009), this was also due to personnel changes. The new head of the BMZ Heidemarie Wieczorek-Zeul put more emphasis on climate change and its security implications than her predecessors (Bohnet 2015) and already in 2000 stressed that climate change 'threatens the foundations of human life' (Wiezcorek-Zeul 2000). Consequently, the BMZ increasingly began to consider the effects of climate change for human security and partly also its potential to fuel violent conflicts. At a conference on disaster preparedness and coping mechanisms (Goldammer 2000, p. 9), BMZ spokesperson Horst Müller claimed that 'environmental degradation and resource scarcity together with population growth' lie at the core of many conflicts (Müller 2000, p. 17). He furthermore underlined the BMZ's emphasis on the promotion of ecological sustainable development, conflict prevention and disaster preparedness (Müller 2000, p. 17).

Thus, a particular focus of the German development policy was on the promotion of sustainable development and on 'global structural policy' (Ziai 2010, p. 136). The basic idea was that improving the living conditions of poor populations around the world and lowering their vulnerability towards climate change would help to prevent violent conflicts. Thus, the influential 2004 report on Civil Crisis Prevention, Conflict Resolution and Peacebuilding matched the disciplinary and sovereign discourse by identifying the physical effects of climate change as important driver for the destruction of habitats, migration movements and violent conflicts that eventually could threaten entire states and the international system (Bundesregierung Deutschland 2004, p. 12). To prevent this, the report emphasised the importance of global climate treaties and also local environmental cooperation projects (Bundesregierung Deutschland 2004, p. 54). As the name of the report implies, a further result of climate security discourses was the increasing combination of climate, foreign, development and security policies, to prevent problematic situations from arising in the first place and if necessary to deal with them in an integrated manner (Bundesregierung Deutschland 2004, p. 55). In this vein, adaptation measures increasingly gained in importance and the already mentioned 2007 BMU report stated that 'The identification and implementation of appropriate adaptation measures can contribute decisively to timely respond to an increased potential for conflict' (BMU et al. 2007, p. 94).

At the peak of the climate security debates, especially the 2007 WBGU report had a direct influence on German development policy. In 2007, at the commencement of the G8 summit in Heiligendamm, Wieczorek-Zeul called climate change 'the biggest security threat the world faces today' and demanded multi-billion euro investments for climate protection instead of financing new defence equipment (N-TV 2007). She also directly adopted the disciplinary argumentation of the WBGU that climate change would become a threat to millions of people especially in developing countries and would lead to millions of climate refugees. To prevent these problems, Germany had to support those people by providing technological knowhow concerning renewables. Thus, mirroring the 2007 BMU report, Wieczorek-Zeul emphasised that in contrast to 'wars over the access to oil', there will be no 'wars over the access to the sun' (BMU 2007b).

Subsequently, the German Development Agency *Deutsche Gesellschaft für Internationale Zusammenarbeit* (GIZ, formerly GTZ) began to integrate climate security aspects into its planning schemes. In its widely noticed 2008 report on climate change and security (GTZ 2008a), it saw an opportunity for the 'foundation for new international alliances geared towards making adaptation to climate change and (conflict-sensitive) preparation for its adverse effects a key issue of negotiations' (GTZ 2008a, pp. 49–50). The report directly adopted disciplinary and governmental notions by emphasising the importance of disaster risk management schemes that would become 'key elements of structural prevention policy geared to avoiding and containing future crises' (GTZ 2008a, p. 53). It furthermore reiterated the risk-group approach by setting regional priorities based on the recommendations of the GTZ's Crisis Prevention and Conflict Transformation Programme (GTZ 2008a, pp. 53–54). In addition, contributing to the merging of climate, development and security policy, the GIZ strived to further integrate its three sector programmes on Climate Protection, Crisis Prevention and Disaster Risk Management and to 'conduct systematic security analyses and risk assessments in the preparatory phase of projects and in affected countries' (GTZ 2008a, p. 54). In the following years, the understanding of climate change as a security problem and the expansion of adaptation, disaster management and climate education schemes became an important aspect in several projects, programmes and reports of the GIZ (GTZ 2008c; Hannen 2012, p. 17; GIZ 2011).

The 2008 German Adaptation Strategy also explicitly highlighted the relevance of adaptation measures in developing countries and identified several tasks for German development policy (Bundesregierung Deutschland 2008, p. 7). In the same vein, the 2011 Action Plan on Adaptation emphasised the vulnerability of developing countries and the need to support their adaptation efforts. This would benefit not only the people in those countries but, because of the prevention of large-scale migration movements, eventually also industrialised countries (Bundesregierung Deutschland 2011a, p. 36). Reflecting the disciplinary and governmental discourses, the focus was particularly on lowering the vulnerability of affected populations, increasing the coping capacity and risk management capabilities of local administrations, and promoting disasters preparedness and early warning systems. Beyond these bilateral instruments, the Action Plan emphasised the importance of international instruments such as the International Climate Initiative, which supported a range of climate risk management and insurance measures in developing countries (Bundesregierung Deutschland 2011a, pp. 37–38).

The growth of climate security articulations that pointed to threats for human security in the Global South also had a lasting effect on policies of the BMZ itself. In 2007, the Programme of Action on Climate and Development emphasised that 'climate change threatened all of humankind' and also that the poor would be disproportionally affected by its adverse effects (BMZ 2007, p. 1). Hence, combating climate change, reducing poverty and keeping peace should go hand in hand to prevent conflicts around the world (BMZ 2007, p. 1), and the BMZ pledged to increase its climate-related funding from 500 million Euros in 2007 (BMZ 2007, p. 2) to about 1.3 billion in 2011 (Bundesregierung Deutschland 2012, p. 156). In the spirit of keeping the risks at a tolerable level and preventing climate change from becoming an issue of international security, the BMZ sought to establish disaster risk management approaches as a cross-cutting issue in high-risk countries as well as to alleviate emerging crises through the Civil Peace Service (GTZ 2008a, pp. 53–54).

Although the 2009 newly appointed Minister for Development Dirk Niebel (FDP) was less enthusiastic about climate protection than his predecessor, climate security considerations remained important. In 2010, at a conference about climate change as a challenge for security policy organised by the Hans Seidel Foundation, the Department of Peacebuilding and Crisis Prevention of the BMZ underlined the need for a holistic strategy to tackle climate-induced conflicts that involved all relevant ministries from the BMZ to the BMU to the Foreign Ministry (Jones 2010, pp. 5–6). Beyond that, in a 2011 Information Brochure, the BMZ called climate change a 'threat to human rights' (BMZ 2011, p. 16), highlighted the devastating physical threats of climate change and emphasised the crucial role of affordable renewable energies to prevent the worst impacts of climate change (BMZ 2011, p. 4). Thus, climate security discourses were a key driver of the integration of development and (preventive) security policy (Pospisil 2009; Ziai 2010, pp. 142–143; Schmidt 2014). In this context, Chancellor Merkel repeatedly emphasised that there was 'no development without security and no security without development'

(Merkel 2010), which also led to an integration of this mantra into BMZ reports (BMZ 2013a, p. 2).

Subsequently, the BMZ recommended to 'strengthen the adaptive capacity and resilience of states and their societies against external shocks and crisis' such as climate change (BMZ 2013b, p. 10) and to include climate adaptation and disaster management as core tasks of German development policy (BMZ 2013a, p. 16). It hence integrated measures such as climate risk management and the promotion of climate-resilient agricultural methods (BMZ 2015a, 10, 17) into its Catastrophe Risk Management Strategy (BMZ 2015a, p. 9). The overall aim of the German development policy in this respect was to prevent risks from spiralling out of control by increasing the coping capacity and resilience of the most affected groups (BMZ 2015a, p. 26; see also BMZ 2015b, p. 26, 2016). This also involved the promotion of insurance schemes such as the G7 Climate Risk Insurance Initiative, which Germany had initiated during its Presidency of the G7 and which strived to increase the number of people with coverage against changes in weather and climate to about 500 million until 2020 (BMZ 2015a, p. 18).

Foreign Policy: Acting as a Good Global Citizen and Strengthening Climate Diplomacy

While climate change had already played a role in Germany's foreign policy in the 1990s, this was mainly in an environmental context; that is, the Foreign Office (*Auswärtiges Amt*, AA) contributed to shape Germany's behaviour in the UNFCCC negotiations but did not on a broader scale devote resources or adopt specific policies on the issue. This began to change in the early 2000s when the AA took a stronger interest in questions of disaster prevention and early warning systems (Goldammer 2000, pp. 12, 14, 16) and became part of the 2004 Action Plan on Civil Crisis Prevention, Conflict Resolution and Peacebuilding (Bundesregierung Deutschland 2004). Gradually, the focus on the various threats of climate change helped to establish the issue as a vital part of Germany's foreign policy. In its quest to act as a 'good global citizen' and civilian power, Germany's foreign policy had for a long time focused on facilitating a more just global order and on supporting disadvantaged populations around the world (Interview 2015b; Hellmann et al. 2006, pp. 188, 192; Gareis 2005, p. 59). Consequently, the growing number of references to conflicts and threats to national and international security in climate debates caught the attention of foreign policy experts. In this context, climate change ceased to be primarily handled by the BMU and as environmental issue and instead the AA progressively resumed more responsibility, particularly concerning the security aspects (Interview 2014b, 2015a).

Around the peak of the climate security debate in 2007, climate change was integrated into Germany's foreign policy on a broader scale (Interview 2015b). Important triggers were the quickly accelerating global climate security debate, the appointment of the more climate (security) friendly foreign minister Frank-Walter Steinmeier (SPD) in 2005 (Interview 2014b) and the 2007 WBGU report on climate change and security. Thus, in 2007, at the 43rd Munich Security Conference, Steinmeier called climate change a problem of national and international security (Steinmeier 2007a) and at the G8 summit in Heiligendamm emphasised the links between the security implications of climate change and energy security (Steinmeier 2007c). In the same year, the AA organised a conference on climate change as security issue (Auswärtiges Amt 2007) where Steinmeier particularly mentioned the 2007 WBGU report. He stressed that 'the foreign and security policy challenges' of climate change 'are enormous' and necessitated new global mitigation alliances, diplomatic efforts and détente policies, that is, 'preventive environmental diplomacy' to avoid conflictual situations especially in developing countries (Steinmeier 2007b). He concluded that Germany would continue to consider climate change as a core aspect of its foreign policy and that it would soon be discussed by the European Council (Steinmeier 2007b). Subsequently, this led to the publication of the influential EU Commission report on climate change and security in 2008 under German EU presidency (Solana and EU Commission 2008).

It was also in the peak year of 2007 that Germany vigorously stressed the security implications of climate change in the UNSC. In the first UNSC debate on climate change in 2007, Wieczorek-Zeul represented Germany, which exemplifies the close interconnections between (climate) foreign and development policy. Wieczorek-Zeul mainly echoed the disciplinary and governmental discourses by emphasising the vulnerability of poor populations in developing countries (UNSC 2007b, p. 20). Consequently, she demanded that the 'security implications of climate change should receive more attention' (UNSC 2007b, p. 19), especially in the context of conflict prevention (UNSC 2007b, p. 19), and that the world needed a 'global framework of risk management' as well as 'preventive diplomacy' (UNSC 2007b, p. 20).

In 2011, Germany organised a second debate on climate change in the UNSC. In a letter addressed to the Secretary-General (UNSC 2011b), the Permanent Representative of Germany to the UN, Peter Wittig, primarily articulated the disciplinary and sovereign discourse. He stressed that 'the impacts of climate change on peace and security are already tangible' and that climate change was a 'risk multiplier' concerning armed conflict especially in already fragile countries in the Global South (UNSC 2011b, p. 2). To tackle these problems, he suggested an 'integrated approach to conflict prevention' that included all relevant UN organs from the United Nations Development Programme (UNDP), the United Nations Environment Programme (UNEP) to the UNSC (UNSC 2011b, p. 5). In the debate itself, he named climate change a 'driver of conflict' that could destabilise whole regions and once again highlighted the importance of crisis prevention policies (UNSC 2011a, p. 21).

Especially from 2011 onwards, the AA continuously expanded its climate security initiatives (Interview 2015b; Auswärtiges Amt 2011). In close cooperation with adelphi, which had been one of the most active non-governmental proponents of climate security, the AA developed the concept of 'climate diplomacy' (Pieper 2011; adelphi 2012, 2013; Adriázola et al. 2013), which also included an online knowledge base on 'environment, conflict, and cooperation'. The aim of these initiatives was to increase the attention for the security risks of climate change and eventually to build support for a comprehensive climate agreement until 2015 that would be able to significantly reduce the security risks of climate change (adelphi 2012, p. 8). Simultaneously, in cooperation with the Research Group Climate Change and Security (CLISEC) at the University of Hamburg, the AA established the global Climate Security Dialogues, which in particular aimed at bridging the science-policy gap (adelphi 2012, p. 30). Moreover, again in cooperation with adelphi, the AA organised policy briefings on climate diplomacy and the security implications of climate change for German embassy officials, officers from German implementing agencies and representatives of German foundations around the world (adelphi 2012, p. 34; Adriázola et al. 2013, p. 13). Finally, in 2011, the AA established the Federal Foreign Office Climate Fund, which financed several projects around the world that aimed at raising awareness for the implications of climate change (adelphi 2012, p. 54).

In 2013, a BMZ report extended the concept of climate diplomacy to 'climate foreign policy' (BMZ 2013a, p. 20), and at the 50th Munich Security Conference in 2014, Steinmeier once again underlined that climate change was a 'pillar of German foreign policy' (Steinmeier 2014). In 2015, at the G7 Foreign Ministers' Meeting in Lübeck under the German Presidency, the security implications of climate change were a central aspect in the final communiqué (G7 2015). Mirroring the dominant discourses in Germany, the communiqué stressed that in order to prevent the spread of instability and fragile states in the Global South, the world needed to 'mitigate the risks of climate change' and to 'support preparedness and resilience to disasters' (G7 2015). Besides establishing a working group supposed to 'conduct integrated climate and fragility risk assessments' (G7 2015), the G7 commissioned a report on 'A New Climate for Peace: Taking Action on Climate and Fragility Risks' carried out by adelphi, International Alert, the Wilson Center and the Institute for Security Studies of the European Union (G7 et al. 2015). It called climate change '[…] a global threat to security in the 21st century' and urged to '[…] act quickly to limit the future risks to the planet […]' (G7 et al. 2015, p. vii). The report particularly engaged in the governmental discourse by highlighting that the aim was to reduce the overall risk of climate change to a controllable or tolerable level by strengthening the 'resilience of states and societies' (G7 et al. 2015, pp. vii, xi) and by carrying out a 'global risk assessments' (G7 et al. 2015, pp. xv–xix).

Defence Policy: Networked Approaches to Security and Crisis Prevention

The overall impact of climate security discourses on Germany's defence policy and the work of the Ministry of Defence (BMVg) was limited. This was not least due to the fact that many advocates of the sovereign discourse themselves did not see an important role for military actors or solutions in the defence sector (Brauch 2002, p. 25; Eberwein and Chojnacki 2001, p. 10). Instead, the construction of climate change as a threat to human and occasionally national security especially in the Global South facilitated the spread of comprehensive or 'networked' approaches to security that included defence, environmental and developmental aspects. Already in 2008 the then Defence Minister Franz Josef Jung (CDU) emphasised the increasing connections between different sectors and government agencies in order to 'prevent crisis' from materialising in the first place (Bundesministerium der Verteidigung 2008). Beyond that, the 2010 published third Report of the German Government concerning the Action Plan on Civil Crisis Prevention and Peacebuilding further specified this holistic approach and recommended to integrate climate security aspects into the planning schemes of various actors such as the EU, NATO, the OECD, the G8 and development agencies (Bundesregierung Deutschland 2010, pp. 33–35).

During the peak of the climate security debate from 2007 onwards and in the wake of the increased appearance of the sovereign discourse in Germany, the Bundeswehr began to acknowledge the security implications of climate change. However, it only very reluctantly deduced concrete pathways of action for itself from this debate (Interview 2014a, 2015c). In stark contrast to the US, German military officials mostly refrained from talking publicly about climate change and if anything began to discuss the issue internally or in reports that were not primarily targeted at a broader audience. A first report published in 2012 mainly revolved around the security implications of climate change for the Middle East and North Africa and emphasised that 'Increasing resilience is in the first place not a military task but rather a challenge for the state as a whole' (Bundeswehr 2012, p. 7). In this vein, it also critically discussed the problems associated with 'securitising' climate change (Bundeswehr 2012, p. 21). In a second report issued in 2014, the Bundeswehr focused on climate security in the Artic, yet only predicted a marginal probability for military conflict and explicitly highlighted that the climatic changes in this region do not entail a 'future challenge for security policy that is relevant for the Bundeswehr' (Bundeswehr 2014, p. 1).

Key Developments Between 2015 and 2020: Expanding Climate Foreign and Development Policy

In recent years, the main threads of the Germany climate and security debate have remained largely stable. The focus is still primarily on climate change as a broad risk that particularly threatens to undermine the human security of poor populations in the Global South (BMZ 2017, pp. 20, 99, 104; Auswärtiges Amt 2019a, pp. 1–2, 5; BMU 2018, p. 5). Examples are the direct physical effects of climate change such as an increase in extreme weather events (BMZ 2019a, p. 5; Auswärtiges Amt 2017, p. 24), droughts, heatwaves and sea level rise (BMZ 2017, pp. 20, 99; Auswärtiges Amt 2019c, p. 3). Closely linked to this overall narrative, the sovereign discourse and fears that local human security violations could spill over to regional conflicts and instability and eventually threaten international peace and security have become more common (Auswärtiges Amt 2018a, p. 1, 2018b, p. 1; BMVg 2016, p. 42, 2019b, p. 1). Especially the AA and the BMVg now have repeatedly stressed that climate change could increase instability, could lead to geopolitical and resource-based tensions (Auswärtiges Amt 2019c, pp. 4, 10; BMVg 2019a, p. 2) or could become a 'catalyst for conflicts' (Auswärtiges Amt 2019a, p. 2).

In terms of political consequences, the impact on German climate policy has become less prevalent in recent years, not least because climate policy in general had become less central after the adoption of the Paris Agreement (PA) in 2015. As climate performance indices indicate (Climate Action Tracker 2020; CPI 2020), Germany has increasingly lost its forerunner position when it comes to climate change and domestically even runs the risk of failing to reach its 2020 goal under the PA (Der Tagesspiegel 2019). And although the attention for climate change has been reinvigorated by the Fridays for Future demonstrations since 2018, which ultimately led to the adoption of a Climate Package in 2019 (Deutscher Bundestag 2019b), this new policy has been heavily criticised for a lack of ambition (Balser et al. 2020). In all these debates and policies, climate security argumentations feature less prominently than in the past and the attention is on regulatory and economic aspects (Bundesregierung Deutschland 2019).

In stark contrast, climate security argumentations are more influential than ever in German foreign policy (Dröge 2018). In 2019, the AA stated that climate change has become a 'catalyst' for German foreign policy (Auswärtiges Amt 2019c, p. 3) and that fulfilling the international climate goals is a 'new imperative of German foreign policy' (Auswärtiges Amt 2019a, p. 3). It has mentioned the security implications of climate change in central policies such as the 2017 Crisis Prevention Guidelines (Auswärtiges Amt 2017, p. 100), and especially in the wake of its campaign for a turn as non-permanent member of the UNSC 2019–2020 (Auswärtiges Amt 2019c, p. 3), the AA has made clear that it wants to anchor climate security issues into key international fora (Auswärtiges Amt 2019c, p. 4). In this context, the AA together with adelphi also organised the 2019 Berlin Conference on Climate Security and issued a Berlin Call for Action (Auswärtiges Amt 2019a, p. 3). Moreover, at the UN level, Germany has initiated a Group of Friends of Climate Security (Auswärtiges Amt 2019c, pp. 4, 7), supports the UN Climate Security Mechanism (Deutscher Bundestag 2019a, p. 2) and continues its focus on climate diplomacy and on mainstreaming climate security issues into the work of its foreign missions (Auswärtiges Amt 2019b, p. 1, 2019c, pp. 4–5).

Beyond foreign policy, climate security aspects have remained a central theme of Germany's development cooperation efforts and the work of the BMZ and have become even more influential in recent years. In line with the disciplinary and governmental discourses, the focus is on saving the most vulnerable people (BMZ 2019a, p. 5) and on 'comprehensive climate and disaster risk management' (BMZ 2019a, pp. 7–9) with a particular emphasis on prevention and early warning (BMZ 2019e, p. 1, 2019a, p. 15). In this context, the BMZ strives to increase Germany's international climate financing to 4 billion per year until 2020 (BMZ 2019b, p. 8) and supports several adaptation and disaster management programmes such as the Global Adaptation Commission (BMZ 2019d, p. 2), the Global Initiative Disaster Risk Management (BMZ 2019a, p. 7) or the G7-Initiative Climate Risk and Early Warning Systems (BMZ 2019a, p. 15). Moreover, in recent years, the BMZ has expanded its emphasis on risk transfer and climate insurance schemes. The central initiative is the InsuResilience Global Partnership (BMZ 2019c), which wants to establish climate finance and insurance as an important part of comprehensive climate risk management strategies and to go from 'ex-post' to 'ex-ante' disaster support (BMZ 2019e, p. 1, 2019c, p. 1).

Finally, although still much less central than in the US, the German defence sector and the BMVg and the Bundeswehr have become slightly more active, not least due to the increased focus on the implications of climate change for international peace and security in German debates. They still primarily see climate change as an issue to be tackled by civilian actors and international climate policy (BMVg 2019b, p. 1). Yet, they have begun to accept a limited role of the defence sector within the networked approaches to security and the increasing links between development and security (BMVg 2016, p. 42, 2017, p. 2). Beyond mentioning climate change in the important 2016 White Book on Security Policy (BMVg 2016, p. 42), the BMVg has begun to gather expertise on climate security (BMVg 2019b, p. 2) and in 2018 has organised a 'Strategic Forecast: The Artic Dialogue', which also discussed the security impacts of climate change (BMVg 2018b, p. 1). Moreover, there are calls for the Bundeswehr to 'recognize the problem of climate change as a concrete security threat' (BMVg 2019a, p. 2) and to prepare for changed missions scenarios (BMVg 2019b, p. 2). In addition, the BMVg has issued a 'sustainability report' in 2018, hence partly engaging in the narrative of greening the military (BMVg 2018a, 19ff).

Conclusion

Connecting climate change to security conceptions played a key role in Germany from early on and, similar to the US, contributed to setting climate change on the agenda in the first place. In the 1990s, it was especially the disciplinary discourse coupled with a concern for humanity that coined the picture of climate change as security problem. In connection with global justice considerations, this discursive representation saw Germany as obliged to stop this from happening and to support and rescue the most affected populations in the Global South. It contributed to Germany becoming one of the leading proponents of the international negotiations and adopting an ambitious mitigation agenda. Together with a focus on ecological modernisation and the precautionary principle, this led to a range of policies mostly at the intersections between environmental and energy or industrial policy and often aimed at combining environmental protection with economic opportunities. In the early 2000s, the disciplinary discourse kept dominating the debate in Germany and, together with the ambitions of the Red-Green coalition government, further expanded Germany's climate agenda. Towards the second half of the 2000s, the climate security debate in Germany intensified and diversified. It peaked around 2007–2009 when numerous influential reports came out and parliamentary articulations of climate security discourses multiplied. While the representation of climate change as threat to human security prevailed, it was increasingly combined with climate risk articulations and occasionally included references to the sovereign discourse.

This predominant securitisation at that time helped to keep the issue on the political agenda when the not overly environmentally enthusiastic CDU/CSU came out strongest after the election. The main impact throughout the 2000s and 2010s was on German climate policy with an emphasis on mitigation measures. It further strengthened domestic climate policies and also sparked renewed efforts to fulfil Germany's self-proclaimed leadership role concerning the international climate negotiations. While the focus continued to lie on regulatory measures and binding reduction commitments, the growing importance of the governmental discourse contributed to ease German reservations against market solutions to climate change such as the European emission trading system. Together with the sovereign discourse, this also helped to increase the importance of adaptation and resilience measures.

Beyond climate policy, the continuous juxtaposition of poor populations in the Global South as first victims of climate change with resourceful industrialised countries in the Global North had a considerable impact on Germany's development sector. The aim was to support Southern populations but also to monitor and control their behaviour to prevent climate-induced crises, which included a merging of foreign, defence and development policies. Gradually, the spread of climate security argumentations also affected German foreign policy in general and led to a transformation of what climate change meant politically. Thus, particularly from 2007 onwards, a growing number of reports and parliamentary debates began to highlight the 'hard' security implications of climate change such as violent conflict or a destabilisation of international security. This increasingly established climate change as a key foreign policy priority. It also shifted the responsibility for climate policy and the international negotiations from the environmental towards the foreign ministry and coined new political approaches such as 'climate diplomacy' or 'climate foreign policy', which have become even more central towards 2020.

In conclusion, the securitisation of climate change in Germany was highly influential when it comes to redefining what climate change meant politically and legitimising concrete policies and practices. In stark contrast to the US, the German case study shows how the securitisation of climate change can develop differently by constructing the issue as a threat to human security and general risk instead of relying on narrow national security conceptions. Accordingly, it mainly affected climate, development and foreign policies and did not lead to meaningful changes in the defence sector. This form of securitisation certainly is more favourable in terms of abating climate change and preventing human suffering and hence partly supports the idea of the emancipatory qualities of human security in comparison to narrow and military-focused national security conceptions. Nevertheless, the case also exemplifies several problematic effects of such human security-driven securitisations, especially when it comes to perpetuating identities and dependencies (see also Chap. 5).

References

adelphi. (2012). *Climate Diplomacy: Reducing Risks for Security*. Berlin: Adelphi/ German Federal Foreign Office.

adelphi. (2013). *Climate Diplomacy: New Approaches for Foreign Policy*. Berlin: Adelphi/German Federal Foreign Office.

adelphi, & UBA. (2013). *Adaptation to Climate Change for Peace and Stability: Strengthening of Approaches and Instruments as well as Promotion of Processes to Reduce the Security Risks Posed by Climate Change in the Context of Climate Change Adaptation (Climate Change 18)*. Berlin: Adelphi/German Federal Foreign Office.

Adriázola, P., Carius, A., Ruthner, L., Tänzler, D., Thwaites, J., & Wolters, S. (2013). *Climate Diplomacy: New Approaches for Foreign Policy*. Berlin: Adelphi/German Federal Foreign Office.

Arnold, M., Dressel, G., & Viehöver, W. (2012). *Erzählungen im Öffentlichen: Über die Wirkung narrativer Diskurse*. Wiesbaden: VS Verlag für Sozialwissenschaften.

Auswärtiges Amt. (2007). *Sicherheitsrisiko Klimawandel: 17. Forum Globale Fragen*. Berlin, 13. - 14. Juni 2007.

Auswärtiges Amt. (2011). Klima und Sicherheit. Retrieved September 1, 2014, from http://www.auswaertiges-amt.de/DE/Aussenpolitik/GlobaleFragen/Klima/KlimaUndSicherheit_node.html.

Auswärtiges Amt. (2017). *Krisen verhindern, Konflikte bewältigen, Frieden fördern: Leitlinien der Bundesregierung*. Berlin.

Auswärtiges Amt. (2018a). Vereinte Nationen: Deutschland gründet Freundesgruppe "Klima und Sicherheit". Retrieved February 5, 2020, from https://www.auswaertiges-amt.de/de/aussenpolitik/themen/klima/freundesgruppe-klima-und-sicherheit/2125194.

Auswärtiges Amt. (2018b). Klimawandel bedroht internationale Sicherheit. Retrieved September 27, 2018, from https://www.auswaertiges-amt.de/de/aussenpolitik/themen/klima/klima-sicherheit-freundesgruppe-deutschland/2141356.

Auswärtiges Amt. (2019a). *Rede von Außenminister Heiko Maas zur Eröffnung der Berliner Konferenz zu Klima und Sicherheit*. Berlin. Retrieved June 4, 2019, from https://www.auswaertiges-amt.de/de/newsroom/maas-klima-sicherheit-konferenz/2223650.

Auswärtiges Amt. (2019b). Dialog für mehr Klimaschutz. Retrieved February 5, 2020, from https://www.auswaertiges-amt.de/de/aussenpolitik/themen/klima/-/205760?openAccordionId=item-205770-0-panel.

Auswärtiges Amt. (2019c). *Bericht des Auswärtigen Amts zur Klima-Außenpolitik*. Berlin. Retrieved February 4, 2020, from https://www.auswaertiges-amt.de/

Palgrave Studies in Sustainable Business In Association with Future Earth

Sustainability in Business is increasingly becoming the forefront issue for researchers, practitioners and companies the world over. Engaging with this immense challenge, Future Earth is a major international research platform from a range of disciplines, with a common goal to support and achieve global sustainability. This series will define a clear space for the work of Future Earth Finance and Economics Knowledge-Action Network. Publishing key research with a holistic and trans-disciplinary approach, it intends to help reinvent business and economic models for the Anthropocene, geared towards engendering sustainability and creating ecologically conscious organizations.

More information about this series at
http://www.palgrave.com/gp/series/15667

András Ócsai

Ecologically Conscious Organizations

New Business Practices Based on Ecological Commitment

palgrave
macmillan

András Ócsai
Business Ethics Center
Corvinus University of Budapest
Budapest, Hungary

ISSN 2662-1320 ISSN 2662-1339 (electronic)
Palgrave Studies in Sustainable Business In Association with Future Earth
ISBN 978-3-030-60917-7 ISBN 978-3-030-60918-4 (eBook)
https://doi.org/10.1007/978-3-030-60918-4

Cover illustration: kenkuza_shutterstock.com

This Palgrave Macmillan imprint is published by the registered company Springer Nature Switzerland AG
The registered company address is: Gewerbestrasse 11, 6330 Cham, Switzerland

Preface

In an interview in 1986, the late Erwin Chargaff, a professor of biochemistry at Columbia University, responded to the question whether there is a place for ethics in science. He claimed that there could not be. He said that while an American Indian might pray to a tree if they needed to cut it, and might also ask for forgiveness, scientists transgressed ethical boundaries in the seventeenth century when they sacrilegiously started digging into nature. For a while, due to its relative inefficiency, what we call modern science did relatively little harm. The nature of the problematic relationship between ethics and science is a question for the present times, Chargaff claimed, mainly due to the splitting of the atom, nuclear energy, and knowledge about the genetic matter of the cell nucleus. Chargaff emphasized that it was now much too late: science had lost its innocence, and science and ethics were now incompatible. He concluded that anyone who sought to be an ethical individual should study bookkeeping, or something other than science (Chargaff 1987, p. 883).[1]

Not knowing about Chargaff's suggestion, I graduated as an economist in 2002, and worked in finance, accounting, and project management for several multinational companies and in the public sector. However, I always found myself looking for something more than just an increase in salary, and further career prospects. I began asking myself, what is the noble goal of ethical, genuine, value-oriented, human behavior? Where is love and respect for nature, and for non-human beings? Can these values prevail in a business organization at all? At the end of 2009, I came across an article about Buddhist economics, which was a revelation in terms of my understanding of the importance of non-mainstream economics. I decided to study alternative economics, ranging from ecological economics to Buddhist economics, and to conduct research on value-oriented, ecologically conscious business organizations.

Unfortunately, the more than three decades that have passed since Chargaff's warning have shown that not only science but also the economy and business are in deep conflict with ethics and ecology. What can we expect from business at its best in the midst of worsening climate change and threats such as coronavirus and other hitherto unknown viral pandemics, at a time when the unpredictable consequences of what may be ecological breakdown have already significantly increased the suffering of human and non-human beings?

This book is a snapshot of the results of my decade-long search for ecological and human values in the economy and business. My intention has been to show the meaning and significance of ecological consciousness in business, and how ecologically conscious business organizations can become exemplary. In this book, I describe unusual business practices that are based on genuine ecological commitment. I hope that these examples will serve as inspiration for a more ecologically conscious way of living and functioning, and help to catalyze endeavors aimed at creatively

[1]Chargaff, E. (1986, November 6): *Az egyre erősödő bizonytalanság légköre* [An Atmosphere of Growing Insecurity]. Interview by Tibor Szántó. Akadémiai Értesítő / Magyar Tudomány [Academic Bulletin / Hungarian Science], Vol. 94, No. 11, pp. 877–883. http://real-j.mtak.hu/143/1/MATUD_1987.pdf. Accessed 20 July 2020 [in Hungarian].

contributing to the spiritual and material betterment of the world of business, and beyond.

Budapest, Hungary
August 2020

András Ócsai

Acknowledgments

I would hereby like to thank Paul Shrivastava and László Zsolnai, series editors of *Palgrave Studies in Sustainable Business In Association with Future Earth*, and the editorial and production team at Palgrave, particularly Jessica Harrison, Srishti Gupta, and Geetha Chockalingam, for their support and professional assistance.

I am grateful to Corvinus University of Budapest for institutional and financial support during the years it took to research and publish this book. I acknowledge the help of my colleagues at the Business Ethics Center at Corvinus University of Budapest, and that of scholars and friends from the Hungarian and international academic world, including leaders and members of the Transatlantic Doctoral Academy (TADA) on Business, Economics and Ethics, with whom I have engaged in fruitful interaction and collaboration.

I sincerely thank the interviewees involved in the empirical research for sharing details about their values, thoughts, and practices related to ecological consciousness in business. I also wish to express my thanks to my language editor, Simon Milton, for his great work improving the English of the text.

I am exceptionally grateful to László Zsolnai for selfless mentorship and continuous spiritual and intellectual support spanning more than a decade. His genuine friendship has helped me through professional and personal challenges in my life.

For their love, inspiration, support, and patience, I thank all my family members and friends, including those who have already crossed over.

András Ócsai

Praise for *Ecologically Conscious Organizations*

"As we step into the beginning of the third decade of the twenty first century the paradigm of business and management is undergoing a radical transformation that addresses such vital questions as meaning of work and purpose of life. After passing through formidable challenges and turmoils including ethical collapses that have compelled modern business organizations to engage in soul searching, conscientious business leaders worldwide have come to realize that reverence for nature and respect for all life forms is of cardinal importance in shaping the course and culture of business in future that accords primacy to ethics and human values, sustainability and spirituality. In the background of these pathfinding developments both in theory and practice for a more humane and sustainable future for the self, organization, community, society and planet this book on Ecologically Conscious Organizations by András Ócsai is a timely and valuable contribution to the existing body of literature, in this multi-disciplinary and multi-dimensional field of research. A committed academic and thinker of Business Ethics Center, Corvinus University of Budapest and also of the European SPES Forum, Professor Ócsai has not only covered the multiple strands of thought and action in the field of Business and Ecology but also elevated it to a higher level of consciousness beyond intellectual exercise by anchoring the movement on the terra firma of spiritual wisdom and experience. Coming from India, I can see and sense a palpable resonance of his work with the pioneering work of the Nobel Laureate poet and philosopher Rabindranath Tagore towards Nature-inspired education and the signal contribution of Prof. S K Chakraborty, Founder-Convener of Management Centre for Human Values, Indian Institute of Management Calcutta, in the field of Human Values and Indian Ethos in Management. I am sure the book will receive wide acceptance through creation of space for dialogue among those who would be committed to the mission of creating a humanistic, value-based and sustainable business scenario in future."

—Sanjoy Mukherjee, *Professor of business ethics and corporate social responsibility, Indian Institute of Management, Shillong, India*

"Grounded in in-depth systematic analyses of a broad array of inspiring case examples of ecologically conscious business organizations, this book is an invaluable treatise on how to achieve the harmonization of economic and deep ecological aims through a fundamental change in the existential mindset and inherent approaches of business enterprises."

—Eleanor O'Higgins, *Adjunct Associate Professor at the College of Business at UCD Dublin, Ireland, and an Associate at the London School of Economics, UK, coeditor of the Palgrave book Progressive Business Models*

"In this book Ócsai is challenging the dominating concept of business models and he gives a convincing and relevant presentation of unusual business practices that are based on genuine ecological commitment. Ócsai goes beyond ethical theory to show moral and environmental considerations and challenges in practice. This book is essential and stimulating reading for students, scholars and practitioners looking for a guide to an ecological responsible economy."

—Ove Jakobsen, *Professor and Director, Center for Ecological Economics and Ethics, Nord University, Norway*

"Bringing together knowledge and insights from a range of disciplines, this book focuses on key areas as spirituality, consciousness, nature and entrepreneurship. The major implications are considered, and the importance are highlighted. It is a beautiful book, and it will appeal to scholars and students of psychology and business, as well as business leaders wishing to gain valuable insights and inspiration into the range of issues we need to understand and act upon in the current atmosphere of bad practice, uncertainty during a pandemic and a threatening climate change."

—Knut O. J. Ims, *Professor in business ethics, Norwegian School of Economics, Bergen, Norway*

Contents

List of Figures

List of Tables

Part I

Introduction

This book addresses the topic of ecological consciousness in business organizations, with a special focus on value orientation and business models. Ecological consciousness is gaining importance today as the ecological crisis of our era, dominated by global big business, becomes increasingly evident. A large body of scientific literature has investigated the value orientations of businesses, but the topic of ecological consciousness has been much less studied in this context. Moreover, there are hardly any examples of the scrutiny of the relationship between value orientation and ecological consciousness in business.

The book goes beyond previous studies, as (i) it summarizes the literature about ecological consciousness and its appearance in business organizations; (ii) it investigates the value orientations of ecologically conscious businesses through a deeper, previously unused type of qualitative methodology; and (iii) it presents in a systematic, comparative way the value commitments and business models of selected ecologically conscious business organizations.

The following main questions come under microscope: What are the central concepts related to ecologically conscious business and an ecologically conscious economy? What are the key features of these concepts, and what is the relationship between them? What kind of value orientations do ecologically conscious business organizations have? What are the business models on which they build their operations? What are the fundamental goals and raison d'être of ecologically conscious businesses? How do they define success?

1

Ecology and Business

Why is it meaningful to examine ecologically conscious business organizations within the social science context—particularly their value background and business models? The first chapter of this book justifies the value of this goal. It also briefly presents the framework of the research that lies behind the book using a short summary of earlier studies and a discussion of related methodological issues. It then concludes with an overview of the structure of the book.

Let the Facts Speak for Themselves

Scientific research (see, among others, Carrington 2016, Waters et al. 2016) suggests that we are now living in a new era that may be called the Anthropocene[1]; a time when human activities have significantly altered

[1] Earth sciences have not yet officially recognized the Anthropocene epoch, and there is no general agreement about its beginning (some have proposed that the Neolithic Agricultural Revolution more than 12,000 years ago could be considered a fitting start date, while others

A. Ócsai, *Ecologically Conscious Organizations*, Palgrave Studies in Sustainable Business In Association with Future Earth,
https://doi.org/10.1007/978-3-030-60918-4_1

Earth's ecosystems (Crutzen and Stoermer 2000; Steffen and Broadgate et al. 2015a; Steffen and Richardson et al. 2015b). Our biosphere-transforming activities—primarily involving the globalization of societies and economies—have by now reached such an extent that they are incomparably larger and more harmful than those of other species, and are largely irreversible. It is not only our well-being, but also the very survival of our own and other species that is at stake (Pataki and Takács-Sánta 2005; Takács-Sánta 2004). To alleviate the ecological crisis affecting our planet, the spread of ecological consciousness and a praxis that focuses on restoring the natural environment is badly needed. This transformation would enable the economy to function again as a system "embedded" in nature and society (Polanyi 1944).

Owing to the work of excellent scholars and scientific communities (including Rachel Carson [1962], Club of Rome [2020], Mihajlo Mesarovic [Mesarovic and Pestel 1974], Jan Tinbergen [1977], Ernst F. Schumacher [1973], and the Stockholm Resilience Center [SRC 2020]), we have known for almost half a century that we are facing very serious ecological problems. What then, are the signs of the ecological crisis?

The main environmental problems include changes in communities of natural living beings, deforestation, and soil degradation; changes in global bio- and geochemical cycles, contamination, acid rain, smog, eutrophication, and the ozone hole; global climate change; the decline in biodiversity; food scarcity; and fresh water scarcity (Takács-Sánta 2010).

Researchers have been dealing with global climate change and the related trends since the 1970s. It is known that colder and warmer periods have periodically occurred on Earth, but climatic processes seem to have changed because of human activity. First, in 1972 the United Nations Conference on the Human Environment (held in Stockholm) addressed the phenomenon in detail (at which time the United Nations Environment Programme, UNEP, was also created). Following this, in

have traced it back to the Industrial Revolution, and others to the nuclear tests at the end of World War II). It is generally accepted, however, that the Anthropocene is dramatically different from the previous eras, and great attention should be paid to this pheonomenon because increasingly accelerating deteriorative ecological, social, and economic processes have been occurring since the 1950s (the so-called "great acceleration" [Steffen et al. 2015a]) that is transgressing planetary boundaries Steffen et al. (2015b).

1985 at the World Climate Conference in Villach, Austria, climate change was identified as a scientifically grounded fact. At conferences held in the following years (1988 in Toronto, 1990 in Geneva) participants also confirmed the need to reduce the greenhouse gasses emitted into the atmosphere. This was also due to the Brundtland Commission, which functioned from 1984 to 1987, and to the activities of the Intergovernmental Panel on Climate Change (IPCC), established in 1988 (IPCC 2020).

The IPCC examines and summarizes research findings about climate change induced by human activities. It published its *Fifth Assessment Report* in 2014, after issuing reports in 1990, 1996, 2001, and 2007. According to the former report, the terrestrial climate is obviously warming, and the rise in global average temperatures since the mid-twentieth century is due to greenhouse gas emissions of human origin. Climate change is causing global sea levels to rise steadily, the thickness of polar ice sheets to decrease, rainfall to become more intense, and drought periods to become longer. In many regions of the world, these phenomena are decreasing residents' access to safe drinking water. Particularly vulnerable regions (those affected by multiple dimensions of the crisis) are the Polar Regions, Sub-Saharan Africa, small islands, and Asian delta regions. Ecosystems facing an elevated level of danger include coral reefs, colonies of marine shellfish, tundra, coniferous forests in the taiga, as well as mountainous and Mediterranean regions. Approximately 20–30 percent of all plant and animal species are presently threatened with extinction (IPCC 2014).

In 2018, the IPCC published a special report on the impact of global warming of 1.5 °C above pre-industrial levels and related global greenhouse gas emission pathways. It concludes that human activities have already caused approximately 1.0 °C of global warming, but this is likely to reach 1.5 °C between 2030 and 2052 if the temperature increase continues at the current rate. Warming from anthropogenic emissions will persist for centuries to millennia, and will continue to cause further long-term changes in the climate system. The climate-related risks to natural and human systems depend on the magnitude and rate of warming, geographical location, level of development and vulnerability, and on the choices and implementation of adaptation and mitigation options. Limiting risks must involve societal and system transition and transformation that can be enabled by increasing adaptation—and mitigation-focused investment, policy instruments, accelerating technological innovation, and behavioral changes. Strengthening the capacity for climate action of national and sub-national authorities, civil society, the private sector, indigenous people and local communities, and greater international cooperation can support the implementation of ambitious initiatives (IPCC 2018).

The latest annual report from the World Meteorological Organization explains the impact of changing weather and climate on the socio-economic sphere, health, migration, food security, and terrestrial and aquatic life. It highlights that the past five years have been the warmest period since the systematic measurement of temperature began, and that 2010–2019 has been the warmest decade. In addition, since the 1980s each successive decade has been warmer than any preceding one. A record amount of ice has melted in both the Arctic and the Antarctic. The oceans have absorbed around 23 percent of annual CO_2 emissions, moderating the increase in the atmospheric concentration. This process, however, enhances ocean acidification. Additionally, almost every region of the World Ocean suffered at least one heat wave in 2019. This warming will continue as the amount of greenhouse gases increases. These tendencies endanger the living conditions of ocean flora and fauna. The effects of climate change are already being felt as extreme weather conditions, and their consequences are dramatically and rapidly accelerating. We are still on the wrong track regarding keeping the likely temperature change to a maximum of 1.5–2 °C above the pre-industrial level, as called for in the Paris Climate Convention. The latter report again calls on decision-makers to start taking the situation seriously, because we are slowly running out of time to act (WMO 2020).

Although the international conferences of the last three decades have generated increasingly profound results (1992 Rio de Janeiro—Agenda 21, United Nations Framework Convention on Climate Change; 1997 Kyoto Protocol; 2002 Johannesburg—World Summit on Sustainable Development; 2005 Montreal—"rulebook" for implementing the Kyoto Protocol; 2009 Copenhagen—legally non-binding Copenhagen

Accord; 2010 Cancún—Cancun Agreements, idea of setting up a Green Climate Fund and a forest protection program; 2011 Durban—Green Climate Fund; 2012 Doha—The Doha Climate Gateway; 2015 Paris—Paris Agreement; 2016 Marrakesh—call for a transformation to a low-emission global economy; 2017 Bonn—Fiji Momentum for Implementation; 2019 Madrid—The European Green New Deal), these have remained mainly diplomatic in nature so far. There are significant differences between expectations and results, so experts have repeatedly considered the series of conferences to be a failure. Considering that the definition of real and significant commitments, concrete deadlines, and sanctions has often failed—that is, no significant turnaround has happened in greenhouse gas emissions—this criticism is unfortunately valid.

Human activity, as confirmed by the *Millennium Ecosystem Assessment Report* (launched in 2000 by Kofi Annan, Secretary-General of the United Nations), is having a decisive and increasingly negative impact on Earth's ecosystems and "ecosystem services" (such as food, water, disease control, climate regulation, spiritual fulfillment, and aesthetic experience) (MEA 2005). The report summarizes the most important findings in four points: (1) in the past 50 years, humankind has changed ecosystems more rapidly and to a greater extent than ever before. This change is already threatening our ability to meet growing needs for food, drinking water, wood, fiber, and energy. We are also causing significant and increasingly irreversible losses of biodiversity. (2) The modification of ecosystems has contributed to an increase in human welfare and economic development, but at the expense of increasing the related cost—that is, the deterioration of a significant proportion of all ecosystems, a rise in the risk of non-linear change, and an increase in poverty among some groups of people. All of these factors also degrade the services that ecosystems will be able to provide to future generations. (3) The deterioration of ecosystems may significantly continue in the first half of the twenty-first century, representing a real threat to meeting the United Nations Millennium Development Goals. (4) In order to reverse the degradation of ecosystems and to meet our growing needs, significant political, institutional, and practical changes are ineluctable, but the signs of such change are not yet visible (MEA 2005).

The figures quoted in the report are astonishing: 60 percent of the ecosystems under investigation have deteriorated, or their services are being used unsustainably (drinking water, fishing, purification of air and water, regulation of regional and local climate, natural risks, and pest control). In the 30 years following 1950, more of the Earth's surface was turned into arable land than between 1700 and 1850. Today, the amount of cultivated land is equivalent to a quarter of the surface area of the Earth. In the last decades of the twentieth century, 20 percent of all coral reefs disappeared, and the condition of another 20 percent deteriorated, while 35 percent of all mangrove forests have been lost. Since 1960, the amount of water that has been collected and stored behind dams has quadrupled. The amount of water extracted from rivers and lakes has doubled. The amount of biologically reactive nitrogen introduced into terrestrial ecosystems has doubled, and the amount of phosphorus tripled. Since 1750, atmospheric concentrations of carbon dioxide have increased by 32 percent, mainly due to the burning of fossil fuels and changes in land use (60 percent of the increase is attributable to the period after 1959). By 1990, two-thirds of the area of two out of the world's 14 largest biomes (ecological communities) and half of the area of another four had been converted into land primarily for agricultural use. The rate of extinction of species has increased a thousand times over the last few hundred years (MEA 2005).

With similar goals to the Millennium Ecosystem Assessment, the Governing Council of the United Nations Environment Programme launched the Global Environment Outlook project in 1995. This is designed to help meet the reporting requirements of the Agenda 21 action plan, and also describes the state of the natural environment in regular reports. The fifth report, entitled *GEO-5*, published in 2012, confirms that human well-being and development are based on the natural environment, and that changes in the latter affect human safety, health, social relationships, and material needs. Scientific evidence supports the claim that ecosystems are operating at their biophysical boundaries, and in some cases, beyond them. Humankind is causing unprecedented changes in the state of the environment at a global and regional level. The global average temperature of the atmosphere and oceans is increasing at an accelerating rate (while the average temperature of the atmosphere in the twentieth century increased by 0.74 percent, in the twenty-first century it is expected to increase by 1.8–4 percent), while the size of polar ice sheets is decreasing and world sea-level is rising. More than two million people worldwide die prematurely each year due to air pollution. The size of the ozone hole above Antarctica is larger than ever. Land degradation caused by unsustainable land use and climate change is endangering the living conditions of approximately two billion people, mainly in developing countries. The amount of fresh water per capita is declining, and if the trend continues, by 2025 1.8 billion people will live in areas with absolute water scarcity. We continue to exploit aquatic ecosystems in a way that seriously endangers the sustainability of food production and biodiversity. The prevalence and number of individuals of a great majority of known species are steadily declining. More than 16,000 species are now at the brink of extinction, and although temperate forests expanded by 30,000 square km per year between 1990 and 2005, during the same period tropical deforestation resulted in the disappearance of 130,000 square km of rainforest per year (UNEP 2012).

The sixth *Global Environmental Outlook* was released in 2019. This reinforced the claim that the world is now not on track to meet the Sustainable Development Goals, even by 2050. The consequences of damage to the natural environment will increasingly threaten people's health and the likelihood that we can meet the goals of the Paris Agreement. Humankind must drastically scale up protective and restorative activities to improve ecological conditions. If this does not happen, by the middle of the century millions of premature deaths will occur in Africa, the Middle East, and Asia, and anti-microbially resistant pollutants in freshwater will become a major cause of death. Many pressing challenges such as the need to reduce food waste, adopt less meat-intensive diets, tackle plastic pollution in the oceans, manage water scarcity, and cope with the loss of ecosystems underline the fact that the health and prosperity of humanity is interlinked with the state of the natural environment. The report concludes that the new types of thinking involving science, technology, finance, and politics that are needed to move toward a more sustainable form of development already exist. Creating a near-zero-waste economy by 2050, green investment using two percent of countries' GDP, improved governance, land-use planning, green infrastructure, strategic investment into rural areas, the use of big data in environmental statistics, and policy interventions that address entire systems rather than individual issues are relevant examples of this. Sufficient support from public, business, and political leaders and the will to implement such policies and technologies at the right speed and scale are, however, still lacking (UNEP 2019).

The *Living Planet Report*, published by the World Wide Fund for Nature for the eleventh time in 2016, also shows a negative picture of the health of our planet, biodiversity, and the impact of human activity (WWF 2016). The most important finding of this is that the needs of humankind, calculated based on indicators from the Living Planet Index (LPI), the ecological footprint, and the water footprint of production, exceed Earth's capacity to provide them. The LPI, developed in 1997, measures the state of biodiversity on Earth by monitoring the average change in the size of vertebrate populations. Between 1970 and 2012, this global indicator declined by almost 58 percent. Thus such populations now contain, on average, fewer than 50 percent of the individuals they did fifty years ago. The ecological footprint measures the magnitude of the biologically productive land and water-covered area that can produce the renewable resources required for human activity, including the area needed for infrastructure, and for vegetation that can absorb the carbon dioxide that is being generated. The footprint had almost tripled by 2012 compared to its size in 1961, and now exceeds Earth's bio-capacity by 66 percent. This status, known as ecological overshoot, has existed since the 1970s and is primarily attributable to the magnitude of the carbon dioxide footprint, which increased from 43 percent in 1961 to 60 percent. The water footprint of production is another measure of the renewable resources that are required for human needs. This indicates that 71 countries are facing challenges in terms of the amount of water that is used but not cleaned, and the amount of water not returned to the natural cycle. The report states that if the world develops along a "business as usual scenario," by 2020 the resources generated by two Earth-like planets would be needed to meet human needs and absorb the carbon dioxide that is being generated.

The 2018 edition of the Living Planet Report restates the scientific evidence that unsustainable human activity is pushing the natural systems that support life on Earth to the edge. Economic activity depends on natural services and is estimated to be worth approximately 125 trillion USD a year. Modern human societies are ultimately built on what nature provides us with, thus are of invaluable worth in terms of our health, wealth, food, and security. Research demonstrates that human consumption is the driving force behind these unprecedented planetary changes. The still increasing demand for energy, land, and water means that the current rate of species extinction is now 100–1000 times higher than the rate it was for many millennia of Earth's history. The maintenance of biodiversity, a prerequisite for human life on the planet and the life of all species, is seriously endangered. To move beyond "business as usual," governments, business, finance, research, civil society, and individuals have to join forces and make a new global deal for nature and people. Strong leadership and the right political, financial, and consumption-related choices at every level are pivotal to sustaining humanity and nature in harmony (WWF 2018).

As our activities are driven by consumption and economic growth, we are causing significant change to our planet. Nevertheless, at this high cost, have we achieved a level of material and spiritual well-being and happiness that all humans can enjoy? A study by Colin Ash, a British economist, concludes that while economically more developed countries have become much richer over the last fifty years, the average level of happiness of people has not changed (the Easterlin Paradox), because income growth increases happiness only until a certain level of income is reached (around 20,000 USD per year in 2005 prices). Above this level, the quality and richness of interpersonal relationships contributes more to the happiness of people than rising income and consumption (Ash 2007). Angus Deaton and Daniel Kahneman, both Nobel laureates in economics from Princeton University, also raised the question whether money buys happiness. They analyzed the Gallup-Healthways Well-Being Index, a daily survey of 1000 US residents conducted by the Gallup Organization. The authors distinguished between emotional well-being (the emotional quality of an individual's everyday experience) and their life evaluation (thoughts that people have about their lives) as two aspects of subjective well-being. The authors concluded that income and education are more closely associated with life evaluation, but health, care-giving, loneliness, and smoking are relatively stronger predictors of daily emotions. Life evaluations rise continuously with level of income, but not emotional well-being. At a certain level of annual household income (around 75,000 USD) no further development in the frequency and intensity of experiences that make one's life pleasant can be observed (Kahneman and Deaton 2010). The London-based New Economics Foundation created the "Happy Planet Index" in 2006, which shows an even more complex picture. This instrument combines life expectancy, life satisfaction, and ecological footprint in a single indicator (NEF 2009). According to the 2016 data of the Happy Planet Index, which were supplemented by data about income inequality, some countries may be able to improve their aggregate score, but no single country can improve all three factors while keeping its ecological footprint below 100 percent (NEF 2016).

In 2012, the Future Earth program was launched at the United Nations Conference on Sustainable Development, or "Rio+20." It is to build on and exceed more than 30 years of research experience of the World Climate Research Programme, the International Geosphere-Biosphere Programme, DIVERSITAS, and the International Human Dimensions Programme on Global Environmental Change. The program brings together over forty thousand scientists from all over the world. It seeks answers to present challenges using a new approach to addressing the pressure on terrestrial systems and global sustainability. Compared to former attempts, it is more integrative, and solution oriented. It strives to create closer links between research programs and trends, and promotes governments, NGOs, local stakeholders, and companies that actively engage in joint work. For humankind to be able to live in a sustainable and just world, there is a need for a new kind of science that can create and operate an effective global innovation system through linking different disciplines, knowledge systems, and social partners that can help meet the United Nations Sustainable Development Goals of 2015. In addition to global research projects and research initiatives, Future Earth's strength lies in associated, so-called "Knowledge-Action Networks." Promoting active collaboration between social partners and the creation of actionable knowledge are the goals at the heart of these networks. They seek solutions to the main challenges associated with the present social and ecological crisis: emergent risks and extreme events; financial and economic reform; health; the state of the oceans; natural assets; systems of sustainable consumption and production; cities; and the water-energy-food nexus (Future Earth 2020).

In the form of a unique community presently comprised of around 100 prominent scientists, business leaders, and former politicians, the Club of Rome has addressed the critical, complex, and interconnected challenges humankind faces for more than five decades. Since its first major report, the 1972 *The Limits to Growth*, its main goal has been to alert the world to the consequences of the interaction between human systems and the health of the planet, and to actively advocate for paradigm and system shifts. Humanity must divert its path of development away from relying on the exponential consumption and population growth that jeopardizes Earth's climate and life-support systems, and address the clamant needs of impoverished billions worldwide who are suffering from increasing social and economic inequality. Through profound scientific research, holistic policy proposals, and high-level events, the Club of Rome is now concentrating on five key areas through work within five so-called impact hubs: Climate-Planetary Emergency, Reframing Economics, Rethinking Finance, Emerging New Civilization, and Youth Leadership and Intergenerational Dialogue. Human beings are inherently relational and interconnected to one another in space and time, and to nature. We thus must combine our efforts to shift to a new paradigm that can reshape our belief system and change complex economic, financial, and social systems. Such a shift can foster fundamental values that promote human dignity, respect for nature, and help protect the commons beyond current generations (Club of Rome 2020).

Indisputable signs of the ecological crisis prove that prevalent, growth-based types of economic system are unsustainable as they deplete natural resources and destroy their own prospects for survival. As American ecological economist Herman Daly writes, global economic growth, when it transgresses ecological boundaries, is associated with greater costs than benefits, thus makes us poorer, not richer. Even popular efficiency-enhancing endeavors are not appropriate solutions to this problem, because they often cause exactly the opposite effect to that we would like to achieve. Improving the efficiency of utilization of a resource increases the supply of that resource, thereby reducing its price, leading to an increase in its consumption (the so-called "Jevons paradox"), meaning that we run up against ecological limits much earlier. Moreover, through a focus on growth we cannot even achieve another goal of development: a global reduction in poverty. Growth does not serve to meet the basic needs of poorer countries, but rather serves the interests of developed ones (Daly 1999, 2008).

Predominantly, the prevailing business model of companies itself is causing and deepening the ecological crisis (Daly and Cobb 1989; Brenkert 1995; Shrivastava 1995a, 1995b; Ims et al. 2014; Capra and Jakobsen 2017; Jakobsen 2017; Shrivastava et al. 2019, 2020). Because of the pace and magnitude of ecological degradation, making small-scale adjustments is not enough. The ecological transformation of the economy, and the development and application of new and progressive business models are of vital importance. Such progressive business models involve ethical considerations, respect ecological values, and apply transdisciplinary approaches and alternative thinking. The existence of these characteristics in business is indispensable if we are to hope for fundamental change (Ims and Zsolnai 2009; Rockström 2010; Ims et al. 2014; Ims and Pedersen 2015; Storsletten and Jakobsen 2015; Zsolnai 2015; Capra and Jakobsen 2017; Jakobsen 2017; Harangozó et al. 2018; Steffen and Rockström et al. 2018; Shrivastava et al. 2019, 2020).

Social Science Context

In exploring the causes of and solutions to the ecological crisis, besides the necessary and valuable natural science approaches, there is an urgent need for the social sciences. The findings of numerous research organizations, research studies, and the reports of individual scholars and scientific communities underline the fact that without identifying and understanding the human roots and social context of ecological symptoms, we will not be able to improve deteriorating natural conditions, or even mitigate the present crisis (Daly and Cobb 1989; MEA 2005; Ims and Zsolnai 2009; Zsolnai 2015; IPCC 2018; Steffen and Rockström et al. 2018; WWF 2018; UNEP 2019; Club of Rome 2020; Future Earth 2020; WMO 2020).

The aim of this book is to explore the theory and practice of ecologically conscious businesses and to present and analyze their value orientation and business models. The book is based on social science research, and its approach is idiographic, inductive, and qualitative (Babbie 2008; Mills et al. 2010). It is idiographic because it seeks to deepen the understanding of the selected businesses cases. It is inductive as it draws conclusions based on the empirical examination of the units of observation. Finally, it is qualitative because it does not work with a representative sample nor process quantifiable data, but engages in a more detailed exploration and description of the details of each case. The results may be utilized in practice and contribute to a better understanding of the phenomenon of ecological consciousness in business and beyond.

Based on its goals, social science research can be exploratory, descriptive, and explanatory (Babbie 2008). Research can thus target several goals at the same time, which is the case here. Typical of exploratory research, the present book serves to satisfy the author's personal interest and desire for understanding, the wish for immersion in a specific social or economic phenomenon, and the desire to lay the foundations for future research. In adapting a descriptive research approach, the analysis is designed to provide a precise and careful description of the subject of the research.

From the epistemological perspective, the present research can be considered constructivist; the author accepts the foundations of scientific approaches that are rooted in the constructivist philosophy of science. These include, inter alia, questioning the idea of the independence of the external, physical world and social reality from human activity and knowledge, and also questioning the existence of a value-neutral scientific method. According to the constructivist approach, "the purpose of both lay and scientific knowledge construction is to provide useful, adequate, coherent, stable, or meaningful representation of the world in accordance with particular sets of systemic and sociolinguistic rules

Table 1.1 Summary of the basic characteristics of the research

Aspect of Classification	Basic Characteristics of the Research
Purpose:	exploratory, descriptive
Epistemology:	constructivist, interpretative
Methodology:	idiographic, qualitative, inductive
Planned utilization:	applied

Source Author's construction

and constraints in given contexts" (Maréchal 2010, p. 220). Individuals who try to understand the world that surrounds them create subjective meaning from their experiences about different things. This is also true of researchers, who accept and search for a diversity of viewpoints, construct theory during social interactions from subjective meaning determined by historical and cultural norms, and specify the patterns of such meaning. The constructivist scholar recognizes that their own philosophical assumptions, worldview, values, beliefs, and experiences influence the interpretation of meanings that are revealed, and therefore interprets the results of research through clarifying and disclosing previous assumptions and analyses (Creswell 2007) (Table 1.1).

What Does Earlier Research Say?

The value orientations of businesses have been investigated in many cases, but the topic of ecological consciousness has rarely been in the spotlight. There are hardly any examples of the study of both phenomenon together. The literature rarely addresses the issue of ecological consciousness, unlike environmental consciousness (the difference between the two concepts is discussed in the third chapter).

Susanne Kaldschmidt studied sustainability values and the impact of the personal values of top business managers on sustainability strategies. Her use of the concept "awareness" does not correspond to the notion of "consciousness," although her inspiring findings include the claim that the personal values of the executives of companies that perform better in the field of sustainability are reflected in the sustainability strategies of such companies. Kaldschmidt further emphasizes that the extent of the transfer of personal values to company strategies largely depends on what executives and managers consider their own role to be, and the role of the economy in society (Kaldschmidt 2011).

In 2008, a study of value-oriented businesses at the Business Ethics Center of Corvinus University of Budapest found that value-oriented company behavior can be implemented in Hungary's economic conditions, because—according to Robert Frank's (2004) hypothesis—the higher operating costs of the former are compensated by specific economic benefits associated with obtaining the trust and commitment of stakeholders (Zsolnai and Győri 2011).

In 2011, a piece of empirical research entitled Sustainable and Socially Responsible Business Practice was conducted by the Business Ethics Center of Corvinus University of Budapest. Within this research, progressive Hungarian businesses that incorporate ecological considerations into their business models were studied. The study reaffirmed the hypothesis of Robert Frank, demonstrating that the surveyed companies base their ecological orientation on the personal commitment and intrinsic motivation of their leaders. The motivation and definitions of success of these ecologically oriented companies are multidimensional, and the well-being of their stakeholders is just as important to them as their own financial prosperity (Győri and Ócsai 2014).

In relation to the subject of previous research, and building upon the results thereof, the research presented in this book goes beyond earlier research in the following ways: It summarizes the theoretical literature about ecological consciousness and its appearance in business enterprises. The research additionally explores the value orientations of ecologically conscious businesses through a deeper, previously unused approach to qualitative methodology; and applies a systematic, comparative method of presenting and analyzing the business models of the selected businesses that is new to this field (Chesbrough and Rosenbloom 2002).

Goals, Questions, and Methods

Due to its exploratory nature, the research required to examine the value orientation of ecologically conscious businesses requires a qualitative approach. According to the interactive model of qualitative research design, the research process should involve five levels: goals, conceptual frameworks, research questions, methods and validity, and their relationships and coherence (Maxwell 2013).

Three types of research goal can be distinguished. Personal goals are the most comprehensive, followed by the practical goals derived from these, and, ultimately, the intellectual goals underlying specific research questions.

The personal goals of the author that led to the research topic are the following:

- to do meaningful things in life, and contribute to activities and efforts aimed at improving the world;
- based on experiencing some of the problems caused by the prevailing consumption-, growth- and profit-oriented economic model, to help individuals and their business organizations recognize that in chasing only materialistic goals humankind will destroy the Earth and the biosphere that ensures their existence;
- to become a better person and a good researcher, and to pass down knowledge to help fellow human beings develop.

Practical goals are designed to help answer the following questions:

- How can we live a meaningful life while contributing to the cessation and reversal of processes destructive to nature?
- How can we develop an economy that does not destroy, but improves the conditions of nature, benefits society, and contributes to the preservation of the values of humankind?
- How can we best participate in research designed to increase understanding of the operation and underlying values of value-oriented, especially ecologically conscious, businesses?

Intellectual goals are aimed at helping explore the following questions:

What are the concepts employed in ecologically conscious businesses and the economy; what are their main characteristics and links to each other?
What values do ecologically conscious businesses hold?
What are the operating environments and contexts of these businesses like?
Why (raison d'être) and how (business model) do they work; how do such businesses implement their ethical convictions and their value orientation, and how can they cope with challenges related to the economy?
How do they define success, what are the determinants of their survival?
What are the environmentally beneficial impacts of their ecologically conscious behavior?
Do these businesses address the question of ecological transformation; do they consider it to be possible, and if so, how?
Drawing from these practical examples, may recommendations for the ecological transformation of the present economic system be made?
Is it possible to discover new theoretical concepts from an analysis of the operation of the former businesses?

A number of intellectual goals have been formulated, but the scope of the book—in order to maintain the focus of the research—is limited to examining some selected questions (the remaining research goals may be investigated in future research, as outlined in Chapter 6).

Based on the personal, practical, and intellectual goals of the author, the following research questions were developed:

1. What are the central concepts of ecologically conscious businesses and of an ecologically conscious economy? What are the key features of these concepts, and the relations among them?
2. What kind of value orientations do the ecologically conscious businesses under analysis have?

3. What are the fundamental goals and raison d'être of the studied businesses?
4. What are the definitions of success of the studied businesses?
5. Which business models do the businesses use?

The initial goal during the current phase of research into the value orientation of ecologically conscious businesses was not to create generalizable results, but to construct a deeper understanding of the selected cases that involves an exploration of related components and relationships, primarily applying a constructivist epistemological approach and less structured qualitative research methodology (Miles et al. 2014; Cassell and Symon 2004). In line with the topic, units of observation were identified from internationally well-known and Hungarian representative cases of ecologically conscious businesses. The sampling procedure was expert-based purposeful sampling. Semi-structured interviews and company documents and websites served as sources for data collection and recording. Qualitative content analysis, involving inductive logic, thematic analysis, and document analysis, was used to explore the relevant value orientations. The additional analysis of responses to further questions about value orientation were based on responses collected using a so-called Likert scale, along with the deductive analysis of answers concerning business model components, raison d'être, and the definitions of success of businesses. Prior exploration, the explication of value judgments and the assumptions of the researcher, careful documentation, clarity, consistency, triangulation of data sources and analytical methods were employed to increase the reliability and validity of the analysis. The ethical aspects of social science research were incorporated via obtaining informed consent from participants, and dealing with the latter with courtesy, respect, confidentiality, openness and honesty (Mason 2002; Krippendorff 2004; Creswell 2007; Babbie 2008; Maxwell 2013; Miles et al. 2014).

Structure of the Book

In this book we examine the theoretical concepts, value orientation, and business models of ecologically conscious businesses. To justify the topic choice, in Part I we present the external and internal drivers and goals that stimulated the implementation of the research behind the book. The research is located within the social sciences; previous studies of the topic are then outlined. In addition, besides the research questions, methodological issues are briefly summarized.

Part II reviews the basic theoretical concepts contained in the book (business enterprise and business model in Chapter 2, ecological consciousness in Chapter 3, values and value orientation in Chapter 4) and provides a comparative analysis of the main fields of the theoretical background.

Based on the results of empirical research, Part III presents working models of ecologically conscious businesses. After analyzing their value orientations, raison d'être and own definitions of success (Chapter 5), elements of their business models are portrayed in a comparative way (Chapter 6).

Part IV (Chapter 7) explores the current global context of ecologically conscious business and makes some conclusions about the future of business.

References

Ash, C. (2007). Happiness and Economics: Sustainability and Sufficiency: Economic Development in a Buddhist Perspective. *Society and Economy, 29*(2), 201–222. https://doi.org/10.1556/SocEc.29.2007.2.5.

Babbie, E. (2008). *The Basics of Social Research* (4th ed.). Belmont, CA: Thomson Wadsworth.

Brenkert, G. G. (1995). The Environment, The Moralist, The Corporation and its Culture. *Business Ethics Quarterly, 5*(4), 675–697. https://doi.org/10.2307/3857409.

Capra, F., & Jakobsen, O. D. (2017). A Conceptual Framework for Ecological Economics Based on Systemic Principles of Life. *International Journal of*

Social Economics, 44(6), 831–844. https://doi.org/10.1108/IJSE-05-2016-0136.

Carrington, D. (2016, August 29). The Anthropocene Epoch: Scientists Declare Dawn of Human-Influenced Age. *The Guardian*. https://www.theguardian.com/environment/2016/aug/29/declare-anthropocene-epoch-experts-urge-geological-congress-human-impact-earth. Accessed 23 March 2020.

Carson, R. L. (1962). *Silent Spring*. Boston, MA: Houghton Mifflin.

Cassell, C., & Symon, G. (2004). Introduction: The Context of Qualitative Organizational Research. In C. Cassell & G. Symon (Eds.), *Essential Guide to Qualitative Methods in Organizational Research* (pp. 1–10). London: Sage. https://doi.org/10.4135/9781446280119.

Chesbrough, H., & Rosenbloom, R. S. (2002). The Role of the Business Model in Capturing Value from Innovation: Evidence from Xerox Corporation's Technology Spin-Off Companies. *Industrial and Corporate Change, 11*(3), 529–555. https://doi.org/10.1093/icc/11.3.529.

Club of Rome. (2020). http://www.clubofrome.org/. Accessed 20 July 2020.

Creswell, J. W. (2007). *Qualitative Inquiry and Research Design: Choosing Among Five Approaches* (2nd ed.). Thousand Oaks, CA: Sage. https://charlesbickenheuserdotcom.files.wordpress.com/.../creswell_2007_qualitative_inquiry_and_research_design__choosing_among_five_approaches__2nd_edition.pdf. Accessed 4 March 2018.

Crutzen, P. J., & Stoermer, E. F. (2000). The Anthropocene. *IGBP Global Change Newsletter*, No. 41, pp. 17–18.

Daly, H. E. (1999). *Ecological Economics and the Ecology of Economics: Essays in Criticism*. Cheltenham and Northampton: Edward Elgar.

Daly, H. E. (2008). Frugality First. In L. Bouckaert, H. Opdebeeck, & L. Zsolnai (Eds.), *Frugality: Rebalancing Material and Spiritual Values in Economic Life* (Vol. 4, pp. 207–226). Bern: Peter Lang.

Daly, H. E., & Cobb, J. B. (1989). *For the Common Good: Redirecting the Economy toward Community, the Environment, and a Sustainable Future*. London: Green Print.

Frank, R. H. (2004). *What Price the Moral High Ground? Ethical Dilemmas in Competitive Environments*. Princeton, NJ: Princeton University Press.

Future Earth. (2020). http://www.futureearth.org/. Accessed 24 March 2020.

Győri, Zs, & Ócsai, A. (2014). Ecologically-Oriented Enterprises in Hungary. *World Review of Entrepreneurship, Management and Sustainable Development, 10*(1), 52–65. https://doi.org/10.1504/wremsd.2014.058053.

Harangozó, G., Csutora, M., & Kocsis, T. (2018). How Big is Big Enough? Toward a Sustainable Future by Examining Alternatives to the Conventional Economic Growth Paradigm. *Sustainable Development, 26,* 172–181. https://doi.org/10.1002/sd.1728.

Ims, K. J., & Pedersen, L. J. T. (Eds.). (2015). *Business and the Greater Good: Rethinking Business Ethics in an Age of Crisis*. Cheltenham and Northampton: Edward Elgar.

Ims, K. J., & Zsolnai, L. (2009). Holistic Problem Solving. In L. Zsolnai & A. Tencati (Eds.), *The Future International Manager* (pp. 116–129). London: Palgrave Macmillan. https://doi.org/10.1057/9780230274068_7.

Ims, K. J., Pedersen, L. J. T., & Zsolnai, L. (2014). How Economic Incentives May Destroy Social, Ecological and Existential Values: The Case of Executive Compensation. *Journal of Business Ethics, 123,* 353–360. https://doi.org/10.1007/s10551-013-1844-6.

Intergovernmental Panel on Climate Change (IPCC). (2014). *Climate Change 2014: Synthesis Report*. Contribution of Working Groups I, II and III to the Fifth Assessment Report of the Intergovernmental Panel on Climate Change. IPCC, Geneva, 151. https://www.ipcc.ch/site/assets/uploads/2018/02/SYR_AR5_FINAL_full.pdf. Accessed 8 March 2020.

Intergovernmental Panel on Climate Change (IPCC). (2018). *Global Warming of 1.5°C*. An IPCC Special Report on the Impacts of Global Warming of 1.5°C Above Pre-Industrial Levels and Related Global Greenhouse Gas Emission Pathways, in the Context of Strengthening the Global Response to the Threat of Climate Change, Sustainable Development, and Efforts to Eradicate Poverty. In Press. https://www.ipcc.ch/site/assets/uploads/sites/2/2019/06/SR15_Full_Report_High_Res.pdf. Accessed 8 March 2020.

Intergovernmental Panel on Climate Change (IPCC). (2020). http://www.ipcc.ch/. Accessed 8 March 2020.

Jakobsen, O. (2017). *Transformative Ecological Economics: Process Philosophy, Ideology and Utopia*. London and New York, NY: Routledge. https://doi.org/10.4324/9781315205434.

Kaldschmidt, S. (2011). *The Values of Sustainability: The Influence of Leaders' Personal Values on Sustainability Strategies*. Dissertation of the University of St. Gallen, School of Management, Economics, Law, Social Sciences and International Affairs.

Kahneman, D., & Deaton, A. (2010). High Income Improves Evaluation of Life but not Emotional Well-Being. *Proceedings of the National Academy of Sciences September, 107* (38), 16489–16493. https://doi.org/10.1073/pnas.1011492107.

Krippendorff, K (2004). *Content Analysis: An Introduction to its Methodology* (2nd ed.). Thousand Oaks, CA: Sage.

Maréchal, G. (2010). Constructivism. In A. J. Mills, G. Eurepos, & E. Wiebe (Eds.), *Encyclopedia of Case Study Research* (pp. 220–225). Thousand Oaks, CA: Sage. https://doi.org/10.4135/9781412957397.

Mason, J. (2002). *Qualitative Researching* (2nd ed.). London: Sage. www.sxf.uevora.pt/wp-content/uploads/2013/03/Mason_2002.pdf. Accessed 12 April 2020.

Maxwell, J. A. (2013). *Qualitative Research Design: An Interactive Approach* (3rd ed.). London: Sage.

Millennium Ecosystem Assessment (MEA). (2005). *Millennium Ecosystem Assessment: Ecosystems and Human Well-being: Synthesis*. Washington, DC: Island Press.

Mesarovic, M., & Pestel, E. (1974). *Mankind at the Turning Point: The Second Report to the Club of Rome*. New York, NY: Dutton.

Miles, M. B., Huberman, A. M., & Saldana, J. (2014). *Qualitative Data Analysis: A Methods Sourcebook* (3rd ed.). Thousand Oaks, CA: Sage.

Mills, A. J., Eurepos, G., & Wiebe, E. (Eds.) (2010). *Encyclopedia of Case Study Research*. Thousand Oaks, CA: Sage. https://doi.org/10.4135/9781412957397.

New Economics Foundation (NEF). (2009). *The Happy Planet Index 2.0: Why Good Lives Don't have to Cost the Earth*. http://happyplanetindex.org/s/2009-Happy-Planet-Index-report.pdf. Accessed 17 March 2020.

New Economics Foundation (NEF). (2016). *The Happy Planet Index 2016: A Global Index of Sustainable Wellbeing*. http://happyplanetindex.org/s/Briefing-paper-HPI-2016.pdf. Accessed 17 March 2020.

Pataki, Gy., & Takács-Sánta, A. (Eds.) (2005). *Természet és gazdaság: Ökológiai közgazdaságtan szöveggyűjtemény* [Nature and Economy: Key Texts of Ecological Economics]. Budapest: Typotex Kiadó. [in Hungarian].

Polanyi, K. (1944). *The Great Transformation: The political and Economic Origins of Our Time*. Boston, MA: Beacon Press.

Rockström, J. (2010). *Let the Environment Guide Our Development*. http://www.ted.com/talks/johan_rockstrom_let_the_environment_guide_our_development. Accessed 15 March 2020.

Schumacher, E. F. (1973). *Small is Beautiful: A Study of Economics As If People Mattered*. New York, NY: Harper and Row.

Shrivastava, P. (1995a). Ecocentric Management for a Risk Society. *The Academy of Management Review, 20*(1), 118–137. https://doi.org/10.5465/amr.1995.9503271996.

Shrivastava, P. (1995b). Environmental Technologies and Competitive Advantage. *Strategic Management Journal, 16*(Special Issue Summer), 183–200. https://doi.org/10.1002/smj.4250160923.

Shrivastava, P., Zsolnai, L., Wasieleski, D., Stafford-Smith, M., Walker, T., Weber, O., et al. (2019). Finance and Management for the Anthropocene. *Organization & Environment, 32*(1), 26–40. https://doi.org/10.1177/1086026619831451.

Shrivastava, P., Stafford Smith, M., O'Brien, K., & Zsolnai, L. (2020). Perspective: Transforming Sustainability Science to Generative Positive Social and Environmental Change Globally. *One Earth, 2*(4), 329–340. https://doi.org/10.1016/j.oneear.2020.04.010.

Steffen, W., Broadgate, W., Deutsch, L., Gaffney, O., & Ludwig, C. (2015a). The Trajectory of the Anthropocene: The Great Acceleration. *The Anthropocene Review, 2,* 81–98. https://doi.org/10.1177/2053019614564785.

Steffen, W., Richardson, K., Rockström, J., Cornell, S. E., Fetzer, I., Bennett, E. M., Biggs, R., Carpenter, S. R., de Vries, W., de Wit, C. A., Folke, C., Gerten, D., Heinke, J., Mace, G. M., Persson, L. M., Ramanathan, V., Reyers, B., Sörlin, S. (2015b). Planetary Boundaries: Guiding Human Development on a Changing Planet. *Science, 347*(6223), 1259855. https://doi.org/10.1126/science.1259855.

Steffen, W., Rockström, J., Richardson, K., Lenton, T. M., Folke, C., Liverman, D., Summerhayes, C. P., Barnosky, A. D., Cornell, S. E., Crucifix, M., Donges, J. F., Fetzer, I., Lade, S. J., Scheffer, M., Winkelmann, R., Schellnhuber, H. J. (2018). Trajectories of the Earth System in the Anthropocene. *Proceedings of the National Academy of Sciences, 115*(33), 8252–8259. https://doi.org/10.1073/pnas.1810141115.

Stockholm Resilience Center. (2020). https://www.stockholmresilience.org/. Accessed 16 March 2020.

Storsletten, V. M. L., & Jakobsen, O. D. (2015). Development of Leadership Theory in the Perspective of Kierkegaard's Philosophy. *Journal of Business Ethics, 128,* 337–349. https://doi.org/10.1007/s10551-014-2106-y.

Takács-Sánta, A. (2004). The Major Transitions in the History of Human Transformation of the Biosphere. *Human Ecology Review, 11*(1), 51–66. http://www.humanecologyreview.org/pastissues/her111/111takacssanta.pdf. Accessed 3 April 2020.

Takács-Sánta, A. (2010). *Rapa Nui or Tikopia?* Notes of the 2010/2011 Course on Human Ecology Titled "Ecological Crisis and Ways Out". Budapest. [in Hungarian].

Tinbergen, J. (1977). *RIO, Reshaping the International Order: A Report to the Club of Rome*. New York, NY: New American Library.

United Nations Environment Programme (UNEP). (2012). *GEO5 Global Environment Outlook: Environment for the Future We Want*. Nairobi: United Nations Environment Programme. https://wedocs.unep.org/bitstream/handle/20.500.11822/8021/GEO5_report_full_en.pdf?sequence=5&isAllowed=y. Accessed 22 March 2020.

United Nations Environment Programme (UNEP). (2019). GEO6 Global Environment Outlook: Healty Planet, Healthy People. Cambridge, UK: Cambridge University Press. https://wedocs.unep.org/bitstream/handle/20.500.11822/27539/GEO6_2019.pdf?sequence=1&isAllowed=y. Accessed 22 March 2020.

Waters, C. N., Zalasiewicz, J., Summerhayes, C., Barnosky, A. D., Poirier, C., Gałuszka, A., Cearreta, A., Edgeworth, M., Ellis, E. C., Ellis, M., Jeandel, C., Leinfelder, R., McNeill, J. R., deB. Richter, D., Steffen, W., Syvitski, J., Vidas, D., Wagreich, M., Williams, M., Zhisheng, A., Grinevald, J., Odada, E., Oreskes, N., Wolfe, A. P. (2016). The Anthropocene is Functionally and Stratigraphically Distinct from the Holocene. *Science, 351*(6269), aad2622. https://doi.org/10.1126/science.aad2622.

World Meteorological Organization (WMO). (2020). *WMO Statement on the State of the Global Climate in 2019* (WMO-No.1248). Geneva: WMO. https://librarywmo.int/doc_num.php?explnum_id=10211. Accessed 25 March 2020.

World Wide Fund for Nature (WWF). (2016). *Living Planet Report 2016: Risk and Resilience in a New Era*. Gland, Switzerland: WWF International. http://awsassets.panda.org/downloads/lpr_living_planet_report_2016.pdf. Accessed 22 March 2020.

World Wide Fund for Nature (WWF). (2018). *Living Planet Report 2018: Aiming Higher*. In M. Grooten & R. E. A. Almond (Eds.). Gland: WWF. https://s3.amazonaws.com/wwfassets/downloads/lpr2018_full_report_spreads.pdf. Accessed 22 March 2020.

Zsolnai, L. (2015). *Post-Materialistic Business: Spiritual Value-Orientation in Renewing Management*. Houndmills: Palgrave Macmillan.

Zsolnai, L., & Győri, Zs. (2011, September 15–17). *Can Ethics Survive in Competitive Environments?* Empirical Evidence from Values-Driven Businesses, European Business Ethics Network 24th Annual Conference.

Part II

Ecological Consciousness in a Business Context

This part of the book first describes the importance of business models and defines the concepts of "business enterprise" and "business model." It then shows how ecological consciousness differs from environmental consciousness, and what the related approaches—such as deep ecology and ecological economics—are. It also discusses the ethical and spiritual aspects of ecological consciousness, and how these are relevant in the real world. The final chapter of this part of the book investigates how business organizations can implement ecologically conscious ways of functioning, and what the underlying values are that can guide them.

2

The Importance of Business Models

In this book, the term "business" is used to refer to business organizations or business enterprises. In the business and management literature there are no standard, widely accepted definitions for the latter terms, but being motivated primarily by profit appears to be an essential part of them. However, ethical, social, and ecological considerations can also be components of a genuine business enterprise. In fact, business practices can only be justified when they are good for society, and do no harm to nature.

Research into business models has expanded over the past few years. After reviewing the most relevant concepts related to innovative and sustainable business models, this chapter highlights the definition and six key dimensions of business models promoted by Chesbrough and Rosenbloom (2002) as a shared framework for the comparative analysis of ecologically conscious business organizations presented in Part III.

A. Ócsai, *Ecologically Conscious Organizations*, Palgrave Studies in Sustainable Business In Association with Future Earth,
https://doi.org/10.1007/978-3-030-60918-4_2

Business Enterprise

In this book, the term "business" is used to mean business enterprise, although in the literature there are no standardized definitions of "business" and "enterprise." The Business Dictionary defines a business as an organization or economic system in which goods and services are exchanged for each other, or for money. Every business needs some investment and a sufficient number of buyers to whom their products or services can be sold on an ongoing basis for a profit. Businesses may be privately or publicly owned. There are also nonprofit businesses (Business Dictionary 2020a).

In examining the concept of "enterprise" we find that it is defined as a company or an entrepreneurial activity that is particularly characterized by initiative and ingenuity (Business Dictionary 2020b). A "business enterprise" especially is usually defined as an endeavor, the main purpose of which is to make a profit, not just to employ the entrepreneur or others (Business Dictionary 2020c). OECD similarly defines a business enterprise as a business project or venture that is launched in the hope of making a profit. The term usually refers to a business entity that engages in business, and is therefore used as a synonym for the words "company" and "firm" (OECD 1993).

Based on such general definitions, the priority of being motivated by profit appears to be a defining characteristic of a business enterprise. Shane and Venkataraman (2000), who believe that entrepreneurial activity involves finding and exploiting profitable opportunities, support this view. Due to the logic of market competition, enterprises are forced to create market value; that is, to survive they are forced to produce and sell their products and services to their market partners (Bruni and Zamagni 2007). Veblen (1904) also emphasizes the relevance of financial gain as the main goal of business, the usual result of which is the accumulation of wealth. As a critique of capitalism, Veblen also adds that business owners and managers tend to hinder the operation of the industrial system by limiting production to maximize profit, which activity can be identified as the cause of many social problems (for instance, high levels of unemployment) (Veblen 1904).

The functioning of the modern economy is based on a materialistic conception of human activity. According to this perspective, human beings are self-interested individuals who have only materialistic desires and sources of motivation. In economics and business, this understanding is called the "homo oeconomicus" model of man. According to this model, individuals seek to maximize their self-interest and are only interested in material value as expressed in monetary terms. Based on this concept, the materialistic leadership model assumes the presence of an external, monetary source of motivation, and measures success in terms of profit (Zsolnai 2015).

Entrepreneurs who create and maintain businesses, however, look for business opportunities, take risks, and create or reorganize their organizations not only to reap financial success, but also in the hope of achieving professional, market, social, and individual success (Vecsenyi 2003). A business enterprise is "a human activity whose primary purpose, its raison d'être[,] is to satisfy consumer needs by [generating] profit" (Chikán 1997, p. 16). A profit orientation is only one of the necessary conditions (besides independence, risk-taking, and operating in a real market) which mark out an organization as a business enterprise. A desire for profit making is therefore a basic requirement, but this is not enough to explain the functioning of a business enterprise: satisfying consumer needs, meeting the organization's mission, and the specific means of achieving basic goals are the starting point for this (Chikán 1997). Some also consider social enterprises to be business enterprises, the mission of which—achieving social goals—takes priority over profit. Profit that is generated by such organizations cannot be distributed among owners, but is used to meet further targets and support communities (Petheő 2009). Hisrich et al. (2013) draw attention to the importance of factors beyond financial profit, pointing out that an enterprise is a value-creating process that takes risks using time and resources to achieve a financial return and increase independence and personal contentment.

According to Drucker (1954), business can only be justified when it is good for society. In his opinion, the prevailing economic theory of business enterprises and behavior, the principle of maximizing profit—which is just a more complicated way of saying "buying cheap, selling dear"—only explains how business enterprises have worked so far, but cannot explain how they *should* work. The concept of profit maximization in this regard is thus meaningless (Drucker 1973). In fact, it may be claimed that profit is not the cause and goal but the outcome of a successful business operation (Drucker 1973; Kay 2010).

Ethical considerations are an integral part of the concept and operation of business enterprises. There is no business which can avoid producing at least some benefit to society, even if it is primarily interested in maximizing profit. However, a profit motive cannot be the sole target of a business, because this does not provide any guidance about which business enterprises should be set up, nor how, or how to operate them. The existence and activities of business enterprises are actually determined by which social needs they meet with their products and services (Allinson 1998).

Business Model

The literature on business models has expanded significantly over the past decade. Zilahy (2016) provides a good overview of the general definitions and features of business models, especially innovative and sustainable ones. In this section we present the main concepts which are the most relevant to the research topic.

According to Joan Magretta (2002), a business model is basically a story that summarizes answers to the following main questions that confront business: who are the customers, what does value mean to them, how can a profit be made by the business, and based on which economic factors can value be created for customers at a reasonable cost? When creating a new business model, a new story is written, but such new stories are in fact iterations of earlier ones, involving revisions and the reuse of universal human experiences. That is, all new business models are based on the general value chain concept that is the basis of all businesses. The value chain has two main parts: first, the design and production of a product or a service; and second, delivering the product or service to the customer and collecting the reward. A business model can generate novelty by creating a new product or service, or improve a process in a preexisting value chain.

Casadesus-Masanell and Ricart (2011), instead of predefining the elements of the ideal business model—which can hamper the creation of revolutionary business models through fixed assumptions—rather focus on the decisions and choices of the leaders who create the business models. Management decisions about how a company should work are all pervasive and include decisions about remuneration practices, procurement contracts, the location of sites, the degree of vertical integration, and sales and marketing initiatives. These decisions have consequences that affect the value creation logic of the company, and thus stakeholders and the company itself, so when defining the business model such consequences must be taken into account: Companies can make three types of decisions when designing business models. (1) Choices regarding operational policies determine the activities of the company. (2) Asset choices affect what resources and tangible assets the company employs and uses. (3) Governance choices involve the allocation of power and decision-making rights, and establish how to make decisions involving the previous two forms of choices. The consequences of these choices can be flexible. For example, price increases can quickly lead to a drop in sales. However, they can be rigid, too. The characteristics of company culture change only over a longer period of time. Differentiating between the consequences of the latter choices is useful, because they have different impacts on the competitiveness of a company. The outcomes of flexible choices can easily be copied, while those of rigid choices are difficult or take a longer time to implement, so they can generate competitive advantage.

The definition given by Chesbrough and Rosenbloom (2002) understands the business model as a coherent framework that transforms technological features and opportunities as inputs into economic outputs through customers and markets. The former authors define six main elements of the business model.

1. Value propositions: what values the company creates for customers based on the technology they use.
2. Market segment: which customers benefit, and for what purpose, from the given technology, and how the company generates revenue from this.

3. Cost structure and profit potential: taking into account the given value propositions and the structure of the value chain, at what cost and with which expectations about profit may value be captured.
4. Structure of the value chain: what the structure of the value creation process is, and what resources are required for its operation.
5. Position within the value network: what place does the company occupy within the network that links suppliers to customers, and who are its competitors and partners.
6. Competitive strategy: with what strategy can the company build and maintain its competitive advantage.

Together, these six components serve, in addition to creating value for customers, to generate a return on financial investment that enables a business model to be implemented and a business to further develop.

References

Allinson, R. E. (1998). Ethical Values as Part of the Definition of Business Enterprise and Part of the Internal Structure of the Business Organization. *Journal of Business Ethics, 17,* 1015–1028. https://doi.org/10.1023/A:1006063114697.

Bruni, L., & Zamagni, S. (2007). *Civil Economy: Efficiency, Equity, Public Happiness*. Bern: Peter Lang. https://doi.org/10.1108/17465680810907323.

Business Dictionary. (2020a). *Business*. http://www.businessdictionary.com/definition/business.html. Accessed 26 March 2020.

Business Dictionary. (2020b). *Enterprise*. http://www.businessdictionary.com/definition/enterprise.html. Accessed 26 March 2020.

Business Dictionary. (2020c). *Business Enterprise*. http://www.businessdictionary.com/definition/business-enterprise.html. Accessed 26 March 2020.

Casadesus-Masanell, R., & Ricart, J. E. (2011). How to Design a Winning Business Model. *Harvard Business Review, 89*(1/2), 100–107. https://hbr.org/2011/01/how-to-design-a-winning-business-model#. Accessed 26 March 2020.

Chesbrough, H., & Rosenbloom, R. S. (2002). The Role of the Business Model in Capturing Value from Innovation: Evidence from Xerox Corporation's

Technology Spin-Off Companies. *Industrial and Corporate Change, 11*(3), 529–555. https://doi.org/10.1093/icc/11.3.529.

Chikán, A. (1997). *Vállalatgazdaságtan* [*Business Economics*]. Budapest: Aula Kiadó [in Hungarian].

Drucker, P. F. (1954). *The Practice of Management*. New York: Harper.

Drucker, P. F. (1973). *Management: Tasks, Responsibilities, Practices*. New York: Harper – Row.

Hisrich, R. D., Peters, M. P., & Shepherd, D. A. (2013). *Entrepreneurship* (9th ed.). New York: McGraw-Hill/Irwin.

Kay, J. A. (2010). *Obliquity: Why Our Goals Are Best Achieved Indirectly*. London: Profile Books.

Magretta, J. (2002). Why Business Models Matter. *Harvard Business Review, 80*(5), 86–92. https://hbr.org/2002/05/why-business-models-matter#. Accessed 26 March 2020.

OECD – Organization for Economic Co-operation and Development. (1993). *Business enterprise in Glossary of Industrial Organisation Economics and Competition Law*. Compiled by R. S. Khemani & D. M. Shapiro, commissioned by the Directorate for Financial, Fiscal and Enterprise Affairs, OECD. https://stats.oecd.org/glossary/detail.asp?ID=3207. Accessed 26 March 2020.

Petheő, A. I. (2009). *A vállalati társadalmi felelősségen túl: A szociális vállalkozás* [*Beyond Corporate Social Responsibility: The Social Enterprise*]. Ph.D. dissertation, Small Business Development Center, Corvinus University of Budapest, Budapest [in Hungarian].

Shane, S., & Venkataraman, S. (2000). The Promise of Entrepreneurship as a Field of Research. *The Academy of Management Review, 25*(1), 217–226. https://doi.org/10.5465/amr.2000.2791611.

Veblen, T. (1904). *The Theory of Business Enterprise*. New York: Charles Scribner's Sons. www.businessbuildersbanquet.com/software/veblen2.pdf. Accessed 26 March 2020.

Vecsenyi, J. (2003). *Vállalkozás – Az ötlettől az újrakezdésig* [*Business—From Idea to Start Over*]. Budapest: Aula Kiadó [in Hungarian].

Zilahy, G. (2016). Sustainable Business Models—What Do Management Theories Say? *Vezetéstudomány* [*Budapest Management Review*], *XLVII*(10), 62–72. http://unipub.lib.uni-corvinus.hu/2497/1/VT2016n10p62.pdf. Accessed 26 March 2020.

Zsolnai, L. (2015). *Post-Materialistic Business: Spiritual Value-Orientation in Renewing Management*. Houndmills: Palgrave Macmillan. https://doi.org/10.1057/9781137525987.

3
Beyond Environmental Consciousness

This chapter first examines the concepts of "ecology" and "human ecology" as types of the transdisciplinary study of man-made environmental problems. It then presents the evolution of the study of and different approaches to "consciousness" in Eastern and Western traditions. The chapter shows how "ecological consciousness" began to be distinguished from "environmental consciousness" from the late 1960s onwards. Environmental consciousness understands environmental problems too narrowly, seeing them mainly as technical problems, and thus does not recognize the root causes of the environmental crisis.

Ecological problems can only be solved if the relationship between nature, society, and individuals is well understood, and if we recognize that environmental problems are the inevitable consequence of false

Based on the permission of the publisher, some parts of this chapter are revisions of one of the author's journal articles formerly published in Hungarian: Ócsai, A. (2018). A vállalkozások ökológiai tudatossága [Ecological Consciousness of Businesses]. *Vezetéstudomány* [Budapest Management Review], *XLIX* (9), 2–19. https://doi.org/10.14267/veztud.2018.09.01.

A. Ócsai, *Ecologically Conscious Organizations*, Palgrave Studies in Sustainable Business In Association with Future Earth,
https://doi.org/10.1007/978-3-030-60918-4_3

ontological and epistemological assumptions and the instrumental rationality embodied in the social and cultural institutions of modernity (modern capitalism, industrial technology, individualism, and mechanistic science). The chapter ends with a summary of further relevant concepts related to the topic (inter alia, deep ecology and ecosophy, ecological economics, and the circular economy), different ethical and spiritual ideas (ecological conscience, land ethics, Gaian ethics, moral disengagement strategies, and the principle of responsibility), and comments on the relevance of some selected worldviews (Christianity, Gandhi's philosophy, and Buddhist economics).

Defining Ecology

The origins of ecology as a science can be traced back to the beginnings of natural science (Benson 2000; Egerton 2001). The term "ecology" was used as early as in 1858 in a letter by Henry David Thoreau (Harding and Bode 1958), but with no explanation of its content. The first definition is attributed to Ernst Haeckel, who in 1866 defined ecology as a comprehensive science that involves examining the relationships between living organisms and their external environment and the organic and inorganic living conditions of organisms (Friederichs 1958). In the science that further developed following the turn of the twentieth century, ecology was defined by Charles Elton (1927) as scientific natural history, while Odum (1963) considered it to be the study of the structures and functions of nature. According to McNaughton and Wolf (1973), ecology is the scientific study of the interactions between organisms and their environment. Whittaker (1975) states that ecology is the examination of living systems and their environmental relationships. Krebs (2001) writes about ecology as the "science of interactions" that determines the distribution and mass of organisms. Such definitions are not uniform, and the different trends involve differences in methodology. Hungarian ecologist Pál Juhász-Nagy has primarily been concerned with clarifying the theoretical foundation of the concept and field of ecology with scientific precision, and the development of mathematical modeling as the study methodology (Juhász-Nagy 1984, 1986, 1993; Pásztor and Oborny 2007; Hahn 2014).

In this book, we understand ecology as "synbiology," which involves examination of the level of organization located above the individual. The supraindividual levels include populations, communities/associations, biomes, and the biosphere. Ecologists study the regularity and types of interaction between organisms and their environment, between populations of living organisms (and systems of populations) and the environment, and the conditions limiting or supporting their existence, and effects on their productivity (Hortobágyi and Simon 2000).

"Human ecology" can be considered the transdisciplinary study of man-made environmental problems. The basic questions human ecology addresses, which overlap with many classical disciplines in the natural and social sciences, are what changes human activities cause in nature; what the social causes and consequences of these changes are; and how it is possible to mitigate ecological changes and adapt to them (Takács-Sánta 2009).

Western and Eastern Approach to Consciousness

According to the Merriam-Webster online dictionary (2020), consciousness is: (1) (a) the state or quality of consciousness, especially being conscious of something within oneself; (b) being conscious of an external object, state, or fact; (c) concern, attention to a social, political cause; (2) the state of being characterized by sensation, emotion, will and thought; (3) the totality of conscious states of an individual; (4) the normal state of conscious life; (5) the higher level of mental life when the individual is conscious, as opposed to unconscious processes.

It is clear from the complexity of the word that no uniform definition of the concept of consciousness has yet emerged. Consciousness is also approached in different ways in the Eastern and Western traditions (Velmans 2009). In the following, we briefly summarize the historical evolution of consciousness in Western thinking, and some approaches to consciousness in the Eastern traditions.

In Western philosophy, the issue of consciousness has been a subject of interest since ancient times. Plato dealt with questions concerning human nature, the soul, and mind, and Aristotle examined the nature of perception and consciousness. In the early modern times, the focus of René Descartes and his followers was the relationship between mind and body. The claim they made was that the mind can be distinguished from the body, and that all mental states can be considered a form of thinking, which requires consciousness. At the end of the seventeenth century, John Locke had similar views about the relationship between thinking and consciousness, but Gottfried Leibniz recognized the existence of unconscious perceptions and distinguished between external consciousness (perception) and self-consciousness (apperception). A century later, the decisive work of Immanuel Kant proved that understanding consciousness requires the acceptance of the existence and study of the mental structures and unconscious mental states and mechanisms that are prerequisites for conscious experience. Since the end of the nineteenth century, consciousness research has taken many directions. The examination of consciousness through introspective methods was pursued in psychology, and the work of Sigmund Freud increased the general acceptance of the existence of unconscious mental states and processes. In philosophy, phenomenology deals with the issue of consciousness, primarily through the work of Edmund Husserl and Martin Heidegger. With the development of brain research, neuropsychology, and neuroscience in the 1980s, interdisciplinary interest in consciousness research strengthened, and dedicated journals, book series, and conferences emerged to generate a deeper understanding of it (Gennaro, n. d.).

Modern medical and psychological research, through psychological experiments and the analysis of cases of altered states of consciousness, examines whether human and animal consciousness is objectively measurable, and seeks to understand its underlying neurological and psychological mechanisms (Koch 2004). "Integrated information theory" aims to scientifically explain consciousness and examine whether specific physical systems are conscious, and, if so, what degree of consciousness they have. The theory states that consciousness is present to varying degrees in biological organizations, ranging from the simplest organisms through non-mammalian species to the human fetus, to premature babies and patients with cerebral cortex injury. Digital computers and forms of artificial intelligence, however, are not conscious (Tononi and Koch 2015).

The Mind and Life Institute, established by the fourteenth Dalai Lama, neuroscientist Francisco Varela, and lawyer and entrepreneur Adam Engle, aims at bridging Western scientific thinking and the contemplative practices inspired by Buddhism. Over the past thirty years, dozens of scientific studies, journal articles, and conferences have pointed out—based on objective observation and analysis—that modern science needs to be complemented by analytical methods based on contemplative practices and self-observation to generate truly "humane" results that can reduce human suffering (Mind and Life Institute 2020).

Contemplative science examines the cognitive, neurological, physiological, behavioral, and social manifestations of the mind in contemplative consciousness. It is inspired by Buddhist values, but its main purpose is to combine science, philosophy, and religion through empirical methods without religious bias (Wallace 2007).

Buddhist philosophy tells us that there is no constant, unchanging soul or ego, and consciousness cannot exist independently of matter, detection, perception, and aspiration. Moreover, they do not exist independently, but only exist, they are formed according to our senses and objects (form, sound, fragrance, taste, tactile things, and thoughts) (Rahula 1974). Thus, our consciousness is made up of seeds that were sown by our own past actions and those of our family and society. In our consciousness, our thoughts, words, and actions arise every moment and create our bodies, minds, and our world, so it is up to us how we shape our consciousness (Nhat Hanh 2008). The fourteenth Dalai Lama also draws attention to the importance of causal relationships in the Buddhist conception of consciousness (2005, quoted by Marques 2012). He also states that consciousness is one of the prominent elements of spiritual development, and that during our lifetime we are capable of spiritual development, having a higher level of experience, and achieving the ultimate state of omniscience through increasing the continuity of our consciousness (identified as the Buddha nature found in every human being) (Dalai Lama 1995 quoted by Marques 2012).

According to the Upanishads, the ancient Indian teachings, consciousness is the essence of an immanent being, the Atman, which can actually be identified with Brahman; with the pure, transcendent, free consciousness that allows nature and man to exist (Sen 2008 quoted by Velmans 2009). Consciousness is described in the Vedas as a principle that is present in matter, in life, and in the mind from the outset after it has crystallized from a finer, universal, and infinite state into a coarser form (Chakraborty 1995). Consciousness is present in every being, is constant, independent from space and time, and is the source of all physical and social reality. Its highest physical manifestation, however, is embodied in humanity, whose members are capable of self-reflection and self-realization; that is, of attaining unity with all other beings, individual consciousness, and the source of all, Universal Consciousness (Pruzan 2016). Moreover, for humans, grasping the power of individual consciousness and attaining the original state of Consciousness is not just an opportunity, but a task that should be accomplished (Chakraborty 1995).

Ecological Consciousness and Its Ethical Aspects

In this section, we examine the topic of ecological consciousness in detail, starting from the origin of the concept, through its appearance in the literature, to a presentation of its ethical implications.

From Ecological Consciousness to the Circular Economy

The origin of ecological consciousness as a phenomenon can be traced back to Paleolithic gatherer-hunter-fisher societies. Archeological and comparative ethnographic studies confirm that the psychological representation of the "natural" state of the human soul was approximately the same on every continent. In such ancient societies, ecological consciousness included the identification of the “self” with nature and respect for life. The universe was considered a holy being, and humans a part of the great unity of nature and dependent on it, not dominant over it. Every being was seen as part of a greater whole; and humankind interconnected with every animate and inanimate being. Ecological consciousness was also characteristic of early agricultural societies and thus part of the mindset of the first great civilizations. In Ancient Greece, the Pythagoreans believed that humans were a microcosm of the universe; the body of a human was made up of the same elements as the universe, and their soul was derived from its soul. Plato and the Stoics also considered the cosmos to be a living creature, together with all living beings, and by nature similar to them. The universe lived and had a soul and consciousness, from which the human soul and intellect originated. In the East, the teachings of Hinduism and Taoism about humanity were similar (Hughes 1991). However, as the biosphere-transforming activity of humankind went through six major leaps (the use of fire, the appearance of language, the development of agriculture, the emergence of civilizations, the European conquests, and the technical-scientific-energetic revolution; Takács-Sánta 2004), each increasingly extensive, humans gradually became physically, psychologically, and spiritually separated from non-human nature. Ecological consciousness weakened at the community level, and the state of the natural environment deteriorated (Hughes 1991).

Since the 1980s, there has been increasing interest in ecological consciousness. The concept began to be distinguished from environmental consciousness at the end of the 1960s by environmental activists and philosophers, who believed that conventional nature conservation movements were interpreting environmental problems too narrowly. This philosophical viewpoint, initially proclaimed by few, has since become the perspective adopted by environmental movements over the last few decades, and has led to profound changes in society, cultural, political and economic institutions by redefining the concepts of individual, society, and nature (Christopher 1999).

In the formulation of Christopher (1999), ecological consciousness is concern for environmental problems fueled by the knowledge and experience of the degradation of nature. Ecological consciousness also implies that modern thinking is based on false ontological and epistemological presuppositions that prevent modern people from correctly perceiving and responding to environmental problems.

According to ecologist thinkers, conventional environmental consciousness degrades environmental problems into technical problems; it is therefore unable to help with understanding their root causes. Environmental problems can only be solved if concern for the devastation of nature is accompanied by an "ecological" understanding of the relationship between nature, society, and the individual (Christopher 1999: p. 358). At the center of ecological consciousness is the belief that environmental problems are the inevitable consequence of an instrumental rationality embodied in the social and cultural institutions of modernity (modern capitalism, industrial technology, individualism, and mechanistic science); thus without new social and cultural institutions based on a new rationality, global disaster is unavoidable (Christopher 1999: p. 361).

Beyond feeling the increasing ecological threat and experiencing related anxiety, ecological consciousness also implies that by formulating a new metaphysical framework of rationality and a new understanding of the nature of the self and its place in the universe we can give meaning to the threat. The steps of this process are the following. (1) Replacing the atomistic concept of the self with a holistic approach (understanding that nature and the individual exist together in ecosystems). (2) Developing an ethical system that allows escape from anthropocentrism in line with the new ontological framework. (3) Recognizing that the realization of an extended self-understanding is possible through a pre-rational process involving changes in individuals' feelings and relationships to the world. (4) Searching for an ecological rationality: combining the previous steps, thereby creating a viable way to re-rationalize the sociocultural relations that support the actualization and empowerment of ecological consciousness (Christopher 1999: pp. 362–363).

According to Hamilton (2010), the development of ecological consciousness depends on the appearance of a new kind of sense of self and a new relationship of man to the natural environment. In the case of rich, consumption-based societies, one must understand how the personal identity of individuals is created by consumption and the institutions that serve it, and how people act when influenced by these factors. Instead of confronting consumers with information about environmental degradation, it may be more effective to show them whether a consumer lifestyle can really make them happy.

Deep ecology, developed by Norwegian philosopher Arne Naess, can be considered both a philosophical and an operative approach (Naess 1989). Naess criticizes environmental thinking, which focuses on short-term problems (i.e. "shallow" ecology), because it assumes that environmental pollution and resource depletion can be solved through technology. In contrast, Naess believes that deep ecology, which involves exploring the roots of problems, is needed, along with a redefinition of predominantly reductionist, anthropocentric ideology, and, within it, the individual. This should involve reconstructing our own self-image, and radically changing our mindset. Based on the teachings of many prominent spiritual predecessors (Aristotle, Buddha, Saint Francis, Spinoza, Thoreau, Aldo Leopold, and Heidegger), the approach seeks to address fundamental questions such as how can we cope with today's ideological struggles, what is the place of man in nature, and how can we care for future generations. Solutions may be found by applying the pluralistic view of ecophilosophy, which involves the consideration of ecological principles (complexity, diversity, symbiosis, and unity) and a personal approach called "ecosophy." On this basis—by viewing the ecology and the economy as of equal importance—every human being must be able to develop their individual worldviews, values, and norms, to act upon them, and take responsibility (Ims 2015).

White (2011) derives the concept of ecological consciousness from the work of researchers who deal with deep ecology and ecosophy. The concept has been addressed—often using different terms, including "ecological self," "ideal state of being," "self-realization," and "ecological sensibility"—by different scholars since its appearance in the 1970s (Beck 1995; Bragg 1996; Christopher 1999; Devall 1988; Devall and Sessions 1985; Drengson 1989, Hill et al. 2004; Leff 1978; Milbraith 1989; Morris 2002; O'Sullivan and Taylor 2004; Uhl 2004). In summarizing the results, White defines ecological consciousness as deep awareness of individuals' biological, ecological, emotional, and spiritual relationships with non-human beings. It is one's personal identification with non-human nature, respect for the intrinsic value of non-human nature, and deep concern for the destruction of our natural environment—and handling this through rational understanding and an active, emotional, psycho-spiritual attitude. It is the intention of "self-realization," which includes commitment to non-human beings. It is the understanding that the global environmental crisis is caused by the prevailing state of—globalized, industrial, consumer—modernity, and a disconnected consciousness. It is the motivation to cope with global and personal concerns, and to understand concepts that are beyond humanity through experientially and emotionally connecting to the natural environment (White 2011: p. 42).

Ecological economics is related to ecological consciousness. This discipline, formed by the thoughts of Karl Polanyi (1944), Kenneth Boulding (1966), Nicholas Georgescu-Roegen (1971), Ernst F. Schumacher (1973), Herman Daly (Daly 1999; Daly and Farley 2004), and Robert Costanza (Costanza et al. 1997), characteristically differs from environmental economics, which is part of mainstream economics. According to ecological economics, the economy is part of the ecosystem, not the other way around. Nature, society, and economy are interconnected in dynamic unity. In ecological economics, a holistic, transdisciplinary, and organic perspective, intergenerational equity, a long time horizon for natural processes, the carrying capacity of the Earth, and the issue of sustainable development play an important role (Daly and Farley 2004; Capra and Jakobsen 2017; Jakobsen 2017). Ecological economics must reinforce its ethical background by relying on Hans Jonas's imperative of responsibility. The main elements of this are that humans are members, not masters of the community of life on Earth. Moreover, that the Earth and its living systems are not merely natural resources, but beings intrinsically worthy of respect and care. Careful use and the fair sharing of low-entropy sources and sink capacities that support living conditions are required (Brown and Timmermann 2015).

A new kind of economics that is capable of dealing with the current challenges of the world should be based on an organic worldview that relies on the unity of economy, nature, and society. A shift from the competition-based, mainstream economic paradigm with its mechanistic, instrumental and reductionist worldview is required that involves change at structural and individual levels so that consumption and lifestyle habits and values can become supportive of solidarity and sustainability. This requires circularity, respect for pluralistic values, and transparent cooperation based on communication (Ims and Jakobsen 2006; Ingebrigtsen and Jakobsen 2007; Jakobsen 2017). This mindset is the basis of the circular economy, in which production, distribution, consumption, and recycling take place in much closer unity through cycles. Instead of one-off consumption events, the goal is to increase the lifetime of products, reduce the amount of waste, and increase commodity efficiency (for example, through repair, reuse, and recycling). In addition to global and national processes, local solutions and local currencies will play a key role in this (Ingebirgsten and Jakobsen 2007; European Commission 2015, 2018; Jakobsen 2017).

Ethical Aspects of Ecological Consciousness

Nobel laureate poet and artist Rabindranath Tagore always had deep respect for the natural environment. He was particularly concerned about deforestation, and probably the first person in India to initiate a festival for tree planting. This occurred at the end of the 1920s (Sen 1997). His desire to sustain nature stems from a central preoccupation with a harmony of relationships based on care and concern. With a perspective rooted in traditional Indian ethical values, non-dualism, oneness, and the interconnection of all beings, Tagore highlighted the importance of ecological stability and the sustainable development of the whole universe (Sarin 2014).

Tagore argued that civilization has become dissociated from the context of the cosmos and that dehumanizing, mainstream materialistic attitudes have caused humanity to lose their harmony and identification with nature. The internal causes of ecological imbalance are basic vices such as greed, jealousy, lust, and pride manifested in commercialism, consumerism, and unbridled competition. Tagore holds that the cosmic unity between man and nature is attainable through developing a deep kinship with natural entities. The latter should not be considered mere physical objects, but God-given gifts that complete us and provide us with sources of bliss and peace that lead to the emergence of our best values. Tagore regards the world as the expression of beauty, which he shows us in many of his creative masterpieces, such as in a poem entitled "When I See the World as a Song." Tagore states that beauty creates harmony in relationships that lead to soul consciousness, and the expansion of aesthetic, moral, and spiritual consciousness. One's whole existence is interlinked, thus through identification with nature we can rise above our narrow ego and realize oneness within our soul, where the key to cosmic consciousness is (Sarin 2014; Mukherjee 2018).

In an essay entitled "The Ecological Conscience," American naturalist Aldo Leopold writes that the conservation of nature must arise from what is ethically and aesthetically correct. Only those things are right that preserve the integrity, stability, and beauty of a natural community, including soil, water, flora and fauna (Leopold 1949).

If we accept the need for an ecological conscience, economic gains cannot be an excuse for "unsocial land use" or ecological "atrocities." It is advisable to encourage proper land use through social rewards that are proportionate to the importance to society of the former (Leopold 1949). The basis of preservation must be "land ethics," which perceives humanity as a part of ecosystems, and expands the scope of ethics to the Earth. This new kind of ecological consciousness means that humans must become members of the ecological community through transforming themselves from masters of "life's commonwealth" into equals with other species (Feng Lin and Fyles 2015).

Experiencing the ecological destruction caused by the economic, industrial, and military activities of the twentieth century and the enormous damage to the natural and human life of the Vietnam War, American Trappist monk and philosopher Thomas Merton agreed with Aldo Leopold that there is a need to strengthen ecological conscience. Without this, the ruthless and mindless exploitation of nature and man will continue. Behind the ecological tragedies caused by business and wars, we find aggression, fear, and the desire for domination of the modern Western man cloaked in clichés, hidden behind the mask of individuals as "defenders of freedom and democracy." The psychological roots of this are dehumanization and alienation. Man confuses the value of objects and abstractions (commodities, money, and property) with life itself, and appreciates immediate profit more than anything else. Developing an ecological conscience requires that we recognize our responsibilities to other members of the terrestrial community, and respect all forms of life (Merton 1968).

A very important representative of ecological consciousness is the English geochemist-ecologist James Lovelock. Based on the experience of research into global ecological change triggered by human activity and a program that searched for Martian life that was launched in the 1960s, Lovelock developed, with the American biologist Lynn Margulis, so-called Gaia theory. According to this theory, terrestrial animate and inanimate systems are closely related and form a self-regulating (homeostatic) system. Named after the Greek goddess Mother Earth, "Gaia" refers to the complex, self-regulating system of Earth's biosphere, atmosphere, water, and land that create the optimal physical-chemical environment for life on earth (Lovelock 2009). In light of this, Gaian ethics requires that economic activities do not undermine "the intactness and integrity of the global patterns and mechanisms of the Earth" (refraining from species extermination, greenhouse gas emissions, and the carbon dioxide emissions that cause global warming) (Zsolnai 2001).

According to Stanford psychologist Albert Bandura (2007), if we really want to be responsible caretakers for nature and preserve it for future generations, then we have to make it harder to apply moral disengagement strategies to those who commit ecologically destructive activities. With the help of psychological strategies, one can disassociate one's conscience in order to continue to undertake activities harmful to nature and avoid facing their negative consequences. Activities that cause global ecological and human damage (such as deforestation, desertification, global warming, melting ice sheets, rising sea levels, extreme weather conditions, soil degradation, biodiversity loss, etc.) are often "justified" by reference to a higher social, national, or economic purpose. Morally disengaged actors may attempt to prove the acceptability of their own practices with "exonerative comparisons," and activities can be hidden behind "euphemistic language." People also can reduce their accountability through the "displacement and diffusion of responsibility"; disregard, minimize, or dispute the harmful effects they cause; dehumanize and disparage victims; or discredit the "messengers of harmful effects." These psychosocial mechanisms work on an individual and social level, so it is particularly important that we reduce or eliminate the factors inhibiting the conservation of the natural environment and the future, and activate our ecological conscience.

Having experienced the dramatic change in the nature, spatial, and temporal scope of human activity, German-American philosopher Hans Jonas drew attention to the need for a new kind of ethics, and to the importance of the principle of responsibility. Humans are causing cumulative and irreversible damage to living nature, thus it is an ethical requirement to care and behave responsibly toward natural beings, which all have intrinsic value. We must live and act so that we do not endanger the future of life on Earth. It is a duty of humanity to adopt a caring attitude based on a non-reciprocal attitude toward future generations and nature. Human responsibility to preserve the living conditions of future generations and natural beings is of a substantive nature (Jonas 1984; Zsolnai 2014a). Economic decision-makers also have an ethical responsibility toward their stakeholders, depending on whether they are linked contractually (suppliers, employees, and consumers) or naturally (local communities, natural environment, and future generations) (Zsolnai 2000).

On the basis of his extensive anthropological, biological, linguistic, psychological, cybernetic, ecological, and philosophical research, British-American biologist Gregory Bateson concluded that the duality found in modern philosophical and scientific thinking (mind–body, man–nature, society–ecology, biological–emotional, God–nature) is basically false. The world is made up of systems involving individuals, societies, and ecosystems, which are part of a supreme, comprehensive system. Many call this God; Bateson refers to it as "Mind." Consciousness is the bridge between the individual, society, and ecology, but the limited consciousness developed through Western epistemology is not compatible with the universal Mind. The reason for this is that Western thinking is too purpose-driven, and thus narrows attention, perception, and the receptiveness of consciousness. It reinforces the misconception that humanity can control all systems and customize them to suit its own needs, thereby unbalancing natural systems and destroying self-correcting mechanisms. As a solution, Bateson proposes the practice of humility and the acceptance of natural cybernetic (self-controlled) systems. Thinking and cognition has to be developed into a more complete form of consciousness that synthesizes emotions and subconscious factors. To achieve this, the following factors are needed: extending love from individuals to societies and ecosystems; practicing arts, poetry, and music to expand the capacity of the mind; developing the relationship between humans and animals and nature; and religious practice, or spirituality. A healthy ecology of human civilization involves a unified system of nature and human civilization, in which civilization is as flexible as nature, and is open even to slow changes in its basic characteristics (Bateson 1972).

Ecological Consciousness and Spirituality

We have seen above how often the subject of spirituality appears in the study of ecological consciousness. Now, let us look at the topic of ecological consciousness through the lenses of spirituality and see how some worldviews approach ecological consciousness.

Spirituality: The Search for the Meaning of Life

Modern economic activity that focuses on maximizing profit and growth is threatening the integrity and diversity of natural ecosystems, along with the autonomy and culture of local communities and the living conditions of future generations. To preserve nature and meet real human needs, a moderate and careful economy is needed. This requires the internal motivation of economic actors to serve the public good, and to measure their success in other categories of value besides money. The background and motivational basis for this can be spirituality, because, without this, ethical business initiatives and environmental measures can become ineffective and often lead to the opposite outcomes to those which were intended (Zsolnai 2015).

The well-being of human communities, natural ecosystems, and future generations, therefore, requires dedicated, genuine care, which can come from an experience of unity with other beings and the universal source of creation. Spiritual experiences can play a prominent role in the development of the personality and ethical behavior of decision-makers, and contribute to a deeper understanding of different decision-making situations and systems, thereby helping create new categories, and overcome existing paradigms (Zsolnai 2014b). Spirituality can be a significant factor in sustainable lifestyles, as spiritually inspired people may pursue more pro-environmental activities and live a happier and more satisfied life than other individuals (Csutora and Zsóka 2014).

Spirituality has long been linked exclusively to religion, but today it clearly exceeds the framework of institutionalized religions. It consists of the following main components: connection to the inner "I"; the search for universal values that help the individual transcend their self-centered aspirations; deep compassion for all living beings; and connecting with the source of life (Bouckaert and Zsolnai 2012). According to the definition developed by the European SPES Institute (2020), spirituality is the search for the meaning of life—in various sociocultural forms—that connects people with every living being and with God or Ultimate Reality.

The Economic Application of Spirituality

Spiritual-based leadership can play a key role in the economic application of spirituality. Spirituality can help guide leaders as an internal source in their search for meaning, purpose, and fulfillment both in the business world and in the world of their consciousness and conscience. Their external action and internal reflection support each other. Rationality and spirituality become reinforcing viewpoints (Pruzan 2011). However, it is not enough to transform managerial education and management practices. The mindset and consciousness of managers must be changed, and their internal guidance and conscience strengthened (Pruzan 2015a).

Coming to know the essence of being, the experience of the true Self, is possible for all leaders. This will then be reflected in the leader's decisions in the external business world. However, the acceptance, development, and promotion of the concept and practice of spiritual-based leadership depend first and foremost on to what extent leaders become conscious of their own spirituality—their inborn divine nature and state of perfection (Pruzan 2015b).

According to Luk Bouckaert, spiritual-based leadership can be considered a kind of transformative leadership. The purpose of transformative leadership is to support the development of individuals and organizations in the following ways: increasing development of a common vision and values, cooperation, open communication, trust, intrinsic motivation, empowerment, and charismatic-, serving- and value-driven management. The four main types of transformative leadership can be displayed in Table 3.1 consisting of a vertical and a horizontal axis. The two extreme values of the vertical axis are market-driven leadership (M) and spirit-centered leadership (S). On the horizontal axis, there are hierarchical (H) and democratic (D) versions of leadership. In the table divided into four parts by the two axes (Market leadership [M + H], CSR Leadership [M + D], Aristocratic leadership [S + H], Spiritual-based leadership [S + D]) leaders can be practically described as being of mixed types, but some features are characteristically predominant. The main characteristics of

Table 3.1 Types of transformative leadership

Market

Hierarchy

Market leadership (M + H)	CSR leadership (M + D)
Aristocratic leadership (S + H)	Spiritual-based leadership (S + D)

Democracy

Spirit

Source Author's construction based on Bouckaert (2016)

spiritual-based leadership are that every human being has the ability to lead and follow, so the operation of an organization and leadership are based on co-creativity and co-responsibility. Spiritual-based leadership empowers participants and supports social relationships based on trust, reciprocity, and interconnectedness (Bouckaert 2016).

Lozano and Ribera (2004) find that spirituality and leadership practice interact. Spirituality manifests itself in leadership when it is developed by leaders through effort and vigilance. Leaders then receive feedback during their exercise of leadership that influences their own conception of life, human characteristics, and spirituality. Spirituality can thus lead to a qualitative change in leaders, and through them also in organizations, which can ultimately affect every member of the organization. In a spiritually based organization containing spiritually driven individuals, the following positive effects can be observed: The personal values and abilities of the members of the organization develop, and professional and ethical responsibility strengthens. Relationships among members, the quality of products, and organizational processes improve. The expression of, development of, and experience of values are enhanced. Relations with internal and external stakeholders are also strengthened (Lozano and Ribera 2004).

During the last few decades, many scholars have argued that a paradigm shift in mainstream management thinking and education is necessary. New insights from the humanities (including the arts, literature, and music) and spirituality have been used to call for the establishment of a better connection between management-type learning and life for the comprehensive and integral development of the individual. Rabindranath Tagore's alternative higher education system at his university Visva Bharati in Santiniketan, West Bengal, India, could serve as an example of how to transform our organizations and ourselves for the better. Based on Tagore's call and practice of humanism, openness and otherness, and learning in an ambience of nature, the university aims at promoting holistic human development and the natural expression of students' full potential. Inspired by Tagore's conception, the integration of liberal arts into management education may also increase the potential of managers by emphasizing the importance of wisdom and character. A liberal arts education can strengthen personal values and interdisciplinary and cross-cultural perspectives, adaptive learning, critical thinking, creativity, human caring, respect for nature and all creation, and reverence for life. Such an approach can help executives to create value for all stakeholders, to aspire to non-materialistic goals, and to live a meaningful and soulful life that is sensitive toward nature and the spirit (Mukherjee 2018).

Worldviews and Ecological Consciousness

In this section, we will examine how ecological consciousness appears in some of the important worldviews and systems of thought of our time: Christianity, Gandhi's philosophy, and Buddhist economics.

Inspired by the teachings of Saint Francis of Assisi, the patron saint of ecology, Pope Francis's encyclical "Praise be to you" (*Laudato si'*) presents the ecological position of the Roman Catholic Church. Pope Francis calls attention to the significance of the ecological crisis and connects it with the imperative of social justice. The encyclical describes the ecological and social crisis of today (environmental pollution, climate change, drinking water scarcity, loss of biodiversity, a decline in human quality of life and the degradation of societies, and global inequality) with scientific precision, and presents the teachings of Saint Francis of Assisi as inspiration for finding solutions. Concern for vulnerable people and natural beings, peace, joy, selflessness, a feeling of openness toward ourselves and the world, and material renunciation that fosters spiritual wealth are the most important components of what is called "integral ecology."

Integral ecology is a transdisciplinary approach that respects both nature and humanity, and is oriented by humane values. Ecology and the economy, the natural and social sciences, the arts and education must work together to replace utilitarian (individualist, endless-growth-based, competition-, consumption- and unregulated market-based) modern thinking with the ideal of "ecological citizenship" based on equity, justice, moderate consumption, recognition of the intrinsic value of nature, and a holistic worldview. The basis of individual and community ecological transformation should be ecological spirituality, in which humanity recognizes, through generous care, sensitivity, gratitude, and self-sacrificing work, that they are not separate from the world but live in a universal community with all other creatures (Pope Francis 2015).

To put these ideals into practice and to address economic challenges, starting with the thoughts and economic activities of young scholars and activists, Pope Francis convened in Autumn 2020 the international meeting "The Economy of Francesco. Young people, a pact, the future and Assisi 2020" (after discussion with the Bishop of Assisi, Domenico Sorrentino, and the economist Luigino Bruni). Lead by genuine concern for the pathological state of the world and care for the poor, the former invited young economists and entrepreneurs to develop and implement a new economic model based on communion, gratitude, and reciprocity that would promote an alternative to the dominant capitalist lifestyle (The Economy of Francesco 2020).

Gandhi's philosophy was based on the teachings of Hinduism, Jainism, and Buddhism, and was inspired by writing by John Ruskin, entitled *Unto This Last*. The ancient ethical principle of non-harming (*ahimsa*) became Gandhi's creed. Non-harming was a central element of his pursuit of Truth as the ultimate reality of existence. Gandhi showed a personal example of this when he achieved his political goals not by force but by passive resistance to the oppressive power of Britain.

Gandhi used non-harming in all aspects of his life—in his work, politics, and in economic, religious, ethical, and ecological issues. His main objective was to alleviate Indian poverty, which he intended to do with the help of self-sufficient, small-scale, decentralized, ecologically sustainable farms that could locally meet basic human needs. He was not the enemy of economic development, capital, machinery, or the market, but he stood against unlimited industrialization, profit-making companies, and the market distribution of resources (Kovács and Ócsai 2015). His ecological views were also based on non-harming: he claimed that animals also have rights; that we can realize the purpose of our lives, understanding ourselves only if we consider that we are one with all beings; and that selfless service leads to the understanding that God is present in all beings (Timmerman 2015).

Gandhi's conception had a great influence on the English economist Ernst Friedrich Schumacher, who developed the concept of Buddhist economics in the 1970s (Kovács and Ócsai 2015). The earliest teachings of Buddhism do not deal specifically with nature, because environmental pollution was not a problem in around the fifth century BC. The main issue was how to eliminate human suffering. The environmental attitude of Buddhism can be considered ecological because it advocates for the development of environmental virtues (non-harming, compassion, preferring the small scale, gratitude, contentment, simplicity, wisdom, and consciousness) that sensitize us to ecological problems. The main purpose of a Buddhist lifestyle and strategy is to eliminate suffering, which can be achieved by following the Noble Eightfold Path (Kovács 2015). This contrasts with the principles of modern Western economics (profit maximization, the satisfaction of desires and introduction of markets, the instrumental use of the world, and ethics based on self-interest). The main elements of Buddhist economics are the minimization of suffering, the simplification of desires, the practice of non-harming, and genuine care and generosity through which production and consumption can be transformed more sustainable (Zsolnai 2014c).

References

Bandura, A. (2007). Impeding Ecological Sustainability through Selective Moral Disengagement. *International Journal of Innovation and Sustainable Development, 2*(1), 8–35. https://doi.org/10.1504/ijisd.2007.016056.

Bateson, G. (1972). *Steps to an Ecology of Mind: Collected Essays in Anthropology, Psychiatry, Evolution, and Epistemology*. London: Jason Aronson Inc. nomadicartsfestival.com/wp.../02/Gregory-Bateson-Ecology-of-Mind.pdf. Accessed 28 March 2020.

Beck, U. (1995). *Ecological politics in an Age of Risk*. Cambridge: Polity Press.

Benson, K. R. (2000). The Emergence of Ecology from Natural History. *Endeavour, 24*(2), 59–62. https://doi.org/10.1016/s0160-9327(99)01260-0.

Bouckaert, L. (2016). Why do we Need a Spiritual-Based Theory of Leadership? In M. Chatterji & L. Zsolnai (Eds.), *Ethical Leadership: Indian and European Spiritual Approaches* (pp. 17–33). London: Palgrave Macmillan. https://doi.org/10.1057/978-1-137-60194-0_2

Bouckaert, L., & Zsolnai, L. (2012). Spirituality and Business: An Interdisciplinary Overview. *Society and Economy, 34*(3), 489–514. https://doi.org/10.1556/socec.34.2012.3.8.
Boulding, K. E. (1966). The Economics of the Coming Spaceship Earth. In H. Jarrett (Ed.), *Environmental Quality in a Growing Economy* (pp. 3–14). Baltimore, MD: Resources for the Future/Johns Hopkins University Press.
Bragg, E. (1996). Towards Ecological Self: Deep Ecology Meets Constructionist Self-Theory. *Journal of Environmental Psychology, 16,* 93–108. https://doi.org/10.1006/jevp.1996.0008.
Brown, P. G., & Timmermann, P. (Eds.) (2015). *Ecological Economics for the Anthropocene: An Emerging Paradigm.* New York, NY: Columbia University Press. https://doi.org/10.7312/brow17342.
Capra, F., & Jakobsen, O. D. (2017). A Conceptual Framework for Ecological Economics Based on Systemic Principles of Life. *International Journal of Social Economics, 44*(6), 831–844. https://doi.org/10.1108/IJSE-05-2016-0136.
Chakraborty, S. K. (1995). *Ethics in Management: Vedantic Perspectives.* Oxford: Oxford University Press.
Christopher, M. (1999). An Exploration of the "Reflex" in Reflexive Modernity: The Rational and Prerational Social Causes of the Affinity for Ecological Consciousness. *Organization & Environment, 12*(4), 357–400. https://doi.org/10.1177/1086026699124001.
Costanza, R., Perrings, C., & Cleveland, C. J. (Eds.) (1997). *The Development of Ecological Economics.* Cheltenham: Elgar.
Csutora, M., & Zsóka, Á. (2014). May Spirituality Lead to Reduced Ecological Footprint? Conceptual Framework and Empirical Analysis. *World Review of Entrepreneurship, Management and Sustainable Development, 10*(1), 88–105. https://doi.org/10.1504/wremsd.2014.058056.
Dalai Lama, H. H. (1995). *The World of Tibetan Buddhism.* Boston, MA: Wisdom Publications.
Dalai Lama, H. H. (2005). *The Universe in a Single Atom: The Convergence of Science and Spirituality.* New York, NY: Broadway Books.
Daly, H. E. (1999). *Ecological Economics and the Ecology of Economics: Essays in Criticism.* Cheltenham and Northampton: Edward Elgar.
Daly, H. E., & Farley, J. (2004). *Ecological Economics: Principles and Applications.* London: Island Press.
Devall, B. (1988). *Simple in Means, Rich in Ends: Practicing Deep Ecology.* Layton, UT: Gibbs M. Smith.

Devall, B., & Sessions, G. (1985). *Deep Ecology: Living as if Nature Mattered.* Layton, UT: Gibbs M. Smith.

Drengson, A. (1989). *Beyond Environmental Crisis: From Technocrat to Planetary Person.* New York, NY: Peter Lang.

Egerton, F. N. (2001). A History of the Ecological Sciences: Early Greek Origins. *Bulletin of the Ecological Society of America, 82*(1), 93–97. http://esapubs.org/bulletin/current/history_list/history_part1.pdf Accessed 26 March 2020.

Elton, C. S. (1927). *Animal Ecology.* London: Sidgwick and Jackson.

European Commission. (2015). *Closing the Loop: Commission Adopts Ambitious New Circular Economy Package to Boost Competitiveness, Create Jobs and Generate Sustainable Growth.* Press release, Brussels, 2 December 2015, http://europa.eu/rapid/press-release_IP-15–6203_en.htm. Accessed 27 March 2020.

European Commission. (2018). *Circular Economy: Implementation of the Circular Economy Action Plan.* 2018 Circular Economy Package. http://ec.europa.eu/environment/circular-economy/index_en.htm.Accessed 27 March 2020.

European SPES Institute. (2020). *Definition of Spirituality.* http://www.eurospes.org/content/our-mission-spiritual-based-humanism. Accessed 28 February 2020.

Feng Lin, Q., & Fyles, J. W. (2015). Following in Aldo Leopold's Footsteps. Humans-in-Ecosystem and Implications for Ecosystem Health. In P. G. Brown & P. Timmermann (Eds.), *Ecological Economics for the Anthropocene: An Emerging Paradigm* (pp. 208–232). New York, NY: Columbia University Press. https://doi.org/10.7312/brow17342-008.

Friederichs, K. (1958). A Definition of Ecology and Some Thoughts About Basic Concepts. *Ecology, 39*(1), 154–159. https://doi.org/10.2307/1929981.

Georgescu-Roegen, N. (1971). *The Entropy Law and the Economic Process.* Cambridge: Harvard University Press.

Gennaro, R. J. (n.d). *Consciousness: The Internet Encyclopedia of Philosophy.* ISSN 2161–0002. http://www.iep.utm.edu/consciou/ Accessed 27 March 2020.

Hahn, I. (2014). *Növényökológia* [Plant Ecology]. Department of Plant Taxology, Ecology and Theoretical Biology, Eötvös Loránd University, Budapest, Projected Slides of the 2014 Lecture on Plant Ecology. http://ramet.elte.hu/~ramet/oktatas/Novenyokol_kornyezettan/2014_novenyokologia.pdf. Accessed 26 March 2020. [in Hungarian].

Hamilton, C. (2010). Consumerism, Self-creation and Prospects for a New Ecological Consciousness. *Journal of Cleaner Production, 18*(6), 571–575. https://doi.org/10.1016/j.jclepro.2009.09.013.
Harding, W., & Bode, C. (Eds.) (1958). *The Correspondence of Henry David Thoreau*. New York, NY: New York University Press. https://www.walden.org/wp-content/uploads/2016//Correspondence.pdf. Accessed 2 April 2020.
Hill, S., Wilson, S., & Watson, K. (2004). Learning Ecology: A New Approach to Learning and Transforming Ecological Consciousness. In E. O'Sullivan & M. Taylor (Eds.), *Learning Toward an Ecological Consciousness: Selected Transformative Practices* (pp. 47–64). New York, NY: Palgrave Macmillan. https://doi.org/10.1007/-1-349-73178-7_4.
Hortobágyi, T., & Simon, T. (Eds.) (2000). *Növényföldrajz, Társulástan és Ökológia* [Plant Geography, Synbiology and Ecology]. Budapest: Tankönyvkiadó. [in Hungarian].
Hughes, J. D. (1991). The Psychology of Environmentalism Healing Self and Nature. *Trumpeter, 8*(3). http://trumpeter.athabascau.ca/index.php/trumpet/article/view/458/759. Accessed 27 March 2020.
Ims, K. J. (2015). Deep Ecology and Personal Responsibility. In L. Zsolnai (Ed.), *The Spiritual Dimension of Business Ethics and Sustainability Management* (pp. 47–56). New York, NY: Springer. https://doi.org/10.1007/978-3-319-11677-8_5.
Ims, K. J., & Jakobsen, O. D. (2006). Cooperation and Competition in the Context of Organic and Mechanic Worldviews—A Theoretical and Case based Discussion. *Journal of Business Ethics, 66*(1), 19–32. https://doi.org/10.1007/s10551-006-9055-z.
Ingebrigtsen, S., & Jakobsen, O. (2007). *Circulation Economics: Theory and Practice*. Bern: Peter Lang.
Jakobsen, O. (2017). *Transformative Ecological Economics: Process Philosophy, Ideology and Utopia*. London and New York, NY: Routledge. https://doi.org//9781315205434.
Jonas, H. (1984). *The Imperative of Responsibility: In Search of an Ethics for the Technological Age*. Chicago, IL: The University of Chicago Press.
Juhász-Nagy, P. (1984). *Beszélgetések az ökológiáról* [Conversations about Ecology]. Budapest: Mezőgazdasági Kiadó. [in Hungarian].
Juhász-Nagy, P. (1986). *Egy operatív ökológia hiánya, szükséglete és feladatai* [The Lack, the Need and the Tasks of an Operative Ecology]. Budapest: Akadémiai Kiadó. [in Hungarian].
Juhász-Nagy, P. (1993). *Az eltűnő sokféleség* [Biological Diversity in Destruction]. Budapest: Scientia Kiadó. [in Hungarian].

Koch, C. (2004). *The Quest for Consciousness: A Neurobiological Approach*. Englewood, CO: Roberts and Company.
Kovács, G. (2015). Buddhist Spiritual Orientation to Nature and Sustainability. In L. Zsolnai (Ed.), The *Spiritual Dimension of Business Ethics and Sustainability Management* (pp. 57–67). New York, NY: Springer. https://doi.org/10.1007/978-3-319-11677-8_6.
Kovács, G., & Ócsai, A. (2015). The Spirit of Non-violene and Peace in Business. In L. Bouckaert & M. Chatterji (Eds.), *Business, Ethics and Peace* (pp. 231–245). Bingley: Emerald Group Publishing. https://doi.org/10.1108/s1572-832320150000024014.
Krebs, C. J. (2001). *Ecology: The Experimental Analysis of Distribution and Abundance*. San Francisco, CA: Benjamin Cummings.
Leff, H. (1978). *Experience, Environment, and Human Potentials*. New York, NY: Oxford University Press.
Leopold, A. (1949). The Ecological Conscience. In A. Leopold (Ed.), *A Sand County Almanach*. Oxford: Oxford University Press. http://www.public.iastate.edu/~fridolph/leopold.html. Accessed 2 April 2020
Lovelock, J. (2009). *The Vanishing Face of Gaia: A Final Warning*. London: Allen Lane.
Lozano, J. M., & Ribera, R. (2004). A New Chance for Management, A New Challenge for Spirituality. In L. Zsolnai (Ed.), *Spirituality and Ethics in Management* (pp. 175–185). Boston, MA, Dordrecht and London: Kluwer Academic Publishers. https://doi.org/10.1007/1-4020-2365-0_15.
Marques, J. (2012). Consciousness at Work: A Review of Some Important Values, Discussed from a Buddhist Perspective. *Journal of Business Ethics, 105*(1), 27–40. https://doi.org/10.1007/s10551-011-0932-8.
McNaughton, S. J., & Wolf, L. L. (1973). *General Ecology*. New York, NY: Holt, Rinehart and Winston.
Merriam-Webster Dictionary and Thesaurus (2020). *Consciousness*. https://www.merriam-webster.com/dictionary/consciousness. Accessed 27 March 2020.
Merton, T. (1968). *The Ecological Conscience*. (kivonat) The Catholic Worker. June 1968. http://www.democraticunderground.com/discuss/duboard.php?az=view_all&address=115x180150. Accessed 2 April 2020.
Milbraith, L. (1989). *Envisioning a Sustainable Society*. Albany, NY: State of New York Press.
Mind & Life Institute. (2020). *Mission*. https://www.mindandlife.org/mission. Accessed 27 March 2020.

Morris, M. (2002). Ecological Consciousness and Curriculum. *Journal of Curriculum Studies, 34*(5), 571–587. https://doi.org/10.1080/0108187.

Mukherjee, S. (2018). Management and Liberal Arts: A Transformational Odyssey with Rabindranath Tagore. In L. Bouckaert, K. J. Ims, & P. Rona (Eds.), *Art, Spirituality and Economy: Virtues and Economics 2* (pp. 139–156). Cham: Springer. https://doi.org/10.1007/978-3-319-75064-4_13.

Naess, A. (1989). *Ecology, Community and Lifestyle: Outline of an Ecosophy.* Cambridge: Cambridge University Press. https://doi.org/10.1017/cbo978 0511525599.

Nhat Hanh, T. (2008). *The Heart of the Buddha's Teaching. Transforming Suffering into Peace, Joy, and Liberation.* E-book. London: Random House. https://terebess.hu/zen/mesterek/Thich%20Nhat%20Hanh%20-%20The% 20Heart%20of%20Buddha%27s%20Teaching.pdf. Accessed 2 April 2020.

Odum, E. P. (1963). *Ecology*. New York, NY: Holt, Rinehart and Winston.

O'Sullivan, E. V., & Taylor, M. M. (Eds.) (2004). *Learning toward an Ecological Consciousness: Selected Transformative Practices.* New York, NY: Palgrave Macmillan. https://doi.org/10.1007/978-1-349-73178-7.

Pásztor, E., & Oborny, B. (Eds.) (2007). *Ökológia* [Ecology]. Budapest: Nemzeti Tankönyvkiadó. [in Hungarian].

Polanyi, K. (1944). *The Great Transformation: The Political and Economic Origins of Our Time*. Boston, MA: Beacon Press.

Pope Francis. (2015). *Encyclical letter Laudato Si':On Care for our Common Home.* http://w2.vatican.va/content/dam/francesco/pdf/encyclicals/docume nts/papa-francesco_20150524_enciclica-laudato-si_en.pdf. Accessed 3 April 2020.

Pruzan, P. (2011). Spirituality as the Basis for Leadership. In L. Bouckaert & L. Zsolnai (Eds.), *The Palgrave Handbook of Spirituality and Business* (pp. 287–295). London: Palgrave. https://doi.org/10.1057/9780230321458_35.

Pruzan, P. (2015a). Spiritual-Based Leadership: A Paradox of Pragmatism. In L. Zsolnai (Ed.), *The Spiritual Dimension of Business Ethics and Sustainability Management* (pp. 169–172). New York, NY: Springer. https://doi.org/10. 1007/978-3-319-11677-8_14.

Pruzan, P. (2015b). Further Reflections on Spirituality and Spiritual-based Leadership. In L. Zsolnai (Ed.), *The Spiritual Dimension of Business Ethics and Sustainability Management* (pp. 191–198). New York, NY: Springer. https://doi.org/10/978-3-319-11677-8_19.

Pruzan, P. (2016). Science and Spirituality. In P. Pruzan, (Ed.), *Research Methodology: The Aims, Practices and Ethics of Science*. Switzerland: Springer. Unpublished Chapter.

Rahula, W. (1974). What the Buddha Taught (2nd and enlarged ed.). New York, NY: Grove Press. https://web.ics.purdue.edu/~buddhism/docs/Bhante_Walpola_Rahula-What_the_Buddha_Taught.pdf. Accessed 4 April 2020.

Sarin, I. (2014). Artistic Disclosure of Nature. Heidegger and Tagore. *Existenz: An International Journal in Philosophy, Religion, Politics, and the Arts, 9*(1), 45–52. https://existenz.us/images/read-pdf.gif. Accessed 2 April 2020.

Schumacher, E. F. (1973). *Small is Beautiful: A Study of Economics As if People Mattered*. New York, NY: Harper and Row.

Sen, A. (1997). Tagore and His India. *The New York Review of Books, 44*(11), 55–63. https://www.nybooks.com/articles/1997/06/26/tagore-and-his-india/. Accessed 10 April 2020.

Sen, S. (2008). The Vedic-Upanisadic concept of Brahman (The Highest God). In A. Eshleman (Ed.), *Readings in Philosophy of Religion: East Meets West* (pp. 43–51). Malden, MA: Blackwell.

Takács-Sánta, A. (2004). The Major Transitions in the History of Human Transformation of the Biosphere. *Human Ecology Review, 11*(1), 51–66. http://www.humanecologyreview.org/pastissues/her111/111takacssanta.pdf. Accessed 3 April 2020.

Takács-Sánta, A. (2009). *Környezeti problémák: A generalisták és a társadalmi dimenzió hiánya* [Environmental Problems: Lack of Generalists and The Social Dimension]. Magyar Tudomány [Hungarian Science]. 2009/1. pp. 62-66. [in Hungarian]

The Economy of Francesco. (2020). https://francescoeconomy.org/who-we-are/. Accessed 3 April 2020.

Timmerman, P. (2015). The Ethics of Re-Embedding Economics in the Real: Case Studies. In P. G. Brown & P. Timmermann (Eds.), *Ecological Economics for the Anthropocene: An Emerging Paradigm* (pp. 21–65). New York, NY: Columbia University Press. https://doi.org/10.7312/brow17342-002.

Tononi, G., & Koch, C. (2015). Consciousness: Here, There and Everywhere? Philosophical Transactions of the Royal Society. *Biological Sciences, 370*(1668), 20140167. https://doi.org/10.1098/rstb.2014.0167.

Uhl, C. (2004). *Developing Ecological Consciousness: Path to a Sustainable World*. Lanham, MD: Rowman & Littlefield Publishers.

Velmans, M. (2009). How to Define Consciousness-and How Not to Define Consciousness. *Journal of Consciousness Studies, 6*(5), 139–156. cogprints.org/6453/1/How_to_define_consciousness.pdf. Accessed 1 April 2020.

Wallace, B. A. (2007). Principles of Contemplative Science. In B. A. Wallace (Ed.), *Contemplative Science: Where Buddhism and Neuroscience Converge*. New York, NY: Columbia University Press.

White, P. R. (2011). A Phenomenological Self-Inquiry into Ecological Consciousness. *Ecopsychology.*, *3*(1), 41–50. https://doi.org/10.1089/eco.201 00054.

Whittaker, R. H. (1975). *Communities and Ecosystems*. New York, NY: Macmillan.

Zsolnai, L. (2000). *A döntéshozatal etikája* [Ethics of Decision Making]. Budapest: Kossuth Kiadó. [in Hungarian].

Zsolnai, L. (2001). *Ökológia, gazdaság, etika* [Ecology, Economy, Ethics]. Budapest: Helikon Kiadó. [in Hungarian].

Zsolnai, L. (2014a). Ethical Decision Making. In L. Zsolnai (Ed.), *Beyond Self: Ethical and Spiritual Dimensions of Economics* (pp. 45–64). Bern: Peter Lang. https://doi.org/10.3726/978-3-0353-0596-8/17.

Zsolnai, L. (2014b). Ethics Needs Spirituality. In L. Zsolnai (Ed.), *Beyond Self: Ethical and Spiritual Dimensions of Economics* (pp. 79–83). Bern: Peter Lang. https://doi.org/10.3726/978-3-0353-0596-8/20.

Zsolnai, L. (2014c). Buddhist Economic Strategy. In L. Zsolnai (Ed.), *Beyond Self. Ethical and Spiritual Dimensions of Economics* (pp. 119–138). Bern: Peter Lang. https://doi.org/10.3726/978-3-0353-0596-8/23.

Zsolnai, L. (2015). Spirituality, Ethics and Sustainability. In L. Zsolnai (Ed.), *The Spiritual Dimension of Business Ethics and Sustainability Management* (pp. 3–11). New York, NY: Springer. https://doi.org/10.1007/978-3-319-11677-8_1.

4

Ecological Consciousness and Value Orientations in Business

This chapter presents different concepts concerning how business organizations may implement ecologically conscious ways of functioning. It covers, inter alia, the "moralist model" of companies, the "ecological worldview," "sustainability-driven businesses," "true business sustainability," "entrepreneurship inspired by nature," "environmental commitment," the "really responsible company," "ecological entrepreneurship," and "ecologically sustainable business." The book suggests considering those business organizations as ecologically conscious which are willing to put ecological aspects at the center of their functioning, and which are able to operate in an alternative way to prevailing materialistic, one-dimensional, and profit-maximizing businesses.

According to business ethicist George Brenkert, the business world operates according to a special "ethics" that is related to the free market system and the nature of business organizations. Based on this, the economy is not able to function in an environmentally responsible way because its primary goal within the framework defined by society and the government is to maximize profits. This is the "economic model" of companies, in contrast to which, according to the "moralist model,"

A. Ócsai, *Ecologically Conscious Organizations*, Palgrave Studies in Sustainable Business In Association with Future Earth,
https://doi.org/10.1007/978-3-030-60918-4_4

companies are moral actors, which—beyond complying with laws—must act morally, socially, and in an environmentally responsible way; that is, minimize damage, and solve environmental problems. To this end, companies should take an approach based on biocentric environmental ethics (Hoffman 1991 quoted by Brenkert 1995). Further, global consciousness that evolves from the environmental conscience of individual companies is needed (Goodpaster 1990 quoted by Brenkert 1995). However, the parameters of the economy and companies (i.e., attitudes and values necessary for a free market enterprise such as bureaucratic business structures, short-term profit-interest, growth constraints, resource utilization, specialization, and price competition) make it difficult for economic actors to behave ethically, and their moral consciousness becomes fragmented. A fundamental change to the economic system and business organizations is therefore needed (strengthening accountability, expanding rationality, democratization, openness, environmental protection, and sustainable development) so that companies and their employees can act morally, and their environmental responsibility can be strengthened (Brenkert 1995).

In terms of environmental problems, the industrial and ecological worldviews are sharply different. As claimed by environmental thinker Timothy O'Riordan, in the case of the ecological worldview, the goal is harmony with nature. Nature is not just a resource, because every species is equal, and natural beings have intrinsic value. The Earth's resources are finite and thus define limits for economic growth. Science and technological development cannot solve all problems; technology has to adapt to nature and man. Instead of striving to satisfy unbounded consumer desires, the emphasis must be on meeting human needs, in line with the idea that people are able and willing to moderate their needs by following a consciously simple lifestyle. The distribution of power should be decentralized and take into account the rights of nature and people (O'Riordan and Cameron 1994 quoted by Zsolnai 2001).

Businesses and Sustainability

Many pieces of research discuss how businesses should support ecological and social processes rather than destroy them. However, most of these studies consider economic actors to be self-interested and profit-maximizing entities that strive to incorporate sustainable development principles into existing business strategies. This limited framework can be widened by reversing the perspective in order to reveal the organizational factors through which entrepreneurs motivated by sustainability-related values can achieve the latter most effectively in competitive market conditions. The qualitative analysis of case studies of successful sustainable enterprises of different sizes and business models operating in four different sectors and continents leads to the conclusion that there are five categories in which the former can be compared with conventional business enterprises: (i) The objective and reason d'être of successful sustainable enterprises is to preserve and strengthen the quality of human and natural resources over the longest possible time; (ii) Sustainable enterprises strive to increase the benefits and synergies of every activity, rather than solely pursuing a strategy of cost-effectiveness; (iii) Sustainable enterprises harmonize their objectives, seeking satisfactory results from multiple goals simultaneously, and do not just aim to maximize the value of a single dimension of their set of goals; (iv) The expected qualitative aspects of the results serve as decision criteria; and (v) They distribute benefits based on merit and non-power relations (Parrish 2010, p. 517).

The values adopted and promoted by sustainability-driven businesses may create balance among individuals, other people, and nature, but can cause tension within the organization. Successful entrepreneurs are distinguished from others by the fact that, besides having a proper value orientation, they can manage emerging organizational tensions by successfully achieving the objectives listed above (Parrish 2010).

Sustainability management is becoming increasingly widespread in large companies, but the impact of such aspirations does not appear to have triggered an overall improvement in the ecological state of our planet. In looking for the reasons for the "big disconnect" between micro-development and macro-degradation, a four-item typology can support the analysis of the sustainability of business models. The starting point is Milton Friedman's concept of today's business attitude of "business-as-usual." Refined shareholder value management models that deal exclusively with ecological and social issues that pose an economic risk or opportunity without changing their core business strategy follow this strategy. At the third level, some business models take into account economic, social, and environmental considerations. This is the "triple bottom line" concept. At the highest level, "truly sustainable business models" extend the scope from making an effort to minimize negative impacts to understanding how to maximize positive impact on critical social and ecological issues. Such enterprises forge their solutions to the most pressing sustainability challenges into business opportunities. They try to exert influence not only at the company level, but also throughout their entire sector and in other sectors. This is done through increasing their transparency, sharing examples of good practice, establishing shared self-regulation practices and standards, building cooperation with non-economic actors, and striving to change the rules of the game. Changing perspective this way is a huge challenge, requiring a fundamentally different strategic attitude from business entities, while a number of very significant barriers—such as the strong value orientation of shareholders toward financial markets and demands for short-term returns—make it very difficult to achieve (Dyllick and Muff 2015).

From Environmental Responsibility to Businesses Inspired by Nature

In the current situation, beyond improving environmental performance, the meaning of which is, unfortunately, not uniformly and comprehensively understood, only new conceptions of ethics and social and environmental responsibility can help ensure environmental protection. This presupposes meeting environmental requirements in the absence of state regulation or norms. The commitment of enterprises to social and environmental issues is evolving from a focus on compliance and from minimizing impact toward making a positive contribution to the ecology of the planet. The basis for such environmental commitment can be social standards (for instance, stakeholder expectations, or state regulations); technological opportunities (successful types of innovation, the nature of relationships within the supply chain); as well as the cognitive and behavioral characteristics of business leaders (environmental commitment, morality, and presuppositions) (Csutora and Kerekes 2004; Kerekes and Wetzker 2007).

Standard environmental management and corporate social responsibility tools aimed at the conservation of natural resources and the reduction of pollution and risk include cleaner production, energy rationalization, eco-efficiency, environmentally friendly products, product life cycle analysis, environmental product labeling, environmental performance assessment, environmental reports, industrial ecology, ecological accounting, environmental management systems, ethical codes, corporate governance standards, sustainability reports, product certificates, and socially responsible investment. Such operational approaches and instruments are necessary but not sufficient for fostering sustainable development. Sustainable development is a strategic issue for the economy and for humankind. "Truly Responsible Companies" can contribute to creating solutions in five key areas: (i) minimizing transport and preferring local markets; (ii) strengthening justice and fairness; (iii) avoiding an excessive focus on profit generation; (iv) achieving and maintaining an optimal size; and (v) supplying products that meet real needs. Such businesses can evolve (modeled as rising up through pyramid-like categories) starting from "normal local businesses" through "value-based organizations" and "entrepreneurs for something else [than just financial return]" into "companies serving their communities." It is also important that individuals (company owners and managers) take responsibility for implementing these radical changes. Even a complete redesign of basic products and strategies or the founding of new companies may be necessary, and much more "real action" than simply communicating about the former is needed (Tóth 2007).

The majority of economic enterprises in modern societies exploit nature and destroy ecosystems because they view them as simple economic resources. Capital and labor are usually considered to be more important factors of production than nature, so business strategies tend to be built on the former. However, nature-inspired, sustainable businesses have a different worldview and innovate products and services that help regenerate natural services, re-connecting people with nature ("regenerative eco-innovation"). The basis of this worldview is a kind of environmental wisdom that recognizes that humankind is a part of nature, and very much depends on it. Nature exists for all species, so limited resources should not be used to satisfy only human needs. Sustainable businesses need to be eco-efficient. They should solve real ecological problems, in parallel with meeting human needs and generating economic value. They should also be eco-effective: i.e., produce products and services that actually serve human needs and improve quality of life (Hofstra 2015).

In order for an entrepreneur inspired by nature to be successful, they should adhere to five principles: (i) predict unexpected threats caused by changes in relation to natural resources and transform these into business opportunities, recognizing the role of nature as teacher and mentor; (ii) take into account the different interests of business, society, and nature; (iii) achieve long-term economic efficiency combined with ethical perspectives and ecological effectiveness; (iv) discover and not exploit nature; and (v) generate business value in a broad sense, including economic, ecological, moral, social, and spiritual value (Hofstra 2015).

Ecological Entrepreneurship

The research area of ecologically sustainable business has been fragmented despite its development since the mid-1990s. There is no uniform definition of ecological entrepreneurship, although many variants of the term are widely used. The most common of these include sustainable entrepreneurship, ecopreneurship, environmental entrepreneurship, and green entrepreneurship. Ecologically sustainable business is the process of identifying, evaluating, and implementing business opportunities that minimize the environmental impact of an enterprise, and thus operating for the benefit of local communities and society as a whole (Gast et al. 2017).

Six main factors characterize the existence and operation of ecologically sustainable enterprises. (1) The drivers of engaging in ecologically sustainable entrepreneurship are opportunity, necessity, and personal values. (2) The drivers of conducting business in an ecologically sustainable way are values, the market, and compliance. For these first two factors, the role of intrinsic motivation, personal values, and volunteering is of paramount importance in the commitment to sustainability. Such enterprises do not primarily pursue economic goals, but seek to change people's mindset and society as a whole, and to create value for the environment and society. The latter outputs refer to Factor (3); that is, the micro-, meso-, and macro-level results of ecologically sustainable businesses. (4) The means of operating in an ecologically sustainable way involve redesigning business practices and networking. The role of networks and cooperating with stakeholders is extremely important, as individual enterprises alone have little chance of influencing public policies, or (5) fully meeting financial and market challenges. Factor (6) is support for ecologically sustainable businesses, of which the role of business schools, universities, and other educational institutions is outstanding. Nowadays, the role of these institutions is changing, and their curricula of business subjects need to be transformed in order to truly fulfill their responsibilities and increase the sustainability knowledge of businesses in relation to setting up ecologically sustainable businesses and making existing ones more sustainable (Gast et al. 2017).

It is an essential goal of societies and economies that humankind creates well-being in a sustainable way. Sustainable development is necessary for mitigating the ecological crisis, which requires social and environmental initiatives that take into account ecological, human, social, and economic factors. The well-being these create should be distributed in a balanced way between meeting current consumption needs and the need for reinvesting into the long-term future. An ecological enterprise is a form of business that meets these requirements, and whose actors develop new practices that create economic assets for local communities while remaining committed to preserving cultural, ecological, and environmental integrity. For their successful operation it is essential to correctly identify the potentially high-value traditional products and practices of local communities and to properly apply legal and regulatory instruments (such as logos and trademarks) that can shape and protect their niche markets. For the survival and development of ecological businesses, it is crucial to be able to successfully involve and mobilize external stakeholders within value networks, through whom more and more links between producers and consumers can be established (Marsden and Smith 2005).

Based on the former paragraphs, those businesses can be considered ecologically conscious which are capable of and willing to operate in alternative ways, and to put ecological considerations at the center of their operations.

Values and Value Orientation

This section presents the most important theories and concepts concerning value, analyzes the issue of value orientation, and discusses the findings of relevant studies.

Values and Value Theories

A number of disciplines—including anthropology, moral philosophy, political science, sociology, theology, social psychology, economic ethics, behavioral economics, management science, and corporate governance—deal with the philosophical, ethical, ideological, social, and aesthetic aspects of value and value orientation. Accordingly, several different definitions of value exist that have evolved over time.

Values are important and lasting beliefs; ideas accepted by members of a society about what is good, what is wrong, what is desirable, and what should be avoided. Values significantly influence the behavior and attitudes of individuals, and serve as comprehensive guidelines for them in different situations (Business Dictionary 2020).

The concept of value has been used since the eighteenth century primarily to express the "value of something"; that is, in the economic sense (for example, in Adam Smith's writings on political economy). Then, in the nineteenth century, as more and more philosophers (including Rudolf Hermann Lotze, Wilhelm Windelband, Friedrich Nietsche, Eduard von Hartmann, Hugo Münsterberg, and Wilbur Marshall Urban) became interested in the problematic of value, the interpretation became increasingly broad. In his book *General Theory of Value*, Ralph Barton Perry considered value as "any object of any interest." He later identified eight areas of importance for values: morality, religion, art, science, economics, politics, law, and custom (Encyclopaedia Britannica 2020).

Several disciplines investigate values (Kaldschmidt 2011). Coherent value theories (for instance, Rokeach 1979; Schwartz 1992) study the nature of values and value systems. In recent decades, various pieces of research have focused on the impact of values on national culture (Kluckhohn and Strodtbeck 1961; Schwartz and Bilsky 1987; Inglehart and Baker 2000; Hofstede and Hofstede 2005; Sagiv and Schwartz 2007) or company culture (Hofstede and Hofstede 2005; Schein 2010). Research has also been conducted on the effects of values on leadership behavior (Agle et al. 1999; Krishnan 2001; Lord and Brown 2001; Klenke 2005; Byrne and Bradley 2007; Resick et al. 2011), on the ethical organizational climate (Grojean et al. 2004), and on ethical decision making (Ford and Richardson 1994; Nonis and Swift 2001; Eberhardt-Toth and Wasieleski 2013; Marsh 2013). The role of values in the operation of social and ecological enterprises (Hemingway and Maclagan 2004; Hemingway 2005; Dixon and Clifford 2007) and environmentally oriented behavior (Karp 1996; Bansal 2003) has also been examined.

There are many differences between value concepts and definitions, and a lack of synthesis (Rohan 2000; Hitlin and Piliavin 2004). Let us review how important value theories define values.

American anthropologist Clyde Kluckhohn called values individual- or group-specific facts that determine human action by defining a combination of tools, goals, and methods (Kluckhohn 1951).

According to social psychologist Milton Rokeach, values are long-lasting beliefs based on whether certain behaviors (instrumental values) or end-states of existence (terminal values) are individually or socially preferred to those which are different from or opposed to them. The emphasis is thus shifted from action to motivation (Rokeach 1973).

Another social psychologist, Shalom Schwartz, suggests that basic values are the goals that we want to achieve in the form of principles that guide our lives. Values are also beliefs that directly affect us, and which overarch our various actions and situations. Values are evaluation criteria that determine how we judge actions, procedures, situations, and people. Differently from norms and attitudes, values can be arranged in a hierarchy. They influence activities on the basis of their relative importance, depending on a given situation (Schwartz 2006).

Geert Hofstede studied people's personal values, researching how each culture influences the cultural values that its members hold dear. Initially four, then six cultural dimensions were differentiated to illustrate the relationship between national and organizational culture: individualism–collectivism; uncertainty avoidance (from weak to strong); power distance (from small to large); masculinity (goal orientation)–femininity (relationship orientation); future orientation (short-term – long-term); and indulgence–restraint (Hofstede and Hofstede 2005).

Edgar Schein's model distinguishes three levels of organizational culture: (1) the level of artifacts and behaviors, which can be identified from outside the organization; (2) the level of declared values and beliefs; and, (3) common underlying assumptions, which are often based on participants' unconscious values or convictions. Thus, the primary sources of organizational culture are the values, convictions, and basic assumptions of the founders, thus the most important task of managers is to recognize and understand these so as to be able to adapt to unavoidable organizational change and meet challenges (Schein 2010).

Value Orientation

Value orientations identify desirable and avoidable goals for individuals, communities, and societies, thereby defining and controlling human behavior (Rokeach 1973).

Anthropologist Florence Kluckhohn and psychologist Fred Strodtbeck studied the phenomenon of value orientation with the aim of fostering understanding of the differences between human societies and their communication with each other. According to value orientation theory, societies face some universal problems, in response to which each culture selects its own solution from a limited number of value-based solutions according to their own preferences and values (Hills 2002). The five basic problems and categories of responses to them are the following: Which aspect of time the latter focus on (past, present, and future); the relation of humans to their natural environment (domination, harmony, and humility); the relationship between individuals (super- and subordination, equality, and individualism); the main motivation for behavior (existence, change, and performance); and what the basic nature of humans is (bad, neutral, good, and mixed) (Kluckhohn and Strodtbeck 1961 quoted by Hills 2002). A sixth value dimension (space: here, there, far) has been proposed but not elaborated in detail, so it has not become part of the theory. Value orientation theory has weaknesses. It is more general than specific, thus it is not well suited to predicting expected behavior in a given situation. It is simplistic, since most behavior cannot be classified into a category. Because of the use of rankings, it is difficult to analyze statistically. Its undisputed merit, however, is that it brings us closer to understanding values. It has inspired many subsequent pieces of research and laid the foundations for the important results of universal value surveys (Rokeach 1979; Schwartz 1992; Hofstede and Hofstede 2005) (Hills 2002). The psychological research of values is also important because it helps us to evaluate the full spectrum of human life, not just the positive and negative aspects of attitudes. In addition, the examination of values is also essential for understanding human thinking, feelings, and behavior, and for undertaking comparative analyses between cultures and groups, and within groups (Hills 2002).

Schwartz (2006) examined the value orientation of national cultures. Culture has been defined as a complex of meanings, beliefs, practices, symbols, norms, and values that are common among members of a society, the most important characteristics of which are prevailing values. The emphasis awarded values expresses the ideals of a given culture; the shared ideas of individuals about what is good and desirable. Values that serve as ideals are therefore the main elements of the cultural value orientation which strengthens the coherence of a given culture. Value orientations gradually adapt to changing external circumstances and internal processes. In every case, their examination focuses on the emphasis on values in deep layers of culture. According to Schwartz's research results, the cultural value orientation of societies can be distinguished according to three main dimensions (the corresponding values are given in parentheses):

(1) the autonomy of the individual (intellectual: broadmindedness, curiosity, or affective: pleasure)—and their embeddedness in society (social order, obedience, respect for tradition);
(2) egalitarianism (social justice, equality)—hierarchy (authority, humility);
(3) harmony with the world (unity with nature, world at peace)—mastery (ambition, daring).

Based on these, Schwartz was able to identify seven transnational cultural groups with specific characteristics differentiable along bipolar dimensions: West European, English-speaking, Latin American, East European, South Asian, Confucian, African, and Middle Eastern.

An Analysis of Ecological Consciousness and the Value Orientation of Businesses

Ecological consciousness and the value orientation of businesses are individually addressed in various ways in the literature, although publications typically do not examine the relationship between them. The following section summarizes how the personal values of individuals and the characteristics of decision making can affect the ethicality of decisions, and what attributes and conditions seem to be needed for corporate social responsibility and value-oriented, ecologically conscious business functioning.

(1) Multiple articles argue that the personal characteristics and values of individual decision makers do affect the ethicality of business decisions.

The value commitments of leaders can be decisive in terms of the corporate social responsibility of a company. The question then arises, is the term "corporate" social responsibility meaningful at all (Hemingway and Maclagan 2004)? The characteristics of individual decision makers include nationality, religion, gender, age, education, occupation, and type of personality. Of these, the impact of personal characteristics (gender, age, nationality, and religion), education and occupation, and the direction of the variable associations is variable. Among individual values, the impact of Machiavellism seems to be the strongest (Ford and Richardson 1994). Important individual factors that determine ethical decision making include individual moral philosophy, gender, age, education, occupation, nationality, and religion. Individual values, indeed, have a positive impact on professional ethics, but length of employment does not necessarily affect the ethical profile of individuals (Loe et al. 2000).

The moral beliefs, attitudes, and values of individuals form personal moral philosophies based on two dimensions: concern for morality, and practical consequences. Situationists strive for the "best" possible outcomes regardless of moral principles. Subjectivists base their judgments on personal values and practical concerns. Absolutists consider actions to be morally acceptable as long as they have positive consequences and comply with ethical rules. Exceptionists follow moral guidelines, albeit allowing some exceptions for practical reasons. Individuals can apply a variety of different personal moral philosophies, so business decisions cannot be uniformly evaluated. Therefore, the latter should not be the aim, but rather examining them from the (former) four perspectives, along with promoting the open discussion of ethical questions, and reaching a deeper understanding of business practices (Forsyth 1992).

The concept of "ethical leadership" varies from one culture to another. There are significant differences, which can be traced back to the different core values of different cultures. However, apart from such differences, the character of leaders, respect for others, following self-interest, and the abuse of power are prominent issues in all cultures, the handling of which requires virtuous behavior and credibility (Resick et al. 2011). Some of the important values for ethical leadership are consciousness, commitment, credibility, and perseverance. The environmental factors involved in decisions are embedded in the value of consciousness (Marsh 2013).

Leaders of large companies perceive stakeholder attributes involving power, legitimacy, and urgency as being of key importance, which affects the degree to which managers prioritize competing stakeholder voices (Agle et al. 1999). Stakeholder attributes considered relevant by sustainability-driven entrepreneurs may consist of legitimacy, philosophy, and impact. The former entrepreneurs tend to work with stakeholders who comply with legitimate societal rules, who share their philosophy and basic values, and who have significant potential impact in terms of generating change in ecological and social areas of the economy (Schlange 2006). The personal values of managers and entrepreneurs do influence the operating- and decision-related processes of companies, which can increase social and environmental performance (Agle et al. 1999; Schlange 2006).

The evaluation of sustainability issues depends on the moral maturity of decision makers and the extent to which they consider sustainability issues to be moral issues. An individual with strong sustainability values is intrinsically motivated to consider sustainability issues as important company strategic issues, and to even act accordingly. The sustainability-related values of an individual determine how morally important they consider sustainable-development-related issues to be (Eberhardt-Toth and Wasieleski 2013).

However, based on the social cognitive theory of self-regulation (Bandura 1991), it is not only values but also personal agency that influence the decisions and actions of managers and entrepreneurs. Even if they have pro-environmental values, they can assess business opportunities that cause harm to the natural environment as attractive and appropriate for exploitation. Decision-makers' cognitive disengagement and violation of their own values seem to be stronger when the former have high entrepreneurial self-efficacy and perceive a state of industry munificence, but resource abundance as low (Shepherd et al. 2013).

Some consider ethical decision making to be no different from other types of decision making. The reason is that decision makers themselves tend not to differentiate in relation to whether they are making ethical or other types of decisions (Elm and Radin 2012). Based on shareholder theory, ethical company behavior toward stakeholders may be possible. The well-being of non-shareholder stakeholders is also compatible with the ethical values described in shareholder theory based on Kantian "duties of rights." To put this into practice, the principles of corporate governance need to be well established, and shareholders should take greater responsibility for the activities of their companies (Mansell 2013).

(2) The characteristics of decision making also affect the ethicality of decisions.

Features of decision-making situations may include reference groups, reward and punishment, codes of conduct, types of ethical problems, organizations, industries, and business competitiveness. Moreover, individuals typically consider themselves to be more ethical than other reference persons. They increasingly identify themselves with the top management's belief system, as they are encouraged by the company reward system. Codes of conduct and an ethical corporate culture have a positive effect on the ethics of individual behavior, but larger company size and higher position have unfavorable effects. Interestingly, the ethicality of the industry overall does not really affect individual decision making, but competitiveness does (Ford and Richardson 1994).

Important organizational factors that influence ethical decision making are codes of ethics, remuneration systems, and corporate culture. A code of ethics mostly expresses the values of top management. The marketing orientation and marketing performance of the company is also an important aspect of ethical decision making (Loe et al. 2000).

(3) What are the characteristics of and conditions required for corporate social responsibility and value-oriented, ecologically conscious functioning?

There are different approaches to corporate social responsibility, such as instrumental, political, integrative, and ethical. These can be distinguished along four dimensions: financial gain, political performance, social needs, and ethical values. The ethical theories that best take into account the natural environment and ecology are stakeholder theory, theories of universal rights, sustainable development, and the public good (Garriga and Melé 2004). Managers' personal value commitments and the changes they make can stimulate corporate social responsibility, not only those corporate policies and business and strategic considerations (image, reputation, stakeholder management, integration into the local community) which are important for the company as a whole (Hemingway and Maclagan 2004).

The strategic, business, and instrumental use of the corporate social responsibility concept does not make it truly sustainable. It is necessary to rethink companies, governance, and organizations, and to reestablish them, therein ensuring the economic system does not destroy, but preserves and restores the natural environment, reestablishes the true meaning of work, and supports valuable innovation and well-being that respects natural boundaries (Milton 2010).

Ecopreneurship is a new approach to creating environmentally and socially responsible and economically viable businesses. Business models based on linking entrepreneurialism and environmentalism may need reciprocal relationships with corporate, community, social, and governmental partners to promote and maintain economic sustainability (Dixon and Clifford 2007). Sustainability-driven entrepreneurs strive to balance the requirements of the triple bottom line and create sustainable values in the most general sense. They can be characterized as having an all-embracing perception that involves integrating the perspectives of their stakeholders, as identified in a comprehensive and future-oriented way (Schlange 2006).

Respect for the natural environment does not appear or is not emphasized in numerous articles. The lack of consideration of the natural environment (which appears indirectly, involving a narrow or short-term focus), is identified in a paper by Resick and co-authors (2011). It is worth examining why ecology is of such little importance to leaders. In Ford and Richardson (1994), Loe et al. (2000), Hemingway and Maclagan (2004), and Marsh (2013), the natural environment and ecology are not identified as particularly prominent parts of ethical action, leadership, and responsible company behavior. Forsyth (1992) describes the constraints imposed by the social environment on the economic system, which should be supplemented by examining the constraints that result from the vulnerability and finite nature of the natural environment. Reflecting on research by Kopfer-Rácz et al. (2013) about the socio-cultural elements of Hungarian SMEs, it is interesting to consider how much the orientation on the short-term material success of Hungarian small enterprises is compatible with the long term, ecological mindset necessary for operating in an ecologically conscious way.

Further examination of the four types of moral beliefs (situationist, subjectivist, absolutist, and exceptionist) that underlie business decisions are needed to more deeply explore the characteristics of ethical decision making and, on the basis of this, generate a better understanding of economic decisions from several perspectives (Forsyth 1992). According to Ford and Richardson (1994), the number of empirical studies about the factors that influence ethical decision making is low (especially in terms of the variables education, age, duration of employment, and income), and some factors have not been studied at all so far (for instance, family status, number of children, and type of career). In addition to studying demographic characteristics, it would be advisable to supplement the previous results with some attitude-related research (i.e., involving work or organizational commitment), decision-making characteristics (in terms of the existence of external regulations, the economic environment, and industry competitiveness), and values at the deep level of company cultures. It may also be useful to develop clearer terminology (such as what ethical behavior means), and to further test existing models. Loe, Ferrell, and Mansfield (2000) suggest that the study of ethical issues related to decision making should be extended to other research areas (such as marketing orientation, quality, and performance); and that longitudinal studies are required to monitor temporal changes, while understanding the broader responses of industry may increase analytical validity and applicability. Further research about the components of intentions and moral character of decision makers also seems to be indispensable.

Based on Mansell's (2013) conclusions, it remains an open question how much the practical design and operation of ethical company policies depend on the moral character and experience of employees; and what the impact of the interests, well-being, and happiness of stakeholders means in practice. In connection with moral character, Eberhardt-Toth and Wasieleski (2013) examined the moral judgment of French financial professionals but not their actual actions. There is no clear correlation between these factors, and the authors consider it to be worth further research. Hemingway and Maclagan (2004) suggest the further study of to what extent individual characteristics (for example, insight, and influence) contribute to the social responsibility of organizations. It is also a methodological question how to separate individual and company values, and how to empirically collect data about interactions between individuals, and individuals and companies.

Elm and Radin's (2012) initial qualitative examination that involved research with decision makers of more value-oriented and profit-oriented businesses helps confirm whether ethical and other types of decision making are actually different. The outcome is also related to the size of the sample, but the fact that the investigated leaders were not selected through representative but purposeful or expert sampling limits the generalizability of Marsh's (2013) study.

Bárth-Fehér (2012), Győri (2012), and Szegedi (2012) are not concerned about how to evaluate the instrumental use of ethics and ethical systems. Milton (2010) leaves open the question how the concept of corporate social responsibility can translate into genuine value commitment. On the topic of corporate social responsibility, Garriga and Melé (2004) propose a more detailed examination of the four dimensions of corporate social responsibility theories (instrumental, political, integrative, and value-based), and possibly the development of a new theory that exceeds the latter in scope.

Finally, based on experience, a strong value orientation is typically found to exist in small businesses, so it may be interesting to compare the value commitments of value-oriented companies with the average socio-cultural attitudes of micro, small-, and medium-sized enterprises (Kopfer-Rácz et al. 2013).

Values and the Value Orientation of Businesses

According to Peter French (1979), companies themselves have a moral personality, while Kenneth Goodpaster (1990, quoted by Brenkert, 1995) draws attention to the existence of their environmental conscience. Patricia Werhane (1985) also views companies as moral actors because their actions are based on the intentions of morally responsible natural persons; that is, the moral inputs of their constituents.

In this book, we identify the values and value orientations of businesses with values revealed during case analyses and personal interviews with the official representatives of such businesses. The results of previous studies (Kaldschmidt 2011; Zsolnai and Győri 2011; Győri and Ócsai 2014) and several pieces of literature (Ford and Richardson 1994; Meglino and Ravlin 1998; Agle et al. 1999; Loe et al. 2000; Grojean et al. 2004; Hemingway and Maclagan 2004; Hemingway 2005; Schlange 2006; Shepherd et al. 2013) support the suitability of this method. These studies confirm that the personal values of founders, owners, and managers of companies have a decisive influence on the values and value orientation of their businesses. However, in order to broaden the focus of the topic, among future research directions it may be interesting to flesh out the values and value orientations of businesses by examining the values and value orientations of other organizational actors, including employees.

References

Agle, B. R., Mitchell, R. K., & Sonnenfeld, J. A. (1999). Who Matters to CEOs? An Investigation of Stakeholder Attributes and Salience, Corporate Performance, and CEO Values. *Academy of Management Journal, 42*(5), 507–525. https://doi.org/10.5465/256973.

Bandura, A. (1991). Social Cognitive Theory of Self-regulation. *Organizational Behavior and Human Decision Processes, 50*(2), 248–287. https://doi.org/10.1016/0749-5978(91)90022-L.

Bansal, P. (2003). From Issues to Actions: The Importance of Individual Concerns and Organizational Values in Responding to Natural Environmental Issues. *Organization Science, 14*(5), 510–527. https://doi.org/10.1287/orsc.14.5.510.16765.

Bárth-Fehér, Sz. (2012). Fenntarthatóság a hazai vállalati gyakorlatban [Sustainability in the practice of Hungarian firms], Vezetéstudomány [Budapest Management Review], Vol. *XLIII, no., 10,* 44–54. [in Hungarian].

Brenkert, G. G. (1995). The Environment, The Moralist, The Corporation and Its Culture. *Business Ethics Quarterly, 5*(4), 675–697. https://doi.org/10.2307/3857409.

Business Dictionary. (2020). Values. http://www.businessdictionary.com/definition/values.html. Accessed 31 March 2020.

Byrne, G. J., & Bradley, F. (2007). Culture's Influence on Leadership Efficiency: How Personal and National Cultures Affect Leadership Style. *Journal of Business Research, 60*(2), 168–175. https://doi.org/10.1016/j.jbusres.2006.10.015

Csutora, M., & Kerekes, S. (2004). A környezetbarát vállalatirányítás eszközei [Instruments of Environmentally Friendly Corporate Governance]. KJK Kerszöv, Budapest [in Hungarian].

Dixon, S. E. A., & Clifford, A. (2007). Ecopreneurship—A New Approach to Managing the Triple Bottom Line. *Journal of Organizational Change Management, 20*(3), 326–345. https://doi.org/10.1108/09534810710740164.

Dyllick, T., & Muff, K. (2015). Clarifying the Meaning of Sustainable Business: Introducing a Typology from Business-as-Usual to True Business Sustainability. *Organization & Environment, 29*(2), 156–174. https://doi.org/10.1177/1086026615575176.

Eberhardt-Toth, E., & Wasieleski, D. M. (2013). A Cognitive Elaboration Model of Sustainability Decision Making: Investigating Financial Managers' Orientation Toward Environmental Issues. *Journal of Business Ethics, 117*(4), 735–751. https://doi.org/10.1007/s10551-013-1715-1.

Elm, D. R., & Radin, T. J. (2012). Ethical Decision Making: Special or No Different? *Journal of Business Ethicsv, 107*(3), 313–329. https://doi.org/10.1007/s10551-011-1041-4.

Encyclopaedia Britannica. (2020). Axiology. https://www.britannica.com/topic/axiology. Accessed 2 April 2020.

Ford, R. C., & Richardson, W. D. (1994). Ethical Decision Making: A Review of the Empirical Literature. *Journal of Business Ethics, 13*(3), 205–221. https://doi.org/10.1007/bf02074820.

Forsyth, D. R. (1992). Judging the Morality of Business Practices: The Influence of Personal Moral Philosophies. *Journal of Business Ethics, 11*(5–6), 461–470. https://doi.org/10.1007/bf00870557.

French, P. A. (1979). The Corporation as a Moral Person. *American Philosophical Quarterly, 16*(3), 207–215. https://www.jstor.org/stable/pdf/20009760.pdf. Accessed 12 April 2020.

Garriga, E., & Melé, D. (2004). Corporate Social Responsibility Theories: Mapping the Territory. *Journal of Business Ethics, 53*(1–2), 51–71. https://doi.org/10.1023/b:busi.0000039399.90587.34.

Gast, J., Gundolf, K., & Cesinger, B. (2017). Doing Business in a Green Way: A Systemic Review of the Ecological Sustainability Entrepreneurship Literature and Future Research Directions. *Journal of Cleaner Production, 147*, 44–56. https://doi.org/10.1016/j.jclepro.2017.01.065.

Goodpaster, K. E. (1990). Can a Corporation Have an Environmental Conscience? In W. M. Hoffman, R. Frederick, & E. S. Petry (Eds.), *Business, Ethics, and the Environment*. New York: Quorum Books.

Grojean, M. W., Resick, C. J., Dickson, M. W., & Smith, D. B. (2004). Leaders, Values, and Organizational Climate: Examining Leadership Strategies for Establishing an Organizational Climate Regarding Ethics. *Journal of Business Ethics, 55*(3), 223–241. https://doi.org/10.1007/s10551-004-1275-5.

Győri, Zs. (2012). Első- és másodfajú etikai kudarcok [First Kind and Second Kind Ethical Failures]. Vezetéstudomány [Budapest Management Review]. Vol. XLIII, No. 10, pp. 56–62. http://unipub.lib.uni-corvinus.hu/945/1/vt_2012n10p56.pdf. Accessed 2 April 2020. [in Hungarian].

Győri, Zs., & Ócsai, A. (2014). Ecologically-oriented Enterprises in Hungary. World Review of Entrepreneurship, Management and Sustainable Development, Vol. 10, No. 1. pp. 52–65. https://doi.org/10.1504/wremsd.2014.058053.

Hemingway, C. A. (2005). Personal Values as a Catalyst for Corporate Social Entrepreneurship. *Journal of Business Ethics, 60*(3), 233–249. https://doi.org/10.1007/s10551-005-0132-5.

Hemingway, C. A., & Maclagan, P. W. (2004). Managers' Personal Values as Drivers of Corporate Social Responsibility. *Journal of Business Ethics, 50*(1), 33–44. https://doi.org/10.1023/b:busi.0000020964.80208.c9.

Hills, M. D. (2002). Kluckhohn and Strodtbeck's Values Orientation Theory. *Online Readings in Psychology and Culture, 4*(4). https://doi.org/10.9707/2307-0919.1040.

Hitlin, S., & Piliavin, J. A. (2004). Values: Reviving a Dormant Concept. *Annual Review of Sociology, 30*(1), 359–393. https://doi.org/10.1146/annurev.soc.30.012703.110640.

Hoffman, W. M. (1991). Business and Environmental Ethics. *Business Ethics Quarterly, 1*(2), 169–184. https://doi.org/10.2307/3857261.

Hofstede, G., Hofstede, G. J. (2005). *Cultures and Organizations: Software of the Mind: Intercultural cooperation and its importance for survival*. McGraw-Hill, New York.

Hofstra, N. (2015). Entrepreneurship Inspired by Nature. In L. Zsolnai (Ed.), *The Spiritual Dimension of Business Ethics and Sustainability Management* (pp. 159–166). New York: Springer. https://doi.org/10.1007/978-3-319-11677-8_13.

Inglehart, R., & Baker, W. E. (2000). Modernization, Cultural Change, and the Persistence of Traditional Values. *American Sociological Review, 65,* 19–51. https://doi.org/10.2307/2657288.

Kaldschmidt, S. (2011). The Values of Sustainability: The Influence of Leaders' Personal Values on Sustainability Strategies. Dissertation of the University of St. Gallen, School of Management, Economics, Law, Social Sciences and International Affairs.

Karp, D. G. (1996). Values and Their Effect on Pro-environmental Behavior. *Environment and Behavior, 28*(1), 111–133. https://doi.org/10.1177/0013916596281006.

Kerekes, S., & Wetzker, K. (2007). Keletre tart a "társadalmilag felelős vállalat" koncepció [The Concept of a "Socially Responsible Company" is Heading East], Harvard Business Manager: Korszerű menedzsment elmélet és megvalósítás [Advanced Management Theory and Implementation] Vol. 9, No. 4, pp. 37–47. [in Hungarian].

Klenke, K. (2005). Corporate Values as Multi-level, Multi-domain Antecedents of Leader Behaviors. *International Journal of Manpower, 26*(1), 50–66. https://doi.org/10.1108/01437720510587271.

Kluckhohn, C. (1951). Values and Value-orientations in the Theory of Action. In T. Parson & E. A. Shils (Eds.) (2013): Toward a General Theory of Action. Harvard Univesrity Press, Cambridge. https://doi.org/10.4159/harvard.9780674863507.c8.

Kluckhohn, F. R., & Strodtbeck, F. L. (1961). *Variations in Value Orientations.* Evanston: Row, Peterson.

Kopfer-Rácz, K., Hofmeister-Tóth, Á., & Sas, D. (2013). A hazai kis- és közepes vállalatok szociokulturális beállítódása a Hofstede-dimenziók mentén [The Socio-cultural Attitudes of the Hungarian Small and Medium

Enterprises along the Hofstede Dimensions], Vezetéstudomány [Budapest Management Review], Vol. XLIV, No. 10, pp. 2–11. http://unipub.lib.uni-corvinus.hu/1353/1/vt_2013n10p2.pdf. Accessed 12 April 2020 [in Hungarian].

Krishnan, V. R. (2001). Values Systems of Transformational Leaders. *Leadership and Organizational Development Journal, 22*(3), 126–132. https://doi.org/10.1108/01437730110389274.

Loe, T. W., Ferrell, L., & Mansfield, P. (2000). A Review of Empirical Studies Assessing Ethical Decision Making in Business. *Journal of Business Ethics, 25*(3), 185–204. https://doi.org/10.1023/a:1006083612239.

Lord, R. G., & Brown, D. J. (2001). Leadership, Values, and Subordinate Self-concepts. *Leadership Quarterly, 12*(2), 133–152. https://doi.org/10.1016/s1048-9843(01)00072-8.

Mansell, S. (2013). Shareholder Theory and Kant's 'Duty of Beneficence'. *Journal of Business Ethics, 117*(3), 583–599. https://doi.org/10.1007/s10551-012-1542-9.

Marsden, T., & Smith, E. (2005). Ecological Entrepreneurship: Sustainable Development in Local Communities Through Quality Food Production and Local Branding. *Geoforum, 36*(4), 440–451. https://doi.org/10.1016/j.geoforum.2004.07.008.

Marsh, C. (2013). Business Executives' Perceptions of Ethical Leadership and Its Development. *Journal of Business Ethics, 114*(3), 565–582. https://doi.org/10.1007/s10551-012-1366-7.

Meglino, B. M., & Ravlin, E. C. (1998). Individual Values in Organizations: Concepts, Controversies, and Research. *Journal of Management, 24,* 351–389. https://doi.org/10.1177/014920639802400304.

Milton, S. (2010). Some Observation on CSR and Strategic Management, Vezetéstudomány [Budapest Management Review]. Vol. XLI, No. 2, pp. 59–67. http://unipub.lib.uni-corvinus.hu/2697/1/vt2010n2p59-67.pdf. Accessed 2 April 2020.

Nonis, S., & Swift, C. O. (2001). Personal Value Profiles and Ethical Business Decisions. *Journal of Education for Business, 76*(5), 251–256. https://doi.org/10.1080/08832320109599644.

O'Riordan, T., & Cameron, J. (1994). The History and Contemporary Significance of the Precautionary Principle. In T. O'Riordan & J. Cameron (Eds.), *Interpreting the Precautionary Principle* (pp. 1–30). London: Earthscan.

Parrish, B. D. (2010). Sustainability-driven Entrepreneurship: Principles of Organization Design. *Journal of Business Venturing, 25*(5), 510–523. https://doi.org/10.1016/j.jbusvent.2009.05.005.

Resick, C. J., Martin, G. S., Keating, M. A., Dickson, M. W., Kwong Kwan, H., & Peng, C. (2011). What Ethical Leadership Means to Me: Asian, American, and European Perspectives. *Journal of Business Ethics, 101*(3), 435–457. https://doi.org/10.1007/s10551-010-0730-8.

Rohan, M. J. (2000). A Rose by Any Name? The Value Construct. *Personality and Social Psychology Review, 4*(3), 255–277. https://doi.org/10.1207/s15327957pspr0403_4.

Rokeach, M. (1973). *The Nature of Human Values*. New York: The Free Press.

Rokeach, M. (1979). *Understanding Human Values: Individual and Societal*. New York: The Free Press.

Sagiv, L., & Schwartz, S. H. (2007). Cultural Values in Organizations: Insights for Europe. *European Journal of International Management, 1*(3), 176–190. https://doi.org/10.1504/ejim.2007.014692.

Schein, E. H. (2010). *Organizational Culture and Leadership*. San Francisco: Jossey-Bass.

Schlange, L. E. (2006). Stakeholder Identification in Sustainability Entrepreneurship: The Role of Managerial and Organisational Cognition. *Greener Management International, 55*(13), 13–32. www.jstor.org/stable/greemanainte.55.13. Accessed 18 April 2020.

Schwartz, S. H. (1992). Universals in the Content and Structure of Values: Theoretical Advances and Empirical Tests in 20 Countries. In M. Zanna (Ed.), *Advances in Experimental Social Psychology* (Vol. 25, pp. 1–65). New York: Academic Press. https://doi.org/10.1016/S0065-2601(08)60281-6.

Schwartz, S. H. (2006). A Theory of Cultural Value Orientations: Explication and Applications. *Comparative Sociology, 5*(2–3), 137–182. https://doi.org/10.1163/156913306778667357.

Schwartz, S. H., & Bilsky, W. (1987). Toward a Universal Psychological Structure of Human Values. *Journal of Personality and Social Psychology, 53*(3), 550–562. https://doi.org/10.1037//0022-3514.53.3.550.

Shepherd, D. A., Patzelt, H., & Baron, R. A. (2013). "I Care About Nature, but…": Disengaging Values in Assessing Opportunities That Cause Harm. *Academy of Management Journal, 56*(5), 1251–1273. https://doi.org/10.5465/amj.2011.0776.

Szegedi, K. (2012). Hazai nagyvállalati etikai kódexek tartalmi elemzése [Content Analysis of Code of Ethics of Hungarian Large Companies]. Vezetéstudomány [Budapest Management Review]. Vol. XLIII, No. 1, pp. 47–55. http://unipub.lib.uni-corvinus.hu/2548/1/vt2012n1ksz-47.pdf. Accessed 3 April 2020 [in Hungarian].

Tóth, G. (2007). A Valóban Felelős Vállalat. A fenntarthatatlan fejlődésről, a vállalatok társadalmi felelősségének (CSR) eszközeiről és a mélyebb stratégiai megközelítésről [The Truly Responsible Company. On Unsustainable Development, Corporate Social Responsibility (CSR) Tools and a Deeper Strategic Approach], KÖVET, Budapest [in Hungarian].

Werhane, P. H. (1985). *Persons, Rights, and Corporations*. New York: Prentice Hall, Englewood Cliffs.

Zsolnai, L. (2001). *Ökológia, gazdaság, etika [Ecology, Economy, Ethics]*. Budapest: Helikon Kiadó. [in Hungarian].

Zsolnai, L., & Győri, Zs. (2011). Can Ethics Survive in Competitive Environments? Empirical Evidence from Values-Driven Businesses, European Business Ethics Network 24th Annual Conference. 15–17 September 2011.

Part III

Working Models of Ecologically Conscious Businesses

This part of the book first introduces the main characteristics of the empirical research and two samples of ecologically conscious business organizations. It then presents and analyzes the value orientations of the selected enterprises, and reveals how they define their raison d'être and what they mean by success. The second chapter in this section reviews the elements of their business models.

5

Ecologically Conscious Business Organizations and Their Value Orientations

The ancient philosophers Plato, Aristotle, and Confucius perceived the essence of human existence as creation; that is—in economic terminology—production, not consumption. However, production may be associated not only with material goods, but also with the creation of higher values (Allinson 2009, 2011). In light of this perspective, how can we relate to the consumer society of today, the "values" it conveys, growth-oriented economies, and the profit-oriented businesses that serve them? Can we create a state of general human well-being and ecological sustainability, in which we can achieve the individual goals that make human existence worthwhile, based on a materialistic conception of the economy? This is hardly possible using the materialistic and mainstream economic and business models with which we have almost completely ruined the biosphere (Zsolnai 2001; Pataki and Takács-Sánta 2005; Magnuson 2007).

What can we do to avoid and even restore the devastation of Earth? We need to make sure that economic activity does not occur for its own sake, but exists to serve people. It should contribute to the creation of real, positive values for society and help reduce and eliminate negative impacts. For this to happen, financial profit and growth should not

A. Ócsai, *Ecologically Conscious Organizations*, Palgrave Studies in Sustainable Business In Association with Future Earth,
https://doi.org/10.1007/978-3-030-60918-4_5

be considered goals, but only means (Allinson 2009). In addition, to preserve the Human-Earth system an ecologically sensitive economy is needed that recognizes the integrity of natural beings and the dignity of their existence. However, this can only be imagined through an ecological turn by which humans become reconciled with themselves and reestablish a covenant with nature (Zsolnai 2001). The necessary ecological turn can be fostered if modern man's distorted attitude and behavior incompatible with sustainability is replaced by a spiritual value orientation that represents unity, compassion, and a demeanor of self-discipline (Lázár 2011).

Earlier research has shown that a positive relationship between a genuine value orientation and economically successful business operations that pay special attention to environmental impacts does exist. Continuing in this direction, this chapter examines the value orientations of some ecologically conscious business organizations to reveal the core values behind their functioning. These businesses, which make ecological considerations the center of their operations, can thus be considered ecologically conscious and capable of and willing to operate in alternative ways to the organizations that currently prevail in our materialistic economic system.

This chapter introduces the main methodological features of the empirical research underlying this book, and presents two samples of ecologically conscious business organizations. One sample consists of eight internationally well-known ecologically conscious enterprises from all over the world, while the other sample contains fourteen well-known ecologically conscious businesses based in Hungary. They are briefly described to familiarize readers with this special set of business organizations.

On Empirical Investigation

In the current phase of research of the value orientations of ecologically conscious businesses, the generalization of results is not the primary goal, but obtaining a deeper understanding of the selected cases. We can examine the characteristics of the chosen businesses and the factors that influence their operations primarily through a constructivist epistemological approach (Creswell 2007; Maréchal 2010) and by applying so-called "less structured qualitative research" methodology (Cassell and Symon 2004; Miles et al. 2014). Further methodological issues in the empirical research process involve the units of observation, the sampling procedure, data collection and analysis, the reliability and validity of the analysis, and ethical questions (Kvale 1994; Hunyadi et al. 1997; Mason 2002; Patton 2002; Krippendorff 2004; Creswell 2007; Babbie 2008; Maxwell 2013; Miles et al. 2014).

The book focuses on the population of ecologically conscious business organizations, of which two samples are examined. The reason for selecting and analyzing two samples is the hypothetical difference between the two subsets, and the (partly) different methods of data collection and analysis that were used. One sample consists of eight internationally well-known ecologically conscious companies from around the world. In these cases, the enterprises were selected as the units of observation. A list of the businesses in the international sample and related data is provided in Table 5.1.

Table 5.1 Main characteristics of ecologically conscious businesses in the international sample

Company name	Seat (country)	Main product/service	Year of foundation
Green Monday	Hong Kong	Plant-based foods, lifestyle	2012
Houdini Sportsware	Sweden	Functional outdoor clothing	1993
Interface	USA	Modular carpet	1973
Natura	Brazil	Natural cosmetics, direct marketing	1969
Organic India	India	Organic teas, herbal products	1997
Patagonia	USA	Outdoor clothing, apparel	1973
Scott Bader	Great Britain	Composites, adhesives, polymers	1921
TerraCycle	USA	Waste collection, recycling	2001

Source Author's construction

The other sample contains fourteen well-known ecologically conscious business organizations based in Hungary. In the case of these companies, the official representatives (owners and/or managers) were the units of observation as we undertook personal interviews with them. A list of the businesses in the Hungarian sample and related data are included in Table 5.2.

Due to the specific nature of the topic, requiring special knowledge, samples were selected through expert or purposeful sampling (Hunyadi et al. 1997; Babbie 2008). They were built up based on personal- and e-mail discussions with distinguished scholars and practitioners from

Table 5.2 Main characteristics of ecologically conscious businesses in the Hungarian sample

Company name	Seat (city in Hungary)	Main product/service	Year of foundation
Skanska Hungary	Budapest	Construction	1987
Nekedterem Hungary	Budapest	Home delivery of local food	2012
Réthy Fashion	Szigetszentmiklós	Fashion design from recycled jeans	2012
Bio-Drog-Berta	Kalocsa	Organic spices, food products	1995
Ecoizm/Zöldbolt	Veszprém	Trade in organic household products	2007
ALTEO Energy	Budapest	Alternative energy production	2008
Hajtás Pajtás	Budapest	Bike courier	1999
Pannonhalmi Abbey Winery	Pannonhalma	Wine and herbal products	2002
NES Shoes	Budapest	Ecological footwear design	2008
Béres Pharmaceuticals	Budapest	Alternative medicine	1993
Kör Architect Studio	Budapest	Architectural design	1989
Pipacs Bakery	Budapest	Organic bread	2014
Pendits Vineyards and Winery	Abaújszántó	Biodynamic wine	1991
MagNet Bank	Budapest	Banking	1995

Source Author's construction

the field (including individuals from the University of Hong Kong; the Norwegian School of Economics, Bergen, Norway; the University of Victoria, Canada; and Corvinus University of Budapest, Eötvös Loránd University, and the University of Pannonia, the latter three from Hungary).

Because of the different characteristics of the companies in the two samples and the availability of the representatives of the Hungary-based enterprises, the two samples were processed using different methods. In the case of the internationally well-known ecologically conscious businesses, the main goal was to depict the value profiles of each company based on publicly available information. With the Hungary-based enterprises, personal contacts were established, and through personal interviews more detailed data sets could be collected and deeply analyzed.

Two types of data collection methods were applied. On the one hand, company websites, publicly available official data, company documents, financial statements, corporate social responsibility or sustainability reports, codes of ethics and conduct, annual reports, information from external sources (media, authorities, customers, and industry peers) were gathered about all the companies in both samples. Semi-structured personal interviews with the owners and managers of the Hungary-based ecologically conscious businesses and one of the international cases (personal conversation with Tom Szaky, founding owner of TerraCycle at a company presentation at the Corvinus University of Budapest, Hungary in 2016) were also conducted (Kvale 1996, 2006; Mason 2002; King 2004; Barlow 2010).

A questionnaire with strictly predefined content and closed questions was not used in the interviews with Hungarian companies, but rather an interview guide (Appendix 1). The related questions were formulated based on the literature related to the research topic, preliminary discussions with experts from the field, and the author's previous knowledge and research experience. The six main topics of the questions were: (1) the characteristics of the business; (2) the interpretation and value orientation of the former regarding ecological consciousness; (3) ecological aspects of the operation of the business; (4) the relationship between competition and ecology; (5) the company's assessment of the potential benefits of ecological consciousness; and (6) business perspectives about ecological consciousness in the Hungarian economy. When a Likert scale was used by an interviewee to answer a question, they were also asked to give a detailed explanation of their response. Thus through the combined use of techniques typical of semi-structured and structured interviews, the advantages of the richness of detail of a qualitative approach could be preserved, but, bearing in mind the limitations of Likert-scale based surveys,[1] answers also became—at least partially—formally comparable and depictable (Barlow 2010; Robbins and Heiberger 2011).

Research into the value orientations of the Hungary-based ecologically conscious businesses was carried out using both deductive and inductive approaches. Analysis of the importance of ecological values[2] followed a deductive logic, as the latter were identified from the theory. Identification of elements of business models and definitions of the business raison d'être and success were similar to a deductive approach because they were designed to map the emergence of preexisting categories. However, transcripts of the interviews, field notes, and documents used as secondary data sources were analyzed using inductive logic through qualitative content analysis (Patton 2002; Krippendorff 2004; Hsieh and Shannon 2005; Bowen 2009; Stan 2010) and thematic analysis (Boyatzis 1998; Fereday and Muir-Cochrane 2006; Lapadat 2010). During the content analysis, starting with texts, patterns that could be used to characterize the value background, motivation, culture, and mindset of the enterprises were sought out. Thirty-six patterns were identified, and aggregate clusters were formed from those patterns more similar to each other.

[1]The limitations of measuring using a Likert scale include, but are not limited to, the following:

1. The Likert scale is an ordinal scale, thus data obtained with the instrument can not be used to perform mathematical-statistical operations (e.g., calculation of arithmetic and geometric means and standard deviations) that can be applied to higher levels of measurement such as interval scales or ratio scales (Hunyadi et al. 1997; Jamieson 2004);
2. Answers to questions may suffer from multiple distortions, such as social desirability, acquiescence, and a centralizing tendency (Cassell and Symon 2004; Babbie 2008; McLeod 2008; Mills et al. 2010).

[2]Thirteen ecological values were predefined through a theoretical literature review. These encompass humility, biodiversity, non-possession, trust, responsibility, holistic approach, temperance, respect for ecosystems, reciprocity, love, natural embeddedness, intrinsic value of natural beings, and openness to mystery.

Qualitative research strives for the greatest possible objectivity, reliability, and validity. In order to increase objectivity, the author's value judgments, assumptions, and goals were explored and presented, and the research process was documented in detail. Data recording and storage were based on the precautionary principle, so that the entire process could be verified and replicated. To increase reliability, different types of triangulation (data source triangulation, and inter-researcher triangulation) were employed. In order to enhance the validity of the research findings, quotations taken from texts of documents and interviews that supported the findings were added, and data analysis was conducted according to both a deductive and inductive approach (Krippendorff 2004; Miles et al. 2014). This qualitative research, which used only two smaller samples, was not designed to be representative, but rather to help with exploring and synthesizing the details of the selected cases. This research strategy can produce new insights of use in further research, and lead to suggestions for business practices.

Consideration of ethical issues was not a separate factor, but a key aspect that was taken into account during every part of the research design and implementation (Maxwell 2013). Interviewees received all the necessary information in time and in sufficient detail to enable them to make a decision about their participation in the research (i.e., to give their informed consent). During the organization and conduct of interviews, interruption of the work of interviewees was minimized. All the information we received was treated confidentially. Throughout the research process, open and honest communication was maintained, and potential information and power asymmetry were carefully considered and avoided (Kvale 2006; Creswell 2007).

The findings of this book should be seen in the light of some of the limitations of the empirical research that underlie them. To answer the research questions, two small samples were used, meaning that it is not possible to consider the research results representative of the total population of ecologically conscious organizations. However, the initial goal during the current phase of research was not to create generalizable results, but to construct a deeper understanding of the selected cases, their components, and relationships. This is why an exploratory research design and a qualitative approach were applied. Capacity and time constraints did not permit the application of longitudinal research, which could have further refined the results of the study. In addition, sampling bias must be acknowledged, as in the international sample only world-renowned companies were studied, but because of local knowledge, even very small businesses were examined in the Hungarian sample. Some of the differences between the ecologically conscious businesses in the international and the Hungarian sample may be explained through this sampling distortion.

Value Orientations of Ecologically Conscious Business Organizations in the International Sample

This section presents the value orientations of ecologically conscious businesses in the international sample using the same structure in each case. A short introduction is followed by a summary of the history of the organization and its core values.

Green Monday

Green Monday is a social start-up group dedicated to tackling climate change and global food security. David Yeung and Francis Ngai founded it on Earth Day in 2012. Two years later, Fast Company Magazine selected it as one of China's 50 Most Innovative Companies. It promotes low-carbon, sustainable lifestyles and works to make the former simple, easy to implement, and disseminate. It functions as a platform that brings together the efforts of companies, restaurants, schools, and members of the public to promote social and environmental responsibility. As Asia's first plant-based grocery store, its Green Common initiative aims to provide its customer community with food that is produced and marketed in a sustainable, innovative, healthy, and ethical way (Green Monday 2020).

History: The basic idea of Green Monday is the recognition that the current system of food production contributes as much to global warming and the degradation of the natural environment as heavy industry, transport, or fossil fuel-based electricity generation (Green Monday 2020). According to a 2006 report by the Food and Agriculture Organization of the United Nations, meat-based food production is responsible for 18 per cent of the world's annual total greenhouse gas emissions (mainly carbon dioxide, methane, and nitrous oxide). This is more than the combined emissions of cars, rail, and aviation (FAO 2006). Accordingly, eating a vegetarian meal is less harmful to the environment, and even having a meat-free day once a week can significantly reduce one's carbon footprint. Green Monday achieved significant success even in its founding year by developing vegetarian menus for Hong Kong's number one fast-food network. In cooperation with twenty leading food service providers, it has launched a school food program for 600,000 students in more than 800 public and higher education institutions. In 2013, it also started offering its services at the University of Hong Kong and Hong Kong International Airport. In 2014, a Green LUCK food aid program was launched, and nearly a quarter of all college students at Columbia University chose the Green Monday menu. In 2015, the food the organization offers became available at Washington University (St. Louis), University of California, Los Angeles (UCLA), a German and an English university, HSBC, a number of restaurant chains, and more than 300,000 students at 39 schools in Albania. It also opened several grocery stores as part of the Green Common business project. During 2016, additional stores were opened, sales channels expanded, and own-brand products were launched. Green Monday also participated in organizing the Earth Day Forum and first TV show on vegetarianism in the US program, while the Green Monday Ventures business services initiative also commenced (Green Monday 2020).

Core values: Green Monday's vision is to be a global player in relation to sustainable public good. Its mission as an innovative and multifaceted social business platform is to guide individuals, communities, and companies to lead a more sustainable, healthy, and conscious lifestyle that supports the fight against climate change, and promotes food safety, health, and animal welfare. Green Monday emphasizes the beneficial effects of vegetarianism, but does not expect people to make a sudden and drastic dietary transition (Green Monday 2020). One reason for this is that the founder, David Yeung, as a practicing Buddhist, seeks to reduce suffering. He encourages people not to strive to fully give up eating meat, but only to start to reduce the amount they consume (Zimberoff 2017). Green Monday sets an example, and is constantly informing its customers and the public that eating meat-free meals on one day a week is an easy and effective way to reduce their carbon footprint and improve their health. The name of the company expresses the fact that Monday, the first day of the working week, fills the body and mind with new energy, giving new impetus to a new beginning. The color green draws attention to the benefits of sustainable living and eating, love of the Earth, and the importance of finding harmony. It also inspires people to practice these principles (Green Monday 2020).

Green Monday's motto is "Make change happen, make green common." The three pillars of its approach are simplicity, ease of implementation, and dissemination. While the issues of climate change and animal welfare are complex, the solutions need to be simple. The message of the relevance of having a meat-free meal once a week is crystal clear, yet it can lead to great results in terms of the health of the planet and people. Green Monday believes that knowledge alone is not enough for successful change, but through influencing people's emotions, a fashionable, viral trend should be created. Using its branding, marketing, and PR experience, Green Monday has managed to build its rapidly growing brand so that customers can perceive its everyday meals as contributing to the mission of a movement that successfully works for the common good. The concept of becoming sustainable ("Go Green") is already supported by most people, but channels and frameworks for joint action are often lacking. For ease of implementation, Green Monday has developed a social business approach that supports a green lifestyle both in terms of demand (e.g., by providing information and conducting awareness-raising campaigns for consumers and partner companies) and the supply side (active market participation in food retail, corporate consulting, and venture capital investment) (Green Monday 2020).

Houdini Sportsware

Houdini Sportsware AB is a Swedish outdoor clothing company founded by Lotta Giornofelice in 1993. Its headquarters is in Nacka, an eastern suburb of Stockholm. It produces and markets a full range of functional clothing in an ethical and sustainable manner. With an annual turnover of almost seventeen million EUR in 2019 and 45 employees, the main goal is to serve its customers who are seeking to actively protect the Earth. Houdini also strives to eliminate all adverse environmental impacts, and achieve a state of sustainability and functionality even from the stage of product design. Its products are now directly available in eighteen countries on three continents (Houdini 2020; Largest Companies 2020).

History: The company's history can be traced back to an idea about a new raw material (stretch wool) that was still unknown in the early 1990s. It was thought up during a New Zealand ski and mountaineering tour. Building on this novelty, the company first started producing functional underwear on a small scale. Houdini's products soon became very popular among climbers. However, the company also attracted attention because it was founded (and is still run by) women, in a male-dominated clothing industry (the current managing director is Eva Karlsson). In the early 2000s, Houdini introduced another innovation: it started to produce underwear and insulated clothing that, in contrast to the norms of the time, was fashionable as well as functional. Despite some negative feedback from other industrial players, customers welcomed the innovation positively. In 2007, Houdini made great strides in sustainability: after years of ongoing consultation with its suppliers, it started manufacturing its first recycled polyester products. Today, it continues to organize its activities around sustainability and in recent years has launched a number of new environmental projects. These include producing clothing from residual fabrics in 2010, second-hand clothing sales and the rental of outerwear since 2012, and the climate-neutral design of its Gothenburg store in 2014, which is bicycle-friendly and promotes sustainability through its operations (the building has been awarded a Green Building Certification called Leadership in Energy and Environmental Design) (Houdini 2020).

Core values: Houdini's core values, sources of motivation, and goals were published in a statement at the end of 2017 called The Houdini Manifesto. The main driving force behind its activities is a deep love of nature and the experiences this provides. Houdini recognizes that nature is of intrinsic value, regardless of human needs, and that human activity has created an unsustainable situation on Earth, in which companies have a significant role to play and therefore a major responsibility. The current system of producing, using, and then degrading products into waste at an accelerating rate is not working. Houdini, therefore, considers its mission to transform its operations into a circular system to be in harmony with natural principles, and it is inspired by nature. The requisite change depends simultaneously on interrelated and closely intertwined natural, social, and economic factors. Houdini fights against overconsumption, for the development and application of sustainable technologies, and for working together with others and sharing knowledge. Houdini constantly monitors its own operations, striving for fuller transparency and continuous improvement in order to meet challenges that arise. Other values include free thinking, encouraging individuality, passion, uncompromising creativity, the desire to explore the world, the importance of communities of friends, joy, long-term commitment, taking into account the interests of future generations, and minimizing the ecological footprint, and having an overall positive impact on the world (Houdini 2018, 2020).

Interface

Atlanta-based Interface, one of the world's largest manufacturers of office and household modular carpet, floor mat, and office fabric, was founded by Ray C. Anderson in 1973. The company manufactures its products on four continents, sells them in more than 100 countries, employs 4100 people, and made a profit after tax of almost 80 million USD in 2019, with sales of more than 1.3 billion USD. The company, which faithfully represents its late founder's commitment to industrial ecology and sustainability, is associated with a number of significant industry innovations, several valuable brands, outstanding performance, and a customer-centric company culture. For more than two decades, Interface has been among the most well-recognized companies in terms of striving for sustainability, along with Unilever and Patagonia, among others. Through its Mission Zero® and Climate Take Back™ programs and attempts to climb up its own "Mount Sustainability" (associated with the goal of sending zero waste to landfill, reducing fossil fuel energy use to zero, as well as eliminating process water use and emitting zero greenhouse gasses), Interface seeks to completely eliminate the negative effects of its operations, reverse negative ecological processes, and maintain its leadership in implementing these changes (GlobeScan 2019; Interface 2019, 2020).

History: Ray Anderson graduated as an industrial and systems engineer in 1956. After a decade and a half of experience in the carpet trade, recognizing the growing demand for free lay carpet tiles, he founded Interface in collaboration with a British company, Carpets International Plc., and American investors. The company, which originally had 15 employees, successfully rose to the challenge of managing steeply rising raw material costs thanks to the use of advanced technology. It has grown rapidly due to its popular products, and had annual sales of 80 million USD at the time of listing in 1983. Interface has expanded internationally over the years and has become an industry leader after numerous product and technological innovations, opening up to newer markets (households, luxury vinyl floors), and having made more than 50 acquisitions (Interface 2020). The deep commitment to sustainability, inspired by Paul Hawken's book *The Ecology of Commerce* (Hawken 1993), has mainly been felt since the mid-1990s. Inspired by the book, Anderson wrote his own book on the environmental impact of Interface production and its business model (Anderson 1998). The next significant step in the unfolding transformation was the Mission Zero® program that was announced in 2006 that had the goal of eliminating all of the company's negative environmental impacts by 2020. The founder's death in 2011 did not change the mission of Interface, which continued to be implemented until 2017 under the leadership of Dan Hendrix and then Jay Gould, and was even strengthened by joining the fight against global warming through the Climate Take Back initiative (Interface 2020).

Core Values: Interface sees itself as a "purpose-driven" company, and its core values are consciousness, compassion, action, and a passionate team that wants to "do things right" in parallel with "doing the right things." Its five value commitments are: (1) to plan better things; (2) to be original and generous; (3) to inspire others; (4) to create unity, and (5) a better future. (1) Interface is a revolutionary and entrepreneurial organization. It is a pioneer in the carpet industry that is working on the renewal of the industrial system by researching unknown, unusual, and unlikely solutions. Its organizational culture is based on discovery, innovation, and experimentation, which can lead to progress, success, and a more livable world. (2) Interface is also a community based on trust, deep relationships, and inclusion, and includes a network of supporters and collaborators. (3) It is determined to stand up boldly, with the help of its innovative products and by setting an example, to make the industrial world realize that it is possible to act for the greatest good, and to encourage companies to change. (4) Interconnectedness is a central element of its activities. Just as a harmonious, functional floor covering is made up of interconnected modules, Interface is also part of a larger whole consisting of interdependent elements, whereby partnerships and collaborations can achieve more together than separately. (5) According to the intellectual heritage of the founder, and in the hope of a livable future, it is understood that the relationship between man and nature must be fundamentally reshaped through the business world. This requires long-term planning, persistent action, and the adoption of guidelines based on natural principles (Interface 2020).

Natura

Founded in 1969, Natura is Brazil's largest cosmetics manufacturer and leader in direct marketing. With the Australian brand Aesop and the English brand The Body Shop, it has been part of a global cosmetics group headquartered in Sao Paulo under the name Natura & Co. since 2017. To this group the iconic American beauty brand Avon joined in January 2020, making it the world's fourth largest pure-play beauty group. Natura has been guided by sustainable development guidelines from the outset, and its customer relations are of paramount importance, so its basic trading strategy has been direct sales. Currently, 1.8 million consulting partners are conveying Natura's values and products to customers in ten countries on three continents. Innovation is at the heart of Natura's sustainable development: the company spent 2.4 per cent of its annual net revenue of over 2.2 billion USD in 2019 on product development, and has continuously launched new products on its main markets. Complemented by similarly committed and strong brands for social and environmental sustainability, Natura & Co. data (which indicate that the brands serve 200 million consumers in more than 100 countries on five continents, have 40,000 employees and 6.3 million consultants and representatives, along with 10 billion USD gross revenue) show that one of Brazil's most emblematic companies has become a truly global organization (Natura 2020a, b; Natura & Co. 2020).

History: The founder of Natura, Antonio Luiz de Cunha Seabra, an economist, worked early in his career on the production of typewriters and calculators before turning to the creation of beauty care products in a small family-owned laboratory in Sao Paulo in the 1960s. Natura was founded in 1969 under the name Natura Cosméticos, and the company's products were sold through a small shop. Remaining in direct contact with his customers, the founder developed his philosophy of beauty care: to connect beauty with well-being and skin care with self-esteem through the production and distribution of skin-care products made of natural ingredients. The by-now well-recognized floral logo of Natura was created in 1970 to symbolize the unity of nature and beauty. By 1977, Natura already had 40 products, and the Natura Brasil magazine was then launched. In 1984, Natura was the first beauty care company in the world to start selling its products in refillable form. In 1995, as a first major step toward social responsibility and recognition of the utmost importance of education for social development, the entire profit of the company's newly launched Crer Para Ver product line was donated to the French organization Entraide Scolaire Amical, which supports schoolchildren. In 2000, the Ekos brand was introduced, whose products stem from Brazil's rich natural treasures. In 2001, Natura inaugurated its headquarters in Cajamar, with the creed of not just dreaming of a better world, but implementing it. In the following years, this idea was gradually put into practice. An ISO 14001 environmental management certificate was obtained in 2004. The testing of products on animals stopped in 2006. In 2007, Natura received a carbon-neutral certificate for its entire production chain. In 2010, Natura became the first Brazilian company to join the WWF Climate Defenders initiative. The Amazonia program of Natura (aimed at preserving the environmental and social values of rainforests) was launched in 2011. In 2013, Natura acquired the Australian brand Aesop, which adheres to similar sustainability principles. In 2014, Natura was the first publicly traded company in the world to attain B Corporation sustainability certification, in line with which the accomplishment of social and environmental goals is a formal and mandatory element of the company's business model. In 2017, The Body Shop, and in 2020, Avon, joined Natura. The company has grown steadily over the decades, from becoming a listed company in Sao Paulo in 2004, starting its expansion in France in 2005, and building a second production facility in Benevides, northern Brazil, in 2012 (Balch 2013; Natura 2020a, b).

Core values: Natura's motto is *Bem Estar Bem* ("well-being well"), which indicates a state in which one feels at harmony and can thus establish an empathetic, successful, and joyful relationship with other people, nature, and wholeness. This is the starting point and the basic philosophy that inspires Natura to be a key player in social progress and to build clear, transparent relationships with all its stakeholders, and to produce and market products and services that help implement these principles. Underlying these goals are six main value commitments. (1) Relation: life is a chain of relationships, everything in the universe is interconnected, and the proper evaluation of relationships is the foundation of human development for peace and solidarity. (2) Improvement: striving for the continuous development of individuals, organizations, and society. (3) Truth: improving the quality of relationships is based on a commitment to truth. (4) Diversity: the diversity of people and perspectives is the source of prosperity and the existence of all beings. (5) Freedom: striving for beauty, people's inner goals must be free from influences and prejudices. (6) Engagement: the company as a living organism is a dynamic set of relationships, the value and lifespan of which is related to its ability to contribute to the sustainable development of society (Natura 2020a, b).

Organic India

Organic India, a world leader in the production and distribution of Ayurvedic Tulsi tea blends and herbal supplements, was founded by Bhavani Lev and Bharat Mitra with some of their friends in 1997 in Lucknow, North India. The founders were inspired by the teachings of the Hindu advaita master Hariwansh Lal Poonja (or, as his followers named him, Papaji). The company took the name Organic India in 2006 and strengthened its mission to operate as an ethical, socially, environmentally, and economically sustainable business that promotes regenerative agriculture and provides livelihoods and training to thousands of impoverished farmers in India's disadvantaged rural, tribal, farming communities. It also aims to combine 5000 years of traditional Indian wisdom with world-class technology to produce bio-certified Tulsi teas, herbals, Psyllium, castor oil products, organic foods and spices, and to promote organic production and sustainable agriculture. After twenty years of development, Organic India had annual turnover of 55 million USD in 2017. Based on an annual growth rate of more than 20 per cent in recent years, it expects to have turnover of over 70 million USD by the end of the fiscal year 2020 (The Economic Times 2017; Organic India 2020a).

History: In the 1990s, the founders of Organic India observed the devastation of nature in India as agricultural industrialization based on chemical pesticides, fertilizers, and genetically engineered seeds spread throughout the country. In parallel with this experience, with the help of Ayurvedic and Western physicians, the founders learned about and researched the positive health effects of herbs, recipes, and herbal products, especially in relation to "Holy Basil," also known as Tulsi. Over the following years they established a specific business model based on responsible economic, social, and environmental principles and fair trade partnerships with farmers in order to help reestablish traditional agricultural wisdom and environmentally regenerate thousands of hectares of farmland and forest. By now, Organic India has managed to contribute to the economic and cultural revitalization of rural communities through 3000 farm partnerships, as well as promoting community health care, education, and inclusive social change for women. It has a subsidiary in the United States certified as a B Corporation, and sells its products in more than 40 countries around the world (Organic India 2020a, b).

Core Values: At the heart of the concept of Organic India is love and the intention of expressing consciousness in action. The main goal is to support true well-being and raise consciousness through a holistic, sustainable business model based on respect for all beings and for Mother Nature. Organic India's mission is to be a trustworthy and innovative market leader in the global market, producing genuine organic products and solutions for healthy and conscious living. Its core values are service to all beings, honesty, total commitment to quality, respect and devotion to Mother Nature, and uncompromising integrity (Organic India 2020a).

Patagonia

Deeply committed to environmental and social sustainability in the sales of outdoor clothing and equipment, Patagonia was founded in 1973 by climber and environmental activist Yvon Chouinard. The California-based company employs approximately 2000 people in more than 50 countries on five continents and generated sales of approximately 800 million USD in 2017 and one billion in 2020. The concept of the company, which aims to produce the best outdoor products without doing unnecessary harm, is that the economy and companies can and should be used to alleviate and remediate the present environmental crisis, and encourage others to do the same. Patagonia recognizes that, as a player in the business world, it is part of the problem because it still uses fossil energy, water, and plastics and emits significant amounts of waste and carbon dioxide as by-products. Patagonia's goal is to make its negative effects transparent, and, as a leading "activist company," through a number of initiatives sufficiently reduce and neutralize the former throughout their supply chain. These initiatives include increasing consumer consciousness, enhancing product durability, monitoring their carbon footprint, embracing environmental movements and campaigns, and promoting business transformation by supporting businesses with similar commitments (Patagonia 2020).

History: Even as a child, founder Yvon Chouinard was in love with the beauty of mountains, open nature, and the depth of personal and community experiences they give us. Chouinard was a freedom-loving young man who spent a lot of time in nature and became an excellent climber. From 1957 to 1965 he was involved in the distribution of his own forged climbing equipment through his company Chouinard Equipment. In 1965, in partnership with an innovative fellow climber, Tom Frost, he further developed his products, resulting in a significant increase in demand for them. Returning from a European tour in 1970, encouraged by the great success of sales of shirts with a collar that he bought back from there (since a collar effectively protects the neck from the skin-rubbing effect of a climbing rope), clothing soon became the company's number one product. However, because of an unfortunate climbing accident and consequent lawsuits, Chouinard had to close his former business and give up the sale of climbing equipment. However, in 1973 he was able to establish a new brand and company, Patagonia, from the formerly successful clothing division. The product range was soon extended to other "quiet" outdoor sports that do not require the use of a combustion engine, such as running, hiking, skiing, surfing, fishing, and non-clothing products (such as backpacks and sleeping bags). Due to Chouinard's personal values, derived from a nature-respecting mountaineering culture and Zen Buddhism, environmental and social sustainability has always been important to Patagonia. Accordingly, it has been committed to improving its sustainability performance from the outset. As a result, since 1996 Patagonia has used only organic cotton for all its cotton products and it co-founded the Fair Labor Association to improve the situation of its workers. In 2002, it appointed a corporate social responsibility manager and began working more intensively with its factories. In 2005, a clothing recycling program called Common Threads was started. In 2007, a project called The Footprint Chronicles was launched, which has since transparently tracked the social and environmental impacts of Patagonia products. The company has been rigorously auditing its raw material suppliers since 2011, and in that year an extended version of the previous clothing recycling program called the Common Threads Initiative was started based on the principles of Reduce, Repair, Reuse, Recycle, and Reimagine. In 2013, Patagonia reinforced the clauses in its Code of Conduct that refer to ethical practices in the supply chain with special emphasis on the requirement of providing living wages that support a decent livelihood. Patagonia's products have been manufactured with Fair Trade Certified Sewing certification since 2014, which indicates that the buyer pays a community development premium above market price to suppliers to improve the quality of life in local communities in the long run. In addition, since 2014 Patagonia has only used down feathers for linings that come from 100 per cent traceable sources that exclude the practice of force-feeding and live-plucking of birds. Since 2016, it has managed to implement ethical wool procurement, whereby the fair treatment of animals, and ethical and sustainable land use are ensured throughout the supply chain. In 2017, Patagonia launched a website called Worn Wear through which its customers can return Patagonia products that have been used but that are still in good condition to the company via their stores or the internet or post office. For doing this, customers receive a credit toward their next purchase, and, after cleaning and repair, the products can be repurchased at a significant discount (Patagonia 2020).

Core Values: The ethos of Patagonia grew out of a climbing culture that, in addition to love and respect for nature, is based on a free entrepreneurial spirit, adventurousness, risk-taking, action, independence, self-motivation, and passion. In addition to these principles, for such intentions to be realized a person should be versatile, flexible enough and open to new things so that they can react to any difficulties that may arise in a timely and appropriate manner. Development and adaptability is essential, so that one never stops asking the question how one can do what they are doing better. Striving to improve the environmental and social sustainability of companies may also be compared to a complex climbing activity (similar to Interface's concept of Mount Sustainability), whereby one needs to know the internal characteristics (strengths and weaknesses) and the external circumstances (opportunities and threats), but most of all respect the limits and constraints of nature. A minimalist style, simplicity, and a focus on utility are needed, because all unnecessary burdens endanger the success of a climb, and even the life of the climber. Nevertheless, according to Chouinard, who was inspired by Zen Buddhism, one should not be afraid: just as a samurai is not afraid of death, neither should one be afraid of failure in business. You need to focus on doing your job well, and then you will achieve results; including in financial terms. It is thus the job of a business leader who is conscious of this not to sit in their office all day, but to be among their employees, to talk to everyone in the right tone, to come up with new ideas from outside the company, and to unite efforts to make changes in the right direction. Of course, it is also the great responsibility of a leader—and correspondingly, their company—that if they have the skills, resources, and opportunities to do good things, they should do them. A leader has the greatest responsibility for the origin of all resources that are utilized, and the state of nature. The overarching core values of Patagonia's commitment are a sense of interconnectedness with the natural environment and all beings, the pursuit of harmony, responsibility, cooperation, the intention to foster stakeholders' internal development and self-realization, and spirituality nourished by the latter (Chouinard 2005; Ooley 2013; Patagonia 2020).

Scott Bader

Scott Bader is a global chemical company with turnover of around 200 million GBP that employs almost 700 people and operates in sixteen countries on five continents. It has six manufacturing sites and fourteen offices, and is a market leader in composite, adhesive, and specialty polymer manufacturing. What makes it special is its self-governing structure. Scott Bader Commonwealth Limited is a common trusteeship; a community-owned company in which employees make decisions and manage assets. It is also a registered charity. Scott Bader does not have a corporate social responsibility policy because CSR is not separate element of its business model, but the basis of how it operates. Since the establishment of the current community ownership and corporate governance system in 1951, the core value of the company has been creating value for society. One of the main guidelines for this endeavor is a long-term vision and plan called 2036 Vision, which also functions as a system that prioritizes social and environmental sustainability elements and monitors and measures their development and impact (Scott Bader 2019, 2020).

History: The company was founded in 1921 in London by Swiss-born Ernest Bader and his wife, Dora Scott. They operated it according to the prevailing capitalist corporate governance principles for three decades. After becoming followers of the Quaker religion in the 1940s, the couple became increasingly preoccupied with the idea that acting more ethically to improve the world should also be emphasized as an economic principle. Their idea was that corporate capital should be distributed and operated in such a way that all employees have a real and equal say in the management of the company. Based on basically Quaker principles, a radical decision was made in 1951 to create the Scott Bader Commonwealth from the company (which had by then moved to Northamptonshire in central England). The organization, run by the employee community, was considered a British company until the mid-1960s, and then began to expand internationally. In 1966, in Amiens, France, a new plant was set up in the form of a joint venture in continental Europe to strengthen the position of the organization in the market for technical, commercial, and industrial composites. It launched an Irish subsidiary in Dunboyne, Eastern Ireland, in the mid-1970s, which has since worked closely with suppliers and customers in product innovation and actively participated in the company's charitable activities. In 1992, Scott Bader acquired Swedish Neidert Trading AB, from which the Nordic subsidiary responsible for supplying Northern Europe was formed. A South African unit was established in 1993 after 30 years of manufacturing Scott Bader products in the country under license, meaning that Scott Bader's history now extends back more than half a century in Africa. International expansion continued in the 1990s and 2000s (1992 USA and Czech Republic; 1999 Croatia; 2001 Germany; 2002 Spain and Dubai; 2007 China; 2013 Canada). In recent years, the company has strengthened its presence in markets in which it has been relatively weak so far, such as in Latin America, while in 2020 an Australian subsidiary was created (Jacob 2012; Scott Bader 2020; Simkin 2020).

Core values: Embedded in the creation and operation of the Commonwealth is the socio-economic vision of the founder, Ernest Bader. According to this, a sustainable and socially responsible enterprise cannot exist solely for its own sake, as it is part of the natural environment, society, and the national and international community, so its responsibilities extend far beyond the factory walls. The concept of joint ownership and "trusteeship" can be seen as a third way compared to capitalist and communist economic systems. Management must be based on consensus rather than instruction, and a company must refuse to participate in industry rivalry, international wars, and the armaments industry. The aim is to provide all stakeholders with the opportunity to develop materially and spiritually, for the members of the working community to be mutually responsible, and for work to be based on democratic participation. Scott Bader's core values can be derived from all these elements and the Quaker principles that inspired the company's transformation in 1951. It defines sustainable development in the same way as the Brundtland Commission. Scott Bader seeks to have a positive effect on the life of all beings affected by its activities. The company intends to develop so as to promote the complete evolution of individuals. It maintains equal opportunities for all to enjoy the benefits of the workplace. Everyone is involved, can participate, and has a say in the affairs of the company. Everyone has an opportunity to participate in social and community activities, but should also take individual responsibility. The leaders of Scott Bader lead by example, and conflict resolution is dialogue-based and nonviolent (WCED 1987; Scott Bader 2019, 2020).

TerraCycle

Hungarian-born Tom Szaky founded TerraCycle in 2001 at Princeton University. The company, which has been headquartered in New Jersey ever since, collects and recycles waste that has been declared unsuitable for recycling. It is now present in 21 countries around the world and had more than 200 employees, annual sales of 27 million USD, and profits of over three million USD in 2019. According to the owner-founder, who uses nature as an example for developing his concepts about waste, the whole idea of "waste" should be eliminated, since there is no waste in nature—rather, there is continuous circulation of materials. TerraCycle is the world leader in the management of waste in terms of the complex material flows of the industrial processes that serve the needs of today's consumer societies. TerraCycle considers this waste to be a usable resource, unlike traditional waste processing companies, and has now collected and recycled more than a billion pieces of garbage, supporting the mostly voluntary work of over 202 million people. For every piece of recycled garbage, it donates two cents to schools and charities around the world, and has raised more than 44 million USD so far (TerraCycle 2020).

Tom Szaky was born in Budapest, Hungary in 1982, but grew up in Canada after moving to Toronto with his parents in 1987. He enrolled in the Department of Psychology and Economics at Princeton University in 2001, but due to the increasing success of his work at TerraCycle he stopped his studies to focus entirely on his business. Szaky's background includes being honored for his community work as early as in his teenage years (he also operated three smaller internet businesses). Along with a classmate, Jon Beyer, Szaky founded TerraCycle in 2001 with loans of a few thousand dollars to recycle food waste from the university kitchens. This food waste was processed and, with the help of earthworms, ripe compost was produced, which was then sold to horticultural businesses and individual buyers. In 2003, the business model won Carrot Capital's Business Plan Challenge competition and the accompanying one million USD award. However, Szaky refused to accept the award because the would-be donors asked him to completely overhaul his business model and leave his colleagues. Returning to the original idea, TerraCycle further developed the product and began selling it in liquid form packaged into used soft drink bottles. Thus, the product came to consist exclusively of previously recycled components; i.e., "waste" (TerraCycle 2018, 2020).

In 2004, in a huge step, TerraCycle's plant fertilizer (named Plant Food) entered the Canadian stores of Home Depot and Walmart, and the company began to grow. From the premises TerraCycle was using for free on campus the company moved to its own office building in New Jersey, which was converted from a disused warehouse. During the reconstruction, local and international urban artists participated with TerraCycle, using only recycled materials and recyclable waste. Since then, the walls have been repainted every few weeks to provide space for exhibitions for emerging young artists. In 2005, TerraCycle organized their first volunteer waste collection team called Bottle Brigade, which collects used soft drink bottles for liquid plant manure. In the same year, the company also began selling their products in Walmart and Home Depot stores in the United States generating annual sales of one million USD by 2006. In 2007, the product range was expanded to offer nine different types of nutrient solutions, and a sponsored waste collection program was launched, according to which the producer of the original product that becomes waste (e.g., empty soft drink boxes, yogurt cups) bears the cost of selective collection and recycling. In 2008, market leader brand Capri Sun became the main sponsor of the Drink Pouch Brigade and launched an innovative marketing campaign in Newsweek magazine's environmental special issue, which resulted in 47,000 people returning 800,000 Target-brand plastic promotional bags. These were then used to make carrying bags—the first time this type of mass upcycling event has ever happened (upcycling refers to transforming "waste" into a product with a different function). In 2009, 25 new types of Brigade programs were launched, and TerraCycle began operations in Brazil, Canada, and the United Kingdom. To raise consumer awareness, the company launched a stand-alone TV reality show series called *Garbage Moguls* and opened their own store in Princeton to market their own products. Their first pop-up store opened in New York in 2010. "Hot Spots" have also been set up in Walmart stores (organic product displays) to present original products that can be considered sources of waste (such as Capri Sun soft drinks) and products with novel functions made from waste (such as backpacks and pen holders). By 2010, more than 50 million empty soft drink pouches had been collected and more than one million USD donated to schools and nonprofits. Due to the increase in the number of employees, TerraCycle also expanded its office in 2010 (TerraCycle 2018, 2020).

In 2011, TerraCycle opened offices in Norway, Spain, Germany, Ireland, Switzerland, Denmark, Belgium, and the Netherlands, and the number of employees increased to over 100. To change consumer awareness, they developed their first online game, Trash Tycoon, in which players have to run a waste recycling business. In 2012, the world's first project to recycle cigarette butts was launched. Two-and-a-half billion butts were collected with the participation of 40 million people, and a further six million USD was donated to charity. In 2013, the company's expansion continued both geographically and in terms of supply. With the newly opened Australian and New Zealand branches, TerraCycle was operating in more than 20 countries and running more than 100 types of program. In collaboration with the city of Vancouver, Canada, the first municipal cigarette waste recycling program has been launched. One of the largest German waste management companies, Landbell GmbH, bought a 25 per cent stake in TerraCycle in three German-speaking countries. In 2014, the sale of collection boxes called Zero Waste Boxes for almost all types of waste commenced. This new waste collection format was also made available in Staples, Office Depot, and Home Depot stores. An office was also opened in Japan, and Canada's largest waste management company, Progressive Waste Solutions, bought a 20 per cent stake in the Canadian subsidiary. Another TV series called *Human Resources* was launched on the Pivot network. In 2015, in cooperation with Unilever, the collection of metal boxes for deodorants was organized in 1600 Drogerie Markt stores in Germany, from which bicycles were made and donated to child welfare organizations. In 2016, a pioneering partnership was formed with French water, energy, and waste management company SUEZ, through which SUEZ acquired a 30 per cent stake in TerraCycle's French, British, Swedish, Belgian, Dutch, and Finnish subsidiaries. In the USA, with Henkel, TerraCycle was the first company in the world to develop a program for recycling industrial adhesive packaging and, together with Bausch + Lomb, for contact lenses and their packaging. They also set up a branch in China. In 2017, Procter & Gamble and SUEZ partnered to create the first fully recyclable shampoo bottle made of marine waste. In Australia, together with L'Oréal, the first playground made of recycled beauty waste was opened. A South Korean subsidiary was also launched in 2018. TerraCycle also developed and launched Loop in 2019. This is a circular shopping system that home-delivers and recollects the world's most favorite brand items in durable, refillable packaging in order to reduce the amount of single-use plastics (TerraCycle 2018, 2020).

Core values: as a genuine social and ecological entrepreneur and committed leader, Tom Szaky's most important core values are honesty, transparency, accountability, enthusiasm, passion for hard work, and creativity. He emphasizes these factors during the operation of Terracycle, because only through adhering to these values can he achieve his goal, which is making a significant change to the whole world, so that the economy can better meet the needs of humanity without generating waste, just as nature does. His philosophy is that everything is possible, and that it is the greatest joy when one can see that others enjoy what they are doing. In the collaborative and caring environment created by the founder, there is a strict work ethic, but also openness and empowerment. All mistakes can be forgiven, unless someone is dishonest with themselves. The company's mission, value commitments, and atmosphere inspire many not only within but outside the company, and both TerraCycle's and Szaky's activities and achievements have already received numerous accolades (CNBC 2012; Parvati Magazine 2017; TerraCycle 2020).

Table 5.3 summarizes the main characteristics of the value orientations of the internationally well-known ecologically conscious business organizations under analysis.

Table 5.3 Main characteristics of the value orientations of ecologically conscious businesses in the international sample

Company name	Vision	Values	Means	Outcomes
Green Monday	Global change through constructing an ecosystem of plant-based food to prevent climate change, food insecurity, food-related illness, planetary devastation, and animal suffering	Sustainability, health and well-being, consciousness, simplicity, feasibility, easy-to-implement and easy-to-share solutions, knowledge sharing, Buddhism	Awareness raising and promoting a sustainable lifestyle, grocery stores for plant-based food, sustainability and CSR consulting, ethical capital investment fund, prestigious advocates, and partner organizations	Increasing vegetarian and "once-a-week" plant-based food consumption, local and global community building, shaping food culture and society, feasible implementation of triple-bottom-line approach

Company name	Vision	Values	Means	Outcomes
Houdini Sportsware	Inspiring and enabling mankind to reconnect to nature through leading a healthier and happier lifestyle in partnership with nature, to evolve as individuals and as societies to improve the situation of present and future generations	Responsibility, sustainability, functionality, regeneration and intrinsic value of nature, passion, consideration of future generations, love, holistic approach, circular systems, cooperation, knowledge sharing, transparency, free thinking, setting an example, leading the change, ecosophy, deep ecology	Ethical and sustainable production of functional outdoor clothing, highly innovative product design based on recycled, biodegradable raw materials, local R&D and production, few suppliers and short supply chain, redefining consumption and customer care, dedication to "recycling, reusing, repair, rental," open-source innovation	Sustainability-based business model with nature as the blueprint, on the road to becoming 100 per cent circular: 100 per cent of products circular by 2022, total value chain circular and using only renewable energy by 2030, then moving beyond net zero, creative innovations, high quality, timeless style

(continued)

Table 5.3 (continued)

Company name	Vision	Values	Means	Outcomes
Interface	Doing right for a positive future with moral courage, showing that sustainability is better for business, transforming itself into an agent of restoration, and being a catalyst for the redesign of global industry	Industrial ecology, purpose-driven and restorative economy, sustainability, consciousness, compassion, action, passionate teams, doing well by doing good, well-being, genuineness, generosity, inspiring others, interconnectedness, striving for a better future, pioneering innovations inspired by nature	Globally producing commercial and household modular flooring, own sustainability advisory group using "The Natural Step" sustainability methodology and the concept of "Mount Sustainability," radical product and process innovation, close cooperation and ethical relationship with members of the value chain	Climate Take Back and Mission Zero programs to become carbon negative by 2040, carbon-neutral products, services, recycled materials to eliminate all negative environmental impacts, cyclical, gentle business model with radical redesign of processes and products mimicking nature, improvement of life of all stakeholders

Company name	Vision	Values	Means	Outcomes
Natura	Becoming the best beauty group for the world through nurturing beauty and relationships that foster a better way of living and doing business, the collaboration of four socially and environmentally conscious brands through local and global expression that promote the well-being of people, animals, and the planet	"Well-being well," harmony, empathy, successful, and joyful relationship with nature, people, and the "Whole," pure, transparent relationships, peace, solidarity, striving for continuous development, truth, diversity, human rights and gender equality, circularity, and regeneration	Production and distribution of innovative, natural, ethical, and sustainable beauty care products from four iconic brands (Natura, Aesop, The Body Shop, and Avon), nature-inspired cyclical production system with true value creation throughout entire partner network, fair relationship with suppliers, caring direct sales and consumer awareness raising	High-quality products and services combining traditional and cutting-edge knowledge, 6+ million representatives worldwide, support for local communities, public education and women's causes, carbon neutral in 2019, Vision 2030 to become net zero CO_2 emitters and zero deforestation in Amazon

(continued)

Table 5.3 (continued)

Company name	Vision	Values	Means	Outcomes
Organic India	Being a trustworthy, innovative global leader of consciousness through a holistic and sustainable business model that promotes True Wellness and respect for all Beings and Mother Nature by providing genuine organic products and services that support more conscious and healthy living	Love, respect for and service and dedication to all Beings and Mother Nature, holistic approach, sustainability, real well-being, increasing consciousness, health, inspiring others, reliability, innovation, honesty, accountability, commitment to quality, integrity, Hinduism	Production of organic teas, herbal preparations, spices, fresh and packaged food, fair and lasting relations with farmers, herb collectors, rural development for communities, cooperation with market-leading handicraft product distributor Fabindia	Genuine organic herbal products available worldwide, holistic business model for the benefit of all stakeholders, restorative agriculture for serving and protecting nature, support for farmers, herb collectors, and their communities in rural India

Company name	Vision	Values	Means	Outcomes
Patagonia	Saving the planet through building the best outdoor clothing products and apparel without causing unnecessary harm and using business to lead society and protect and restore the stability, integrity, and beauty of the web of life, and continuously developing new ways to do things	Love and respect for nature, free entrepreneurial spirit, self-motivation, passion, diversity, openness, fearlessness, doing good and well, respect for limits, minimalist style, simplicity, transparency, innovation, humility, interconnectedness, empowerment, courage, meaning, Buddhism	Retailing of ecological outdoor clothing and apparel, leading "activist company" that aims at increasing consumer consciousness, support for environmental movements and campaigns and businesses, and promoting business transformation	Using business to alleviate environmental crisis and encouraging and supporting others to do the same (campaigns, movies, investment fund), monitoring and reducing carbon footprint, material, product and value chain innovations, participation in political activism and regenerative agriculture
Scott Bader	Creating a world in which humanity thrives without compromising the natural systems it depends on through providing essential technologies that address the challenges of a changing society, and being a trustworthy partner that harnesses the power of chemistry as a force for good	Being a part of natural environment, society, national and international community, mutual responsibility, joint ownership and asset management through democratic participation, material and spiritual development of all stakeholders, circular economy, full personal evolution, equal opportunities, participation, leading by example, dialogue, Quakerism	Production, distribution of composite materials, adhesives, special polymers, fireproof-, smoke- and toxin-resistant materials, gel coatings, trusteeship as a unique form of community ownership and corporate governance system, innovation, production, use of "The Natural Step" sustainability methodology	2036 Vision with three pillars: humanity, ecology, business, development without jeopardizing natural systems, use of safe, ethical and transparent chemical industry for public good, knowledge and profit sharing, circular business model, workforce development and cooperation with the wider society Commonwealth Strategy

(continued)

Table 5.3 (continued)

Company name	Vision	Values	Means	Outcomes
TerraCycle	Being a social enterprise that works to eliminate the idea of waste through offering a range of easy-to-use recycling platforms that help everyone to recycle everything and move away from single-use packaging	Honesty, transparency, accountability, credibility, enthusiasm, hard work, strict work ethics, creation, changing the world, everything is possible, cooperative and caring atmosphere, openness, empowerment, community, meaning, circular economy, eco-capitalism ("People, planet, profit" and "Better, greener, cheaper")	Collecting, recycling, reusing waste formerly considered non-recyclable, national-, brand-funded or customer-funded waste collection programs, close cooperation with production companies associated with global and local brands, continuous scientific research to eliminate more waste streams, partnering with millions of individuals, conscious consumers, civil and business communities collecting waste	Implementing multitude of waste collection programs, supporting 200 + million people to collect billions of pieces of garbage, donations to schools and charities, opening subsidiaries worldwide, broadcasting TV series, developing an online game Trash Tycoon, providing value-added consultancy for media, PR, social media, communication, marketing, enhancing brand/consumer loyalty and awareness raising

Source Author's construction

Value Orientations of Ecologically Conscious Business Organizations in the Hungarian Sample

This section first briefly presents the company profiles and value background of the Hungary-based ecologically conscious business organizations that participated in the sample. During the examination of their value orientations, an exhaustive exploration of the value profile of the individual enterprises was not the primary focus. Through aggregation of the interviews with fourteen businesses, the general value characteristics of the Hungarian sample were analyzed to identify similar motives and patterns.

Skanska Hungary

Sweden-based Skanska is one of the world's leading project development and construction groups, with expertise in construction, residential, and commercial property development. Skanska Property Hungary (https://www.skanska.hu/) is its Hungarian subsidiary, with several decades of experience. Its main goal is building for a better society and creating shareholder value alongside social responsibility, and creating innovative and sustainable solutions for a sustainable future for all stakeholders (employees, customers, and communities). Further core values include commitment to security, ethical and transparent operations, green and community investments, diversity, inclusion, protection of life, cooperation, joint development, and commitment to partners.

Nekedterem Hungary

Nekedterem (https://nekedterem.hu/) was founded to provide consumers with a web shop and home delivery service for high-quality fresh vegetables, fruit, and food products from local small-scale producers in Budapest and its surroundings. Its main goal is buying and selling local, artisanal Hungarian products that also meet special customer dietary needs (organic, sugar-free, gluten-free, GMO-free, lactose-free, paleo, vegan, etc.), minimizing the length of supply chain, using recyclable, biologically degradable packaging materials, and selective waste collection. Core values include offering a personal, reliable, friendly service, fostering direct relationships between producers and consumers, saving time for customers, community building.

Réthy Fashion

Réthy Fashion (http://rethy-fashion.com/) is a small family business founded by fashion designer Éva Réti with the aim of rethinking the relation between fashion and environmental protection through recycling and reusing jeans-based clothing and materials. Its main goal is to create high quality, new, innovative jeans products, combined with environmental awareness raising, the promotion of selective waste collection, social value creation, and the employment of disadvantaged people. Réthy Fashion is engaged in climate protection by fighting against global warming and striving for a circular economy.

Bio-Drog Berta

Bio-Drog Berta Ltd. (https://www.bioberta.hu/english/pages/about_us/) was founded to produce organic vegetables, herbs, and fruits. It integrates the work of 50 families who produce 30 species of plants on 70 hectares of irrigable land. Its core values cover value creation for the community, use of their own talent and skills in a novel way, producing as many innovative products as possible, utilizing the largest proportion of processed plants as possible from an excellent agricultural area, building an international reputation, and supplying products that promote the health of consumers, especially Hungarian ones.

Ecoizm/Zöldbolt

Ecoizm/Zöldbolt (https://www.ecoizm.com/) launched its service to provide consumers with eco-friendly products in one place (on its webshop and in two bricks-and-mortar shops). Its goal is to offer real, eco-friendly alternatives in almost all areas of everyday life that represent small ways of helping to protect the environment. Ecoizm strives for quality, not quantity, and strictly selects the items it trades. Core values include using everyday products that have some significant environmental benefits over substitute products, the continuous monitoring of innovation and the promotion of environmental awareness, use of natural ingredients, minimal use of synthetic chemicals, durability and ease of repair, pollutant- and hazardous-waste-free, recycled, recyclable, and reusable materials, minimal or natural packaging materials, locally available products, fair trade, handmade, and organic products.

ALTEO Energy

ALTEO Plc. (https://alteo.hu/?lang=en) is an energy service and trading company engaged in renewable and natural gas energy production, energy trading and bespoke energy services, and development for small and large companies. It strives to help meet climate protection goals while maintaining an appropriate cost level and security of supply. ALTEO promotes the principles of democratized energetics, customized service, and sustainable business benefits for customers. Its core values are adaptability, innovation, leading by example, efficiency, flexibility, long-term economic, environmental and social value creation, responsible corporate governance, operating ethically, protection of human life and security, equal treatment, right to a healthy environment, environmental responsibility throughout the supply chain, building lasting and mutually beneficial partnerships, and transparency.

Hajtás Pajtás

Hajtás Pajtás (https://hajtaspajtas.hu/) was the first Hungarian bicycle courier enterprise and is still the market leader. It is also engaged in civil activism to promote bicycling, a more livable environment, and a more sustainable urban lifestyle. Its core values include service based on trust, reliability, transparent and law-abiding business operations, sustainability, environmentally conscious activity, the avoidance of unnecessary pollution, and social responsibility.

Pannonhalmi Abbey Winery

Pannonhalmi Abbey Winery (https://apatsagipinceszet.hu/?lang=en) is an enterprise based on a thousand years of tradition and the Benedictine monks' values (a balance between spiritual life and physical activity, love of work, self-sustainability, education, and the dissemination of culture based on the Regula of Saint Benedict) that is attempting to revive viticulture and wine-making traditions. Its goal is also to make the nonprofit activities of the abbey (operating social, educational, and cultural institutions) economically sustainable, and to produce its own altar wine. Core values include a community based on the values of the gospel that functions well in terms of everyday life and has a positive spirit. Monastic and secular participants are organically united as regards shared values including responsibility, sustainability, preservation, and preserving resources for future generations.

NES Shoes

Designer Andrea Szilágyi graduated in Paris and founded her brand, NES Shoes (https://nesshoes.com/home/) to promote environmentally friendly shoes. Focusing on individual and customized orders and driven by sustainable practices intrinsic to the design and message, NES products are created exclusively from natural, biodegradable, vegan, recycled, and/or recyclable raw materials in an increasingly ecological way. Production is now based in Eastern Hungary at a biomass-heated manufacturing plant that also uses solar electricity from panels and strives to source local raw materials. Core values include supporting the self-expression of self-aware women and men through the provision of comfortable and sophisticated yet affordable shoes, ecological consciousness, social and individual responsibility, aesthetics and joy, and forward-looking, ever-evolving ethical manufacturing and design.

Béres Pharmaceuticals

Béres (https://www.beres-international.com/?en/) is one of the most well-known brand names in Hungary. The family-owned and managed Béres Pharmaceuticals was named after the legendary inventor, Dr. József Béres Sr. His invention, Béres Drops, a unique trace element supplement, was created in 1972 to maintain health by improving the functioning of the immune system. Core values of the company, based on the founder family's deeply felt Christian values, include the prevention of disease using natural products, human-centeredness, supporting people in need, education, culture, science, righteousness, reliability, predictability, social, national, and environmental responsibility, high quality products, perseverance, diligence, compassion, commitment, successful organization, team spirit, a transparent and efficient organization, and an attractive working environment. Béres is also engaged in eco-friendly wine production through a subsidiary.

Kör Architect Studio

Kör Architect Studio is a micro enterprise based on the personal engagement and enthusiasm of the architectural designer Attila Ertsey (http://www.ertseyattila.eoldal.hu/). The company designs ecological, sustainable buildings, and is also involved in educational activities within Hungarian higher education. It strives for an ecological approach based on Rudolf Steiner's anthroposophy and commitment to the idea of a self-governing society. It represents a complete social alternative to the prevailing social system by focusing on the freedom of spiritual life, direct democracy, and the solidarity economy. Core values also include healthier economic operations on the basis of alternative economics (not building on the principle of profit or the exploitation of natural and human resources), cooperation, a focus on the human scale, and harmony between human and environment.

Pipacs Bakery

Pipacs Artisan Bakery (https://pipacspekseg.com/) is an organic bakery that supplies healthy, additive-and-preservative-free bread. Core values of the founder-owners Ádám Fülöp and Márton Tóth and their company are human-centeredness, the material and moral appreciation of employees, avoiding the night shifts that are common to the industry, the importance of ecological values, connection to nature, respect for all the elements and community of the value chain, protection of life and health, and promoting human relationships with partners based on trust and mutual responsibility.

Pendits Vineyards and Winery

Pendits (https://www.pendits.com/) is a small family business, founded by Márta Wille-Baumkauff and her husband in Eastern Hungary, that follows the principles and values of organic, biodynamic farming, and crafts wines as a philosophy of life. Its spiritual approach based on anthroposophy means that the owners seek harmony with nature and the organic oneness of man, plants, animals, and habitat, timing work in the vineyard to natural cycles. The economy is seen as an organism that should have self-sustaining ability and autonomy. Core values also include building on regional values, products, and relationships, creating a clean, healthy product of the finest quality, social responsibility, conservation of terroir, sustainable development, and strengthening and utilizing environmental, cultural, and local community values, innovative research, and supporting family farms.

MagNet Bank

MagNet Bank (https://www.netbank.hu/en) is the largest financial institution in 100 per cent Hungarian private ownership, and has operated profitably for over two decades in Hungary. It is a commercial bank that started as a savings cooperative but became a triple-bottom-line focused values-based community bank that strives to promote sustainability (i.e., to be an ethical, sustainable, and impact-based bank), as well as environmental and social responsibility, human relationships based on respect and trust, consciousness, innovation, transparency, and human-centeredness. MagNet Bank is a member of the Global Alliance for Banking on Values and is committed to advancing positive change in the financial sector through sustainable economic, social, and environmental development, with a focus on helping individuals fulfil their potential and building stronger communities.

Radar charts of the enterprises' value profiles were drawn up (Appendix 2) based on the interviewees' responses (on a seven-item Likert scale) to the following question: how important to the operation of their business do they consider each of the thirteen ecological values identified from the literature? The figures show the differences between the businesses.

Figures 5.1 and 5.2, show more clearly what the radar charts indicate. The opinion of the fourteen Hungarian businesses is that the most important ecological values are responsibility, trust, openness to mystery, respect for ecosystems, love, temperance, and reciprocity. Respondents consider humility, a holistic approach, the intrinsic value of natural beings, natural embeddedness, and biodiversity to be important values, but less decisive. The value of non-possession appears to be unimportant to them.

Figure 5.2 shows more accurately the direction and intensity of deviations from the mean value (which takes a value of four, as indicated by the vertical line in the middle of the diagram) than other types of figures (Robbins and Heiberger 2011). In the illustration, the color red indicates a value of 1 ("least important"), while dark green represents 7 ("most important"), and dark gray indicates a lack of response, or the

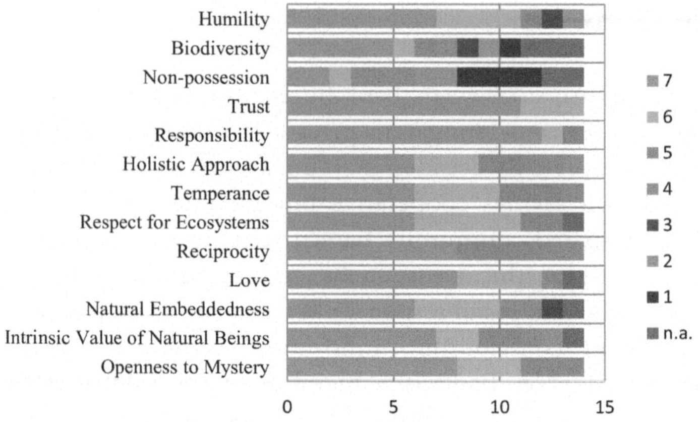

Fig. 5.1 Stacked bar chart of the importance of ecological values for selected Hungarian businesses (*Source* Author's construction)

response "not relevant." In the middle of each colored field, the number of interviewees who chose that particular value is listed.

The importance of the thirteen ecological values was also examined through content analysis of the whole texts of the interviews in order to reveal the presence of these values in the responses of the interviewees. Table 5.4 summarizes the frequency of occurrence of the ecological values identified during the content analysis. The most frequently identifiable values were responsibility, a holistic approach, and temperance. In addition, typical values included respect for ecosystems, trust, and reciprocity. The values of natural embeddedness, biodiversity, openness to mystery, humility, love and the intrinsic value of natural beings occurred rather less often. The value of non-possession was not identifiable using this method either.

The value orientations of the Hungarian businesses were also mapped using an inductive approach, in addition to the deductive method described above. During the content analysis of the transcripts of the interviews, patterns were explored that could characterize the value background, motivation, culture, and way of thinking of the given enterprise through coding and categorizing the meaningful content that could be identified in each paragraph. During the analysis and interpretation of the meaning of 36 patterns that were identified, consolidated sets were

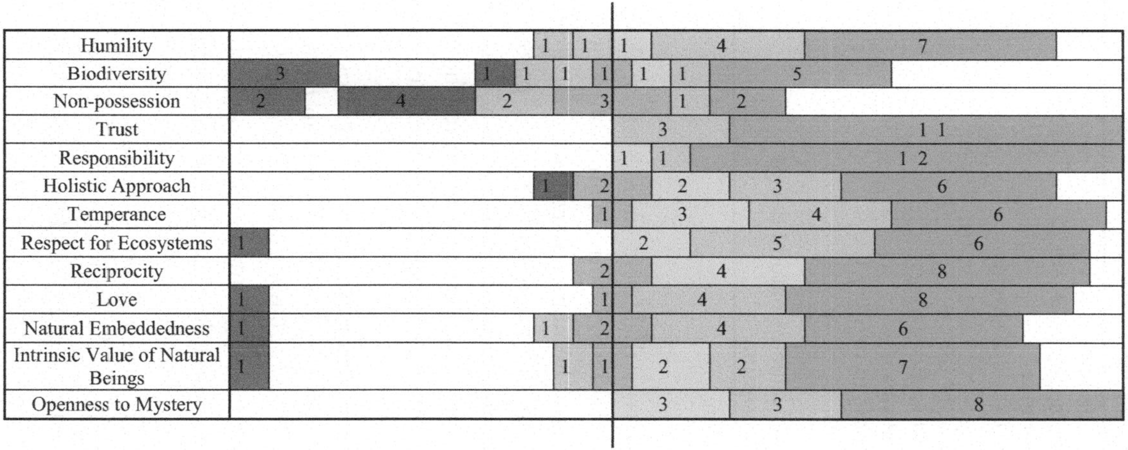

Fig. 5.2 Diverging stacked bar chart of the importance of ecological values for selected Hungarian businesses (*Source* Author's construction)

Table 5.4 Frequency of occurrence of ecological values in the interviews with selected Hungarian businesses

	Ecological value	Frequency
1	Responsibility	57
2	Holistic approach	49
3	Temperance	46
4	Respect for ecosystems	25
5	Trust	22
6	Reciprocity	20
7	Natural embeddedness	13
8	Biodiversity	11
8	Openness to mystery	11
10	Humility	10
11	Love	9
12	Intrinsic value of natural beings	8
13	Non-possession	0

Source Author's construction

formed from those patterns that were more similar. According to the clustering presented in Table 5.5, four main groups of patterns could be defined: namely, general human characteristics, environmental focus, social focus, and economic focus. In Table 5.5, the patterns are listed in order (from top to bottom) pursuant to the frequency of their occurrence (the number after the name of the pattern indicates the number of interviewees out of the sample of fourteen in whose interviews the pattern could be identified).

Based on Table 5.5, the identified patterns are most diverse in terms of human characteristics and most concentrated in terms of an environmental focus. Almost all respondents stressed the need for value creation for social and local communities, human-centeredness, the importance of human relationships, and the role of economic factors (difficulties, challenges, and success factors) and the importance of innovation and quality. Besides the creation of the former groups, an interesting phenomenon is that the lower five patterns in the table can be recognized as overarching the four previously identified groups. These include (1) consciousness, search for meaning, (2) cooperation, avoidance of competition, (3) environmental, social, and economic sustainability, (4) self-sustainability, self-management, autonomy, adaptability, resilience, and (5) organicity.

Table 5.5 Clustering of patterns identified in the interview transcripts in selected Hungarian businesses

Sample of Hungarian ecologically conscious businesses			
Human characteristics	Environmental focus	Societal focus	Economic focus
Commitment 11	Environmental consciousness 13	Social utility, local communities 13	Economic factors, need for entrepreneurial, business knowledge 14
Non-harming, non-violence 11	Education, environmental education 9	Humaneness, human relationships, human-centeredness, human scale 12	Innovation 13
Optimism, hope, gratitude 11	Recycling 5	Health, protection of life 7	Quality 10
Life philosophy, worldview 10		Culture, aesthetics 6	Value-based, ethical operation, transparency 7
Setting an example 10		Mission 6	Certification 5
Authenticity 6		Lack of social demand 6	Labor shortage 4
Dedication, perseverance 5		Tradition 2	Lack of comparative advantage 1
Negative expectation, disappointment, hopelessness 5			
Harmony with nature, connection with nature, interconnectedness 5			
Struggle, fight 4			
Self-confidence 3			
Self-realization, joy 3			

(continued)

Table 5.5 (continued)

Sample of Hungarian ecologically conscious businesses			
Human characteristics	Environmental focus	Societal focus	Economic focus
Respect 3			
Common sense, practicality 2			
Patterns overarching the four groups			
Consciousness, search for meaning 10			
Cooperation, avoidance of competition 9			
Sustainability, triple bottom line 6			
Self-sustainability, self-management, autonomy, adaptability, resilience 5			
Organicity 5			

Source Author's construction

These umbrella patterns may be seen as factors that help synthesize the operations of the businesses.

Summary

For the ecologically conscious business organizations in the international sample, the most important values are circularity, well-being and health, respect for nature and future generations, cooperation and knowledge sharing, passion, sustainability, transparency, responsibility, genuineness, honesty, credibility, accountability, setting an example, regeneration, simplicity, consciousness, and action. Values such as innovations inspired by nature, feasibility, ecology and ecosophy, a holistic approach, doing well by doing good, inspiring others, interconnectedness, diversity, empowerment, free thinking, fostering the material and spiritual development of all stakeholders, love, openness, and meaning are also important to them. Respondents also consider compassion, generosity, humility, respect for limits, solidarity, integrity, harmony, peace, equal opportunities, fearlessness, self-motivation and quality to be relevant, but these values were less emphasized, as identified during the case analyses.

For Hungary-based ecologically conscious business organizations, the following values are the most important: responsibility, holistic approach, trust, openness to mystery, respect for ecosystems, love, temperance, reciprocity, consciousness, cooperation, human-centeredness, commitment, non-harming, optimism, education, knowledge, innovation, and quality. Values such as humility, the intrinsic value of natural beings, natural embeddedness, biodiversity, authenticity, transparency, and aesthetics are important but not decisive for them. Non-possession is a value that they do not consider relevant to them at all.

In brief conclusion, the operation of the ecologically conscious business organizations in the sample under study is overwhelmingly based on similar values, and concerning their value orientations no sharp distinction can be drawn between international and Hungarian cases.

The raison d'être of the international and Hungarian business organizations in the sample is the pursuit of multidimensional (social, ecological, and economic) sustainability and the exercise of authenticity. Their basic goals often include the protection of health, the integrity of people in general, and of nature, preserving the quality of life, and supporting domestic culture, society, economy, and local communities. The businesses strive to produce high quality products and services, and do not pursue growth at all costs but rather organic development, as well as self-realization, and the expression of the personal values of the owners and managers.

Both the international and the Hungarian ecologically conscious businesses interpret success in a multidimensional way. Financial profit is not the only objective, but a means of ensuring the long-term survival of the enterprises. Real success means simultaneously harmonizing, nurturing, preserving, and restoring social and ecological values. An important element of their definitions of success is how well the businesses take into consideration the well-being of the whole stakeholder network, as defined in a wide sense.

Appendices

Appendix 1 Interview Guide for Research on the "Value Orientation of Ecologically Conscious Businesses" in Hungary

1. Characteristics of the business
    1.1. Basic characteristics of the enterprise: foundation, activities, number of employees, annual sales, economic situation.
    1.2. Does the enterprise have a written mission statement, and are there any written documents regarding its value commitment?
    1.3. How do the former relate to the personal value commitment of the founder/manager?
2. The interpretation and value orientation of ecological consciousness
    2.1 What do you understand by the term ecological consciousness?
    2.2 Are you committed to a worldview (religion, spirituality, community, …)? How do the values of this worldview relate to ecological consciousness?
    2.3 As an entrepreneur, do you have any value commitments that are not related to ecology? What is the relation between these values and ecological values?
    2.4. Please rate how important the following ecological values are in your entrepreneurial life!

Humility	0	1	2	3	4	5	6	7
Biodiversity	0	1	2	3	4	5	6	7
Non-possession	0	1	2	3	4	5	6	7
Trust	0	1	2	3	4	5	6	7
Responsibility	0	1	2	3	4	5	6	7
Holistic approach	0	1	2	3	4	5	6	7
Temperance	0	1	2	3	4	5	6	7
Respect for ecosystems	0	1	2	3	4	5	6	7
Reciprocity	0	1	2	3	4	5	6	7
Love	0	1	2	3	4	5	6	7
Natural embeddedness	0	1	2	3	4	5	6	7
Intrinsic value of natural beings	0	1	2	3	4	5	6	7
Openness to mystery	0	1	2	3	4	5	6	7

Other value(s)

0	1	2	3	4	5	6	7
0	1	2	3	4	5	6	7
0	1	2	3	4	5	6	7

3. Ecological aspects of the operation of business

3.1. What does ecology mean in terms of the operation of your enterprise? In what areas does a "green" approach prevail, and how? (for instance, product development, new investments, production, office activities, transportation, etc.)

3.2 In which areas is your ecological performance best, and where do you think further development and progress is needed?

3.3 Do you have certified (environmentally friendly, organic, etc.) products? If so, why was it important to become certified?

3.4 Does the enterprise have an environmental management system, and does it use an environmental management standard (for example, ISO 14000)? If so, what are your experiences with this? If not, do you consider it important to obtain one? If not, why is it not important to you?

3.5 What was the main motivation for using progressive ecological practices? Where did the incentive come from? (for instance, from a manager, owner commitment, external benefits, etc.)

3.6 How important are state regulations or subsidies in stimulating and promoting ecological and environmental activities, and what do you think of these instruments?

3.7 How important are civil organizations, NGOs, and social support in stimulating and promoting ecological and environmental protection activities, and what do you think of these organizations/types of activity? Do you have a relationship with any NGOs? If so, what is their nature (what do they do)?

4. Competition and ecology

4.1. How would you rate the strength of competition in your business sector?

0	1	2	3	4	5
Do not know	Very weak	Weak	Moderate	Strong	Very strong

4.2 How would you evaluate your competitors' ecological consciousness in your business sector?

0	1	2	3	4	5
Do not know	Very weak	Weak	Moderate	Strong	Very strong

4.2 In general, is ecological consciousness an advantage or disadvantage in this business sector?

0	1	2	3	4	5
Do not know	Significant disadvantage	Disadvantage	Indifferent	Advantage	Significant advantage

5. Does the ecological consciousness of a business represent an advantage?

5.1. Does having a business with an ecological consciousness bring advantages to owners and managers?

0	1	2	3	4	5
Do not know	Definitely not	Not	Partly yes, partly no	Yes	Definitely yes

5.2 Does the ecological consciousness of a business influence the performance and motivation of employees?

0	1	2	3	4	5
Do not know	Definitely not	Not	Partly yes, partly no	Yes	Definitely yes

5.3 Do future managers/employees find businesses attractive because of their ecological consciousness?

0	1	2	3	4	5
Do not know	Definitely not	Not	Partly yes, partly no	Yes	Definitely yes

5.4 Does the ecological consciousness of a business contribute to the loyalty of customers?

0	1	2	3	4	5
Do not know	Definitely not	Not	Partly yes, partly no	Yes	Definitely yes

5.5 Does the ecological consciousness of a business have an impact on relationships with suppliers and partners?

0	1	2	3	4	5
Do not know	Definitely not	Not	Partly yes, partly not	Yes	Definitely yes

5.6 On the whole, does the ecological consciousness of a business pay off in the context of the present Hungarian economy?

0	1	2	3	4	5
Do not know	Definitely not	Not	Partly yes, partly no	Yes	Definitely yes

6. Prospects for an ecological consciousness in the Hungarian economy
   - 6.1. How do you rate the chances of an ecological consciousness developing in today's Hungarian economy?
   - 6.2. How can the economy and businesses be made more ecology-oriented?
   - 6.3. What kind of changes do you expect in the next 5-10-25 years with regard to the ecological consciousness of business?
7. Which other ecologically conscious businesses do you know? Who would you recommend that we include in the research?
8. Do you have any other comments or questions about the topics we have discussed?

Appendix 2 The Importance of Ecological Values for Ecologically Conscious Business Organizations in the Hungarian Sample

Radar charts of the value profiles of each business compared to the arithmetic mean and the median of the importance of values.

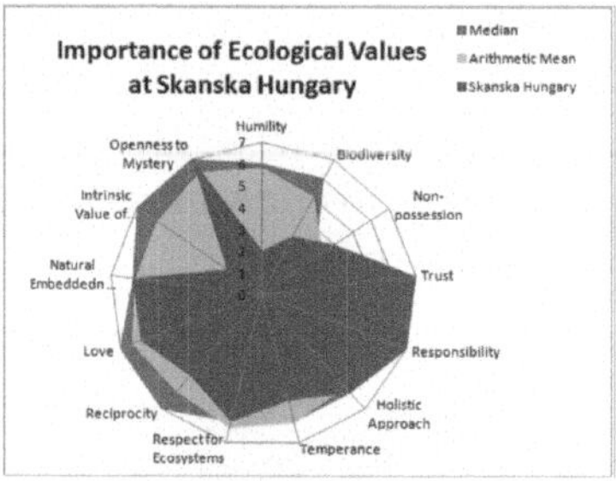

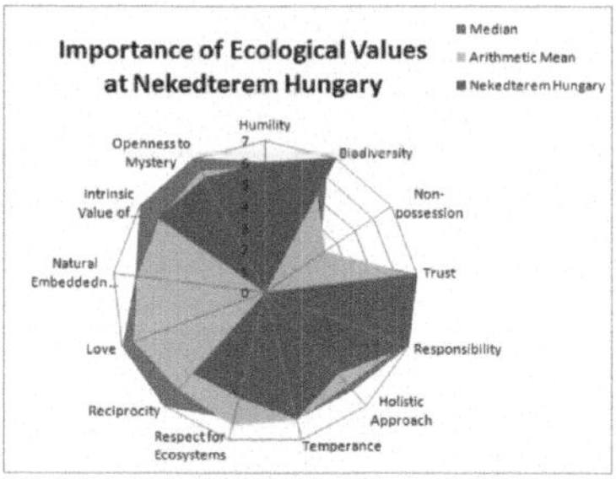

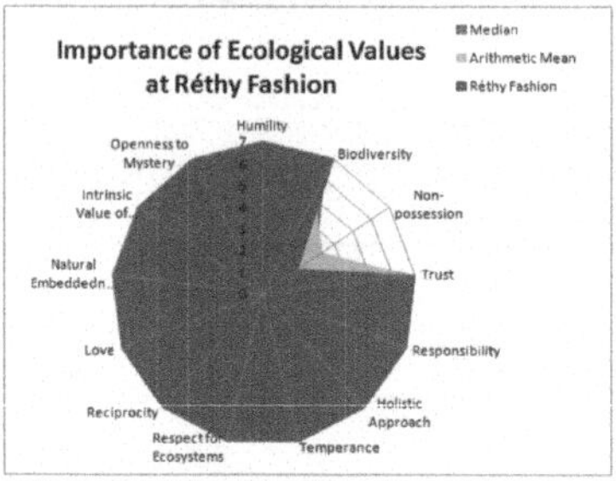

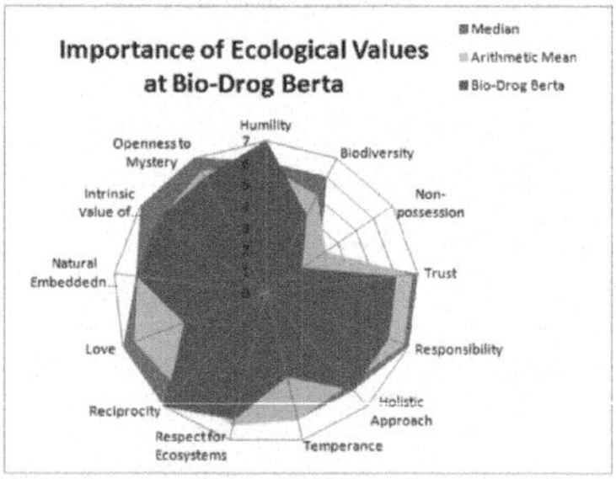

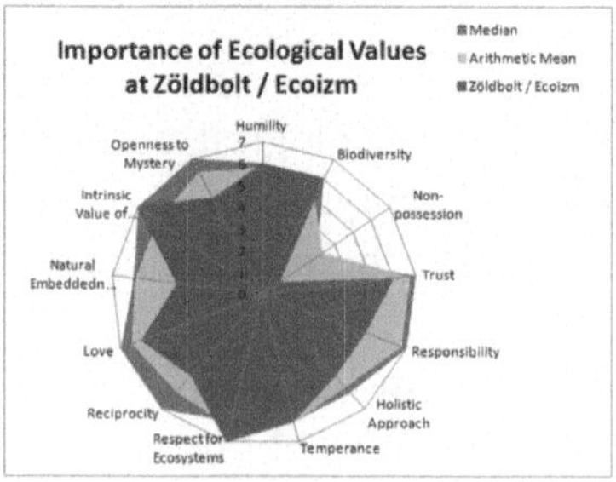

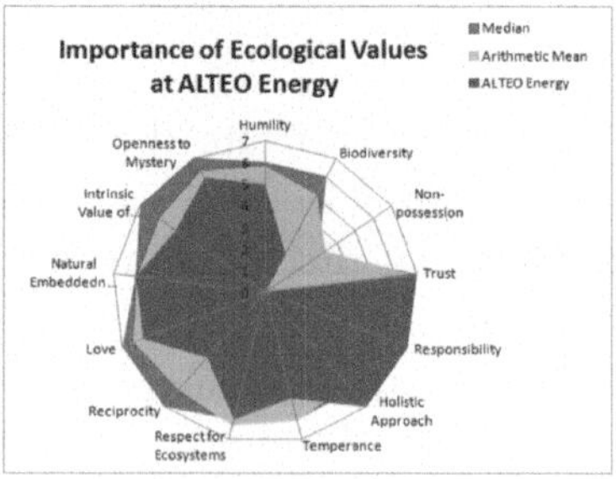

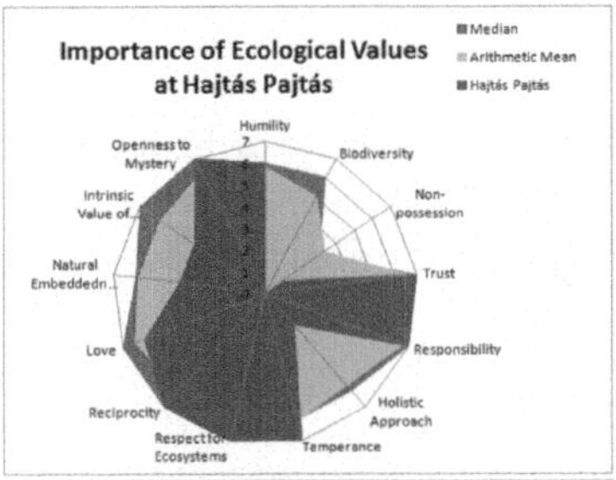
Importance of Ecological Values at Hajtás Pajtás
Median
Arithmetic Mean
Hajtás Pajtás
Humility
Biodiversity
Non-possession
Trust
Responsibility
Holistic Approach
Temperance
Respect for Ecosystems
Reciprocity
Love
Natural Embeddedn...
Intrinsic Value of...
Openness to Mystery

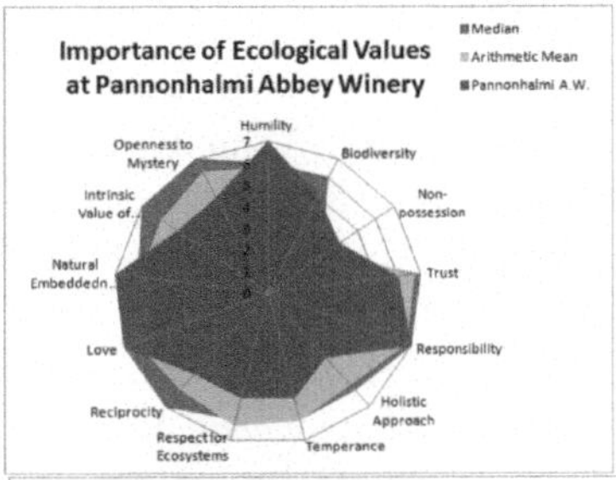
Importance of Ecological Values at Pannonhalmi Abbey Winery
Median
Arithmetic Mean
Pannonhalmi A.W.
Humility
Biodiversity
Non-possession
Trust
Responsibility
Holistic Approach
Temperance
Respect for Ecosystems
Reciprocity
Love
Natural Embeddedn...
Intrinsic Value of...
Openness to Mystery

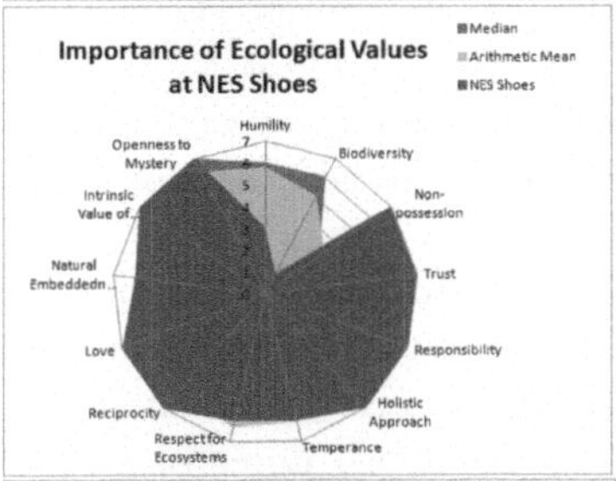
Importance of Ecological Values at NES Shoes
Median
Arithmetic Mean
NES Shoes
Humility
Biodiversity
Non-possession
Trust
Responsibility
Holistic Approach
Temperance
Respect for Ecosystems
Reciprocity
Love
Natural Embeddedn...
Intrinsic Value of...
Openness to Mystery

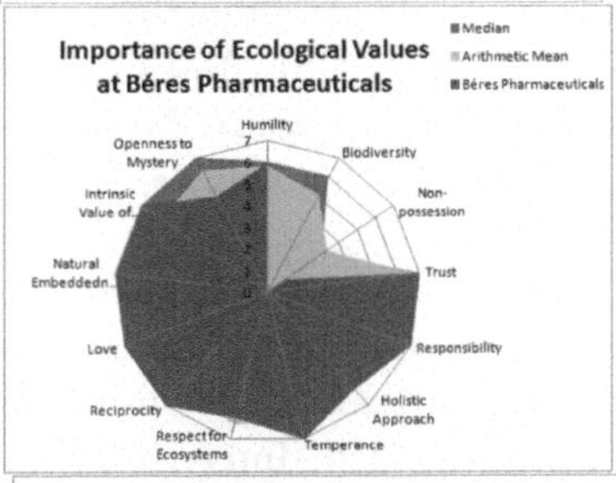
Importance of Ecological Values at Béres Pharmaceuticals
Median
Arithmetic Mean
Béres Pharmaceuticals
Humility
Biodiversity
Non-possession
Trust
Responsibility
Holistic Approach
Temperance
Respect for Ecosystems
Reciprocity
Love
Natural Embeddedn...
Intrinsic Value of...
Openness to Mystery

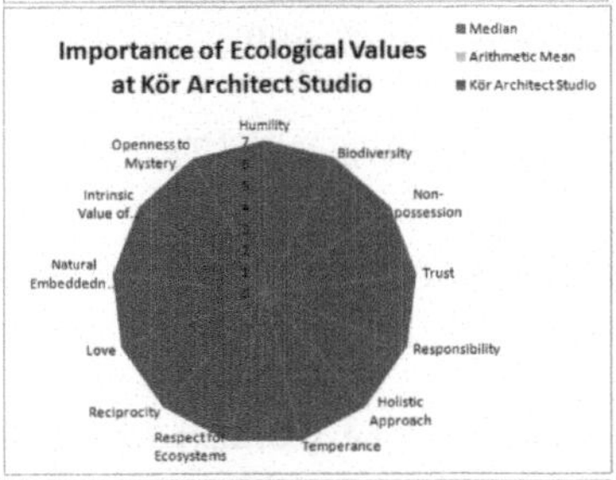
Importance of Ecological Values at Kör Architect Studio
Median
Arithmetic Mean
Kör Architect Studio
Humility
Biodiversity
Non-possession
Trust
Responsibility
Holistic Approach
Temperance
Respect for Ecosystems
Reciprocity
Love
Natural Embeddedn...
Intrinsic Value of...
Openness to Mystery

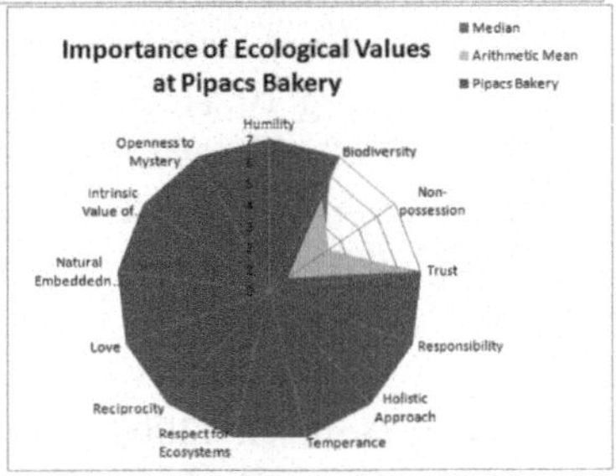
Importance of Ecological Values at Pipacs Bakery
Median
Arithmetic Mean
Pipacs Bakery
Humility
Biodiversity
Non-possession
Trust
Responsibility
Holistic Approach
Temperance
Respect for Ecosystems
Reciprocity
Love
Natural Embeddedn...
Intrinsic Value of...
Openness to Mystery

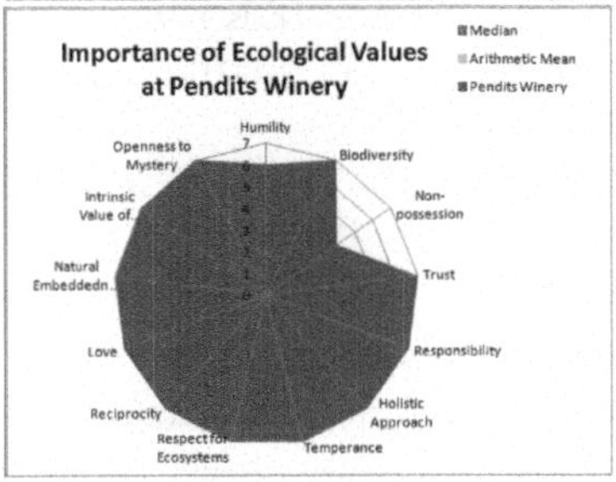
Importance of Ecological Values at Pendits Winery
Median
Arithmetic Mean
Pendits Winery
Humility
Biodiversity
Non-possession
Trust
Responsibility
Holistic Approach
Temperance
Respect for Ecosystems
Reciprocity
Love
Natural Embeddedn...
Intrinsic Value of...
Openness to Mystery

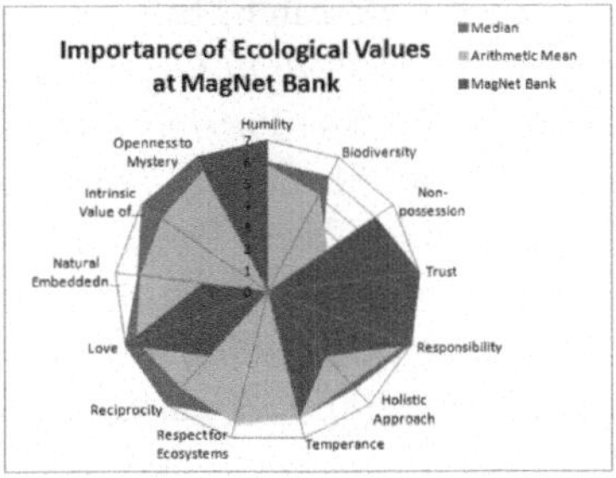
Importance of Ecological Values at MagNet Bank
Median
Arithmetic Mean
MagNet Bank
Humility
Biodiversity
Non-possession
Trust
Responsibility
Holistic Approach
Temperance
Respect for Ecosystems
Reciprocity
Love
Natural Embeddedn...
Intrinsic Value of...
Openness to Mystery

References

Allinson, R. E. (2009). Value Creation as the Foundation of Economics. In L. Zsolnai, Z. Boda, & L. Fekete (Eds.), *Ethical Prospects: Economy, Society and Environment* (Vol. 1, pp. 63–87). Dordrecht: Springer. https://doi.org/10.1007/978-1-4020-9821-5_4.

Allinson, R. E. (2011). The Ethical Producer. In L. Zsolnai (Ed.), *Spirituality and Ethics in Management*. Issues in Business Ethics (Vol. 19, pp. 53–73). Dordrecht: Springer. https://doi.org/10.1007/978-94-007-1153-2_4.

Anderson, R. C. (1998). *Mid-Course Correction: Toward a Sustainable Enterprise: The Interface Model*. Atlanta: The Peregrinzilla Press.

Babbie, E. (2008). *The Basics of Social Research* (4th ed.). Belmont: Thomson Wadsworth.

Balch, O. (2013, March 18). Natura Commits to Sourcing Sustainably from Amazon. *The Guardian*. https://www.theguardian.com/sustainable-business/natura-sourcing-sustainably-from-amazon. Accessed 8 June 2020.

Barlow, C. A. (2010). Interviews. In A. J. Mills, G. Eurepos, & E. Wiebe (Eds.), *Encyclopedia of Case Study Research* (pp. 495–499). Thousand Oaks: Sage. https://doi.org/10.4135/9781412957397.

Boyatzis, R. E. (1998). *Transforming Qualitative Information: Thematic Analysis and Code Development*. London: Sage.

Bowen, G. A. (2009). Document Analysis as a Qualitative Research Method. *Qualitative Research Journal, 9*(2), 27–40. https://doi.org/10.3316/qrj0902027.

Cassell, C., & Symon, G. (2004). Introduction: The Context of Qualitative Organizational Research. In C. Cassell & G. Symon (Eds.), *Essential Guide to Qualitative Methods in Organizational Research* (pp. 1–10). London: Sage. https://doi.org/10.4135/9781446280119.

Chouinard, Y. (2005). *Let My People Go Surfing: The Education of a Reluctant Businessman*. New York: Penguin Group.

CNBC. (2012, June 24). *I Am American Business: Tom Szaky*. CNBC. https://www.cnbc.com/id/100000748. Accessed 10 June 2020.

Creswell, J. W. (2007). *Qualitative Inquiry and Research Design. Choosing Among Five Approaches* (2nd ed.). Thousand Oaks: Sage. https://charlesbickenheuserdotcom.files.wordpress.com/.../creswell_2007_qualitative_inquiry_and_research_design__choosing_among_five_approaches__2nd_edition.pdf. Accessed 4 March 2018.

FAO—Food and Agriculture Organization of the United Nations. (2006). *Livestock a Major Threat to Environment—Remedies Urgently Needed*. http://www.fao.org/newsroom/en/news/2006/1000448/index.html. Accessed 7 June 2020.

Fereday, J., & Muir-Cochrane, E. (2006). Demonstrating Rigor Using Thematic Analysis: A Hybrid Approach of Inductive and Deductive Coding and Theme Development. *International Journal of Qualitative Methods.*, *5*(1), 80–92. https://doi.org/10.1177/160940690600500107.

GlobeScan. (2019). *The 2019 GlobeScan/SustainAbility Leaders Survey* (Report). https://globescan.com/wp-content/uploads/2019/07/GlobeScan-SustainAbility-Leaders-Survey-2019-Report.pdf. Accessed 7 June 2020.

Green Monday. (2020). https://greenmonday.org/en/. Accessed 6 June 2020.

Hawken, P. (1993). *The Ecology of Commerce: A Declaration of Sustainability*. New York: HarperBusiness.

Houdini. (2018). *Planetary Boundaries Assessment 2018: This Is Houdini*. https://api.houdinisportswear.com/storage/2A69199BFCBA925CC9260D61F41301EA566C760FB9A727B5DABB2C330C13D1BC/08df8496f36f49f0bb821fdeafdd775e/pdf/media/e5eec5e201b242e9a2aa14aba9c3b696/Houdini_Planetary_Boundaries_Assessment_2018.pdf. Accessed 6 June 2020.

Houdini. (2020). https://www.houdinisportswear.com/en/. Accessed 6 June 2020.

Hsieh, H.-F., & Shannon, S. E. (2005). Three Approaches to Qualitative Content Analysis. *Qualitative Health Research, 15*(9), 1277–1288. https://doi.org/10.1177/1049732305276687.

Hunyadi, L., Mundruczó, G., & Vita, L. (1997). *Statisztika [Statistics]*. Budapest: Aula Kiadó [in Hungarian].

Interface. (2019). *Annual Report 2019*. https://s22.q4cdn.com/139673446/files/doc_financials/2019/ar/tile20191229_10k_web.pdf. Accessed 7 June 2020.

Interface. (2020). https://www.interface.com/US/en-US/homepage. Accessed 8 June 2020.

Jacob, A. (2012, August 28). Scott Bader—A Long Term Commitment to Composites. *Reinforced Plastics Magazine*. https://www.materialstoday.com/composite-industry/features/scott-bader-a-long-term-commitment-to-composites/. Accessed 9 June 2020.

Jamieson, S. (2004). Likert Scales: How to (Ab)Use Them. *Medical Education, 38*(12), 1217–1218. https://doi.org/10.1111/j.1365-2929.2004.02012.x.

King, N. (2004). Using Interviews in Qualitative Research. In C. Cassell & G. Symon (Eds.), *Essential Guide to Qualitative Methods in Organizational Research* (pp. 11–22). London: Sage. https://doi.org/10.4135/9781446280119.n2.

Krippendorff, K. (2004). *Content Analysis: An Introduction to Its Methodology* (2nd ed.). Thousand Oaks: Sage.

Kvale, S. (1994). Ten Standard Objections to Qualitative Research Interviews. *Journal of Phenomenological Psychology, 25*(2), 147–173. https://doi.org/10.1163/156916294x00016.

Kvale, S. (1996). *Interviews: An Introduction to Qualitative Research Interviewing*. Thousand Oaks: Sage. https://books.google.hu/books/about/InterViews.html?id=lU_QRm-OEDIC&redir_esc=y. Accessed 12 April 2020.

Kvale, S. (2006). Dominance Through Interviews and Dialogues. *Qualitative Inquiry, 12*(3), 480–500. https://doi.org/10.1177/1077800406286235.

Lapadat, J. C. (2010). Thematic Analysis. In A. J. Mills, G. Eurepos, & E. Wiebe (Eds.), *Encyclopedia of Case Study Research* (pp. 925–927). Thousand Oaks: Sage. https://doi.org/10.4135/9781412957397.

Largest Companies. (2020). http://www.largestcompanies.com/company/Houdini-Sportswear-AB-227636. Accessed 7 June 2020.

Lázár, I. (2011). Spirituality and Human Ecosystems. In L. Zsolnai (Ed.), *Spirituality and Ethics in Management*. Issues in Business Ethics (Vol. 19, pp. 95–105). Springer. https://doi.org/10.1007/978-94-007-1153-2_6.

Magnuson, J. C. (2007, August). Pathways to a Mindful Economy. *Society and Economy,* 29(2), 253–284. Sustainability and Sufficiency: Economic Development in a Buddhist Perspective. https://doi.org/10.1556/SocEc.29.2007.2.8.

Maréchal, G. (2010). Constructivism. In A. J. Mills, G. Eurepos, & E. Wiebe (Eds.), *Encyclopedia of Case Study Research* (pp. 220–225). Thousand Oaks: Sage. https://doi.org/10.4135/9781412957397.

Mason, J. (2002). *Qualitative Researching* (2nd ed.). London: Sage. www.sxf.uevora.pt/wp-content/uploads/2013/03/Mason_2002.pdf. Accessed 12 April 2020.

Maxwell, J. A. (2013). *Qualitative Research Design: An Interactive Approach* (3rd ed.). London: Sage.

McLeod, S. A. (2008). *Likert Scale.* https://www.simplypsychology.org/likert-scale.html. Accessed 12 April 2020.

Miles, M. B., Huberman, A. M., & Saldana, J. (2014). *Qualitative Data Analysis: A Methods Sourcebook* (3rd ed.). Thousand Oaks: Sage.

Mills, A. J., Eurepos, G., & Wiebe, E. (Eds.). (2010). *Encyclopaedia of Case Study Research*. Thousand Oaks: Sage. https://doi.org/10.4135/9781412957397.

Natura. (2020a). https://www.naturabrasil.fr/en-us/. Accessed 8 June 2020.

Natura. (2020b). https://www.naturabrasil.com/. Accessed 8 June 2020.

Natura & Co. (2020). *Creating the Best Beauty Group FOR the World* (2019 Report). https://naturaeco.com/report_2019_en.pdf. Accessed 8 June 2020.

Ooley, A. (2013, March 11). *15 Quotes to Inspire Small Business Owners from Yvon Chouinard*. LogoMaker. https://www.logomaker.com/blog/2013/03/11/15-quotes-to-inspire-small-business-owners-from-yvon-chouinard/. Accessed 9 June 2020.

Organic India. (2020a). https://www.organicindia.com/. Accessed 8 June 2020.

Organic India. (2020b). https://organicindiausa.com/. Accessed 8 June 2020.

Parvati Magazine. (2017, April 17). Community: TerraCycle, with Tom Szaky. *Parvati Magazine*. http://parvatimagazine.com/2017/04/community-terracycle-tom-szaky/. Accessed 10 June 2020.

Patagonia. (2020). http://eu.patagonia.com/hu/en/home/. Accessed 9 June 2020.

Pataki, G., & Takács-Sánta, A. (Eds.). (2005). *Természet és gazdaság. Ökológiai közgazdaságtan szöveggyűjtemény [Nature and Economy. Key Texts of Ecological Economics]*. Budapest: Typotex Kiadó [in Hungarian].

Patton, M. Q. (2002). *Qualitative Research and Evaluation Methods* (3rd ed.). Thousand Oaks: Sage.

Robbins, N. B., & Heiberger, R. M. (2011). Plotting Likert and Other Rating Scales. In *JSM Proceedings, Section on Survey Research Methods* (pp. 1058–1066). Alexandria: American Statistical Association. www.montana.edu/msse/Data_analysis/Likert%20Survey%20Graphs.pdf. Accessed 12 April 2020.

Scott Bader. (2019). *Annual Report 2018*. https://www.scottbader.com/wp-content/uploads/Annual-Report-2018.pdf. Accessed 9 June 2020.

Scott Bader. (2020). https://www.scottbader.com/. Accessed 9 June 2020.

Simkin, J. (2020). *Ernest Bader*. http://spartacus-educational.com/JbaderE.htm. Accessed 9 June 2020.

Stan, L. (2010). Content Analysis. In A. J. Mills, G. Eurepos, & E. Wiebe (Eds.), *Encyclopedia of Case Study Research* (pp. 225–230). Thousand Oaks: Sage. https://doi.org/10.4135/9781412957397.

TerraCycle. (2018). https://www.terracycle.com/en-US/. Accessed 5 July 2018.

TerraCycle. (2020). https://www.terracycle.com/en-US/. Accessed 10 June 2020.

The Economic Times. (2017). *Organi India Eyes Rs 500 Crore Turnover by FY20 End*. https://economictimes.indiatimes.com/industry/cons-products/fmcg/organic-india-eyes-rs-500-crore-turnover-by-fy20-end/articleshow/62240529.cms. Accessed 8 June 2020.

WCED—World Commission on Environment and Development. (1987). *Our Common Future*. Oxford, NY: Oxford University Press.

Zimberoff, L. (2017, June 12). *This Guy Is Trying to Break Hong Kong's Meat Addiction*. Bloomberg. https://www.bloomberg.com/news/articles/2017-06-12/this-guy-is-trying-to-cure-hong-kong-s-meat-addiction. Accessed 7 June 2020.

Zsolnai, L. (2001). Ökológia, gazdaság, etika [*Ecology, Economy, Ethics*]. Budapest: Helikon Kiadó [in Hungarian].

6

Comparative Analysis of Ecologically Conscious Business Models

This chapter analyzes the business models of ecologically conscious business organizations in the international and the Hungarian sample. The elements and characteristics of the business models are presented in six main dimensions, as defined by Chesbrough and Rosenbloom (2002). These dimensions include value propositions, market segment, cost structure and profit potential, structure of the value chain, position within the value network, and competitive strategy. This approach makes it possible to compare the results of the case analyses of the international and the Hungarian enterprises.

Ecologically Conscious Business Models in the International Sample

In this section, the business models of ecologically conscious enterprises in the international sample are presented in a uniform way.

A. Ócsai, *Ecologically Conscious Organizations*, Palgrave Studies in Sustainable Business In Association with Future Earth, https://doi.org/10.1007/978-3-030-60918-4_6

Green Monday

Value propositions: Green Monday's activities are aimed at helping combine the seemingly small but easy-to-implement steps of individuals who are looking to change their lifestyles to create global-level change. The method it promotes supports and indicates reductions in the carbon footprint, while leaving individuals free to decide when and how to reduce their meat consumption. For instance, people can stop eating meat for a full day once a week, or every lunchtime. They can also reduce the amount of meat they consume during meals, or omit red meat, which has a high environmental impact, from their diet. An important element of Green Monday's value proposition is that consuming less meat has a significant impact on the planet as well as on humans. The production of plant-based food may have a carbon footprint that is only about 3 per cent of meat production, and use a tenth of the water. It can help reduce global hunger, because using one unit of land for crop production can provide more than ten times as many people with proper nutrients than if that same unit is used to produce meat. A plant-based diet can have the following beneficial effects on human health: it can reduce the risk of lung and kidney problems, breast cancer, strokes, heart attack, cardiovascular disease, infertility, osteoporosis, type-two diabetes, and obesity. Additionally, it increases the intake of antioxidants, flavonoids, fiber, polyphenols and micronutrients, and the number of healthy life years due to the beneficial effects of the former (Green Monday 2020). Founder David Yeung also considers business sustainability to be important and is therefore a proponent of the triple-bottom-line concept. He wants to make his mission-based business economically successful while making favorable environmental and social impacts (Zimberoff 2017).

Market segment: Green Monday is a platform that has both a social and a business segment. The social branch, led by the Green Monday Foundation, engages in awareness-raising, dissemination, and knowledge promotion campaigns about climate change, Earth Day activities, restaurant network development, and food aid campaigns to promote sustainable lifestyles. The business side consists of three main areas. (i) As an active market player, Green Monday intends to expand its sustainable food retail supply through an initiative called Green

Common. Its network of stores sells plant-based food products and sustainable household items, and provides advice on how to prepare plant-based food at home. (ii) Green Monday Solutions offers sustainability and social responsibility consulting (involving the themes of sustainable development, CSR, employee loyalty, marketing and business development, and the introduction of Green Monday products and services) to domestic and international companies (restaurants, event organizers, food producers, supermarkets, and marketing companies). (iii) Green Monday Ventures is a private equity fund for impact investing that supports innovative and ethical business ideas whose business models, operations, products and services address global climate change, food safety, health, or animal welfare issues (AVPN 2016; Green Monday 2020).

Cost structure and profit potential: The business side of Green Monday has become successful and the company has been recognized with a number of business awards since its inception (Green Monday 2020). Partner companies have also benefited from the successful operation of Green Monday and the successful dissemination of a sustainable lifestyle. For example, England-based PizzaExpress, which has 20 stores in Hong Kong, has seen double-digit growth in vegetarian menu sales on Mondays. As a side effect, the company has also boosted its total daily turnover on Mondays. The vegetarian community has also evolved significantly, as proven by the increase in the number of vegetarian restaurants in Hong Kong from 130 to 250 from 2013 to 2017. Success is also reflected in company finances. Green Monday's total retail and reseller sales approached 12 million USD in 2017, and have since increased significantly. The sales and profit outlook is favorable, as the global market for alternative protein rose to about 2.2 billion USD in 2019, compared to a global meat market of 1.7 trillion USD. Market forecasts estimate that the alternative meat market could be worth between 90 billion USD and 470 billion USD by 2040. Green Monday is now present in five countries in Southeast Asia, in the United Kingdom, and in the United States. Its products are available in ten countries and will be in ten more within the year (Zimberoff 2017; Green Monday 2020; Yusof 2020).

Structure of the value chain: according to the concept of Green Monday, it is not enough to shed light on things that should be changed; feasible alternatives must also be provided in order for real change to take place in people's behavior and lifestyles. Important elements of the value chain are therefore research and development and knowledge sharing, implemented in part through the Green It Yourself community program that provides sustainability advice regarding housekeeping and food preparation and organizes community programs (with a focus on recipe databases, cooking tips, cooking competitions, and blogger meetings) (AVPN 2016; Green Monday 2020). The Green Knowledge & Research Center aims to raise awareness about sustainability and share best practices, mainly in the Hong Kong region. Its research interests range from environmental and behavioral change associated with climate change, cultural phenomena, sustainable consumption, innovation, and environmental policy to regional development. It shares its latest results and news with its stakeholders through international events and a monthly newsletter (Green Monday 2020). Its research and development activities also include the development of new plant-based foods and meals. As part of its Right Treat initiative, the type of food innovations are studied that will bring long-term benefits to the planet, humanity, and animals alike, and will not impose additional liabilities and suffering on other beings and future generations. The first major achievement of this program was the development of a mushroom-, rice-, and pea-based pork substitute called Omnipork that took place over a year and a half with Canadian researchers. However, Green Monday does not only innovate by itself. Among others, it financially supports research by the California-based company Beyond Meat, which makes plant-based burgers. Green Monday also supports Perfect Day, a cell-based farming enterprise that produces dairy products from cell cultures, and a food-design-based technology company called Lighter (AVPN 2016; Zimberoff 2017; Mannion 2018; Green Monday 2020).

Food and other products sold in Green Common stores are sourced from strictly sustainable and ethical suppliers. The main sourcing principles are meat-free, healthy, and fair trade. Green Monday prefers local products from Hong Kong's organic producers, social enterprises, small-, and medium-sized enterprises and fair trade organizations (AVPN 2016; Green Monday 2020).

Green Monday Ventures can be seen as a seed investment fund that invests in companies that serve social or environmental causes (reducing the ecological footprint, increasing health and well-being, and supporting individuals and businesses to become sustainable), which are also capable of generating a profit. In addition to financial investments, in many cases Green Monday has also supported companies through appropriate forms of synergy. For California-based Beyond Meat, for example, providing access to Asian commercial channels has represented significant help for their development (AVPN 2016).

Green Monday's mission is actively supported by many renowned and popular personalities, from domestic and international artists and government officials to opinion leaders. Its clients include a number of major financial institutions, including the Bank of China and HSBC, public utility companies like Hong Kong International Airport, real estate management businesses, food companies such as Beyond Meat and Del Monte, retail businesses like Quiksilver and Procter & Gamble, and service companies such as eBay and Google. It also serves more than 1000 restaurants mainly from the region, but also others in the USA, Peru, the UK, Italy, Australia, and Japan, and over 2000 schools in 33 countries. Around 600 domestic universities such as the University of Hong Kong, Chinese University of Hong Kong, Hong Kong Baptist University, and Tsinghua University, and numerous foreign universities like Columbia University, Washington University in St. Louis, University of California Los Angeles, and Wharton School are among its partners. Green Monday also supplies governmental organizations such as the Hong Kong Office of Health, the United States Consulate in Hong Kong and Macao, and the Canadian and Italian Chamber of Commerce (Green Monday 2020).

Green Monday's value chain also includes the Green LUCK Banquet program, which works with event management companies, food aid, and shark rescue organizations to make events healthy, sustainable, and free from food waste. By developing event management culture, the program aims to reduce the carbon footprint of events through providing vegetarian food, eliminating food waste, serving the community (offering leftover food to Hong Kongese families in need), and not harming the world (by reducing waste and protecting sharks by avoiding traditional dishes made from shark fins) (Green Monday 2020).

Position within the value network: Green Monday's social and business branches work together across a wide range of areas. According to its own surveys conducted in cooperation with market researcher Ipsos Hong Kong, the reputation of Green Monday and its mission rose between 2014 and 2016 from 22 to 45 per cent among 15–24-year-olds, and 22 to 33 per cent of the total female population (percentages indicate the proportion of respondents who recognized the company name). In 2020, an independent survey showed that more than 45 per cent of people in Hong Kong know about the company's activities (Green Monday 2020; Yusof 2020). Another significant impact is that, due to Green Monday's activities, by 2016, 1.6 million people, (22 per cent of Hong Kong's total population—where per capita meat consumption is among the highest in the world—had become "flexitarians," meaning people who mostly eat a vegetarian diet but sometimes consume meat (AVPN 2016; Hackett 2019). Figures from 2008 showed that only 5 per cent of people in Hong Kong were flexitarians. By 2020, the proportion of vegans and flexitarians in Hong Kong had risen to 34 per cent. More than three million people worldwide have joined the Green Monday movement through taking a "Pledge to go Green Monday." All these steps and changes have reduced CO_2 emissions by 900,000 tons a year, which would have required 40 million trees to absorb. In addition, the lives of 300 million animals (cattle, pigs, and poultry) have been saved along with 375 billion liters of water—enough to supply more than 500 million people (Green Monday 2020; Yusof 2020).

Green Monday's activities were recognized with significant awards as early as in the year after its foundation. These included the "Best Idea of the Year 2013" award and the East Week "Outstanding Corporate Strategy Award." In 2014, founder David Yeung received an award for being one of the "100 Most Creative People in China" from *Fast Company* magazine, and Green Monday received the magazine's "Top 100 Most Innovative Companies in 2014 China" award. In 2015, Green Monday also won Fast Company's "Top 50 Most Innovative Companies in China" award, the title of one of the "Asian 100 Pioneers" from Purpose Economy, Conscious Company Magazine's "17 Rising Social Entrepreneurs" award, and the "Ten Outstanding Young Persons in Hong Kong" award. In 2018, the chief executive officer was named "Social Entrepreneur of the Year" by the World Economic Forum and Schwab Foundation. Green Monday was awarded "Company of the Year" by the animal rights organization People for the Ethical Treatment of Animals in 2019 (Green Monday 2020; Yusof 2020).

Competitive strategy: When Green Monday was established, there were only a few vegetarian restaurants in Hong Kong. Since its inception, it has not only expanded the supply with its own business portfolio, but also inspired others to launch vegetarian restaurants and food services by taking advantage of growing awareness about sustainable lifestyles and diets. However, Green Monday did not primarily see this as an increase in the strength of the competition, but as the spread of its concept and values, which they see as a positive development, meaning they are able to work with an increasing number of partners. Opportunities for cooperation have been consciously sought from the outset. One of these involved joining the global Meatfree Monday movement, which has made significant strides in reducing meat consumption in countries including the United States and Great Britain through more than 20 worldwide campaigns. Through its corporate partners, Green Monday is involved in many sectors, and has the primary target of connecting different stakeholders throughout the entire value chain and ecosystem of the sustainable food industry (from producers and traders to consumers) and developing mutually beneficial forms of collaboration (AVPN 2016; Green Monday 2020).

Houdini Sportsware

Value propositions: The most important element of the value propositions of Houdini is sustainability, which does not influence the company's operations as a separate functional area, but is an integral part of all its activities. For Houdini, sustainability is not the opposite of good business, but rather a precondition for it. Nature has intrinsic value and provides the blueprint for transforming the current business system into a circular one. Nature is the most important source that makes it possible to earn a living, so Houdini intends to do everything to protect it and improve its current state (Houdini 2020).

Market segment: Houdini is a men's, women's and children's clothing retailer. It specializes in the production and distribution of functional clothing in an ethical and sustainable manner. It offers its products primarily to ecologically conscious customers to whom sustainability is important. In addition to ecological and social principles, it wants to reach all consumers who like outdoor sports through offering high quality and timeless product design (Houdini 2020; Largest Companies 2020).

Cost structure and profit potential: In the financial year ending April 2019, annual sales increased by 11 per cent to 16.8 million EUR, indicating a doubling in the growth rate compared to the previous year. The average growth rate of revenue was around 20 per cent in previous years. Profit had been stable at around 2 per cent before 2017. It then dropped to a loss of 6.6 per cent in 2017, which was likely due to the increase in expenses associated with growing sales, a significant increase in assets (by 33 per cent to 9.5 million EUR), the number of employees (from 37 to 42), and rising current liabilities. Earnings before taxes were less than 1 per cent in 2018 and minus 1 per cent in 2019. The amount of current liabilities has fluctuated, but as turnover has grown steadily, it is now at a healthy ratio. Due to stable demand, brand popularity, industry recognition, and the company's continuous innovation activity, financial indicators can be expected to remain acceptable and turnover and asset expansion can continue on a path of organic development (Largest Companies 2020).

Structure of the value chain: Based on a holistic approach, Houdini seeks to roll out the principle of sustainability along its entire value chain. This effort is manifested in its business model innovations, mainly within the field of the collaborative economy, the goal of which is to decouple growth from the use of resources. Houdini has introduced several services for extending product lifetime, such as "Care," "Repair," "Reuse," and "Rental." In its brick-and-mortar retail stores and an e-commerce channel, through the application of green building design the company inspires, educates, assists, and guides its community of consumers to engage in better product care, provides a repair service, and creates a marketplace for the reuse of second-hand yet fully functional Houdini products. Houdini considers sharing and product-as-service solutions to be alternatives to conventional ways of consuming. Through its rental initiative, Houdini provides access to otherwise expensive products as an alternative to ownership while eliminating unnecessary resource demand for fabrics that may be used less frequently (Krarup et al. 2015; Houdini 2018, 2020).

The principle of sustainability is strictly applied from the product design phase onwards. For each product idea, and in several phases during product development, Houdini utilizes a Designer's Checklist, involving asking the questions: Does this product deserve to exist and is it versatile and durable enough? Is nothing added that is not needed? Will the product last long enough, age with beauty, and be easy to repair? And is there an "end-of-life" solution for the product? Houdini also adheres to strict principles (weather resistance, safety, visibility, and timelessness) when choosing from its own self-developed color set, so that its customers can use their comfortable, functional, and eye-catching products for as long as possible (Houdini 2018, 2020).

The environmental footprint of clothing products largely depends on the raw materials that are used. Since the fall of 2019, Houdini's new products have been made of fabrics that are 100 per cent recycled, recyclable, renewable, biodegradable, or sustainable Bluesign® certified. Sixty-three per cent of that collection was produced in a completely circular way, and Houdini's goal is to be 100 per cent circular by 2022 (Houdini 2020). Polyester-based fabrics are widely used due to their durability, fast drying, moisture repellency, reusability, easy handling, and favorable properties due to the lower energy and water requirements of production. In collaboration with a Japanese company called Teijin, Houdini has been using polyester fibers from a fully closed-loop recycling system (Eco Circle®) for its products since 2007 (Krarup et al. 2015). In the case of natural fabrics, merino wool, which has the most favorable properties in terms of retaining heat and repelling moisture and bacteria of nearly all natural materials, has been used since 2011. For Houdini's wool products, full traceability of wool origin was achieved in 2016, which is an important milestone in its sourcing policy that is associated with strict animal welfare standards. In order to preserve biodegradability and reusability, natural fiber fabrics are blended during manufacture solely with natural fibers (e.g., silk), not synthetic fibers (Houdini 2020). A project called the Houdini Menu was initiated in 2016 to prove that the properties of the company's materials are also favorable in terms of sustainability, during which second-hand Houdini clothes made of 100 per cent renewable organic materials were composted. Using this compost, vegetables and spices were grown that were then served as part of a menu at the world's first vegetarian reception of its kind in April 2017 (Di Ciancia 2017). In the fall of 2018, Houdini also opened a composter for worn-out sportswear in Rosendals Garden, Stockholm, to test on an ongoing basis how its natural clothes can be transformed into healthy soil (Houdini 2020). In the design of materials, Houdini strives to use the simplest and fewest possible ingredients to improve their recyclability. In order to utilize the residue generated during cutting, a project called Scrap Me Up was launched in 2010. During this event, a series of a limited number of pieces was made from materials previously classified as waste (Krarup et al. 2015). Of the various fabric treatment materials (used for improving water repellency, and with antibacterial effects), after eight years of research and experimentation, highly harmful fluorinated substances (PFOA, PFOS) were entirely replaced (by 2018). During these activities, Houdini worked closely and continuously with raw material suppliers. The company co-creates, when necessary, new materials that are not protected by patents. Such open-source innovation thus helps the use of ecological materials to gain ground (Houdini 2018, 2020).

True to its strict environmental principles, Houdini exercises great care in selecting its manufacturing partners and manufacturing processes. The company has consciously decided to form partnerships with a minimal number of suppliers and manufacturers so that environmental- and human-rights-related conditions can be managed (monitored two to five times a year) through long-term, close, and stable forms of cooperation. Almost all Houdini products are now made in Europe (in the Baltics, Poland, and Portugal), with only 3.5 per cent being produced in China. Since 2019, all materials have been recycled, renewable, recyclable, biodegradable, or closed life cycle materials. Fabrics are sourced mainly from Japan, Italy, USA, Lithuania, Taiwan, and Sweden. Houdini's suppliers are not Fairtrade-certified, but, according to the company, its own internal standards are stricter than Fairtrade standards, so Houdini does not plan to join the Fairtrade certification scheme (Krarup et al. 2015; Houdini 2018, 2020).

Attitudes toward customers are also in line with Houdini's set of principles. The company tries to redefine the concept of consumption by reducing overconsumption as much as possible. Since 2012, Houdini has allowed its consumers to try out outerwear products (jackets and trousers) by renting them on a short-term basis. Conscious customers can consider whether it is necessary to buy and possess those products, or if it is sufficient to just rent them for a while. With this approach, fewer products are enough for more customers, while the value of use and enjoyment is not reduced. For a similar reason, Houdini has also launched a second-hand clothing sales service in its stores, which it occasionally links to charitable offerings. Customer appreciation is also increased by providing a lifetime warranty for all Houdini products. If a product is damaged through the customer's own fault, it is replaced at a lower price than a new one. However, after a while products wear out from regular use, and if they can no longer be repaired, they can be dropped off at Houdini stores and partner stores to be included in Houdini's own recycling system (Houdini 2020).

Position within the value network: Houdini is a leading representative and supporter of sustainability principles across its entire value network. It has continuously worked to transform consumption patterns, and its products are increasingly being rented, reused, and repaired. Besides these efforts, based on the Stockholm Resilience Center's concept of Planetary Boundaries, Houdini has been working with the organizations Albaeco and Mistra Future Fashion to develop an open-source, holistic environmental assessment system that will allow any business to measure its performance in line with the above-described concept and the UN Sustainable Development Goals (SDGs). The goal is to eliminate all negative effects in the shortest possible time, which can be achieved through knowledge sharing and innovation involving interdisciplinary collaboration. In line with these aspirations, in 2018 the first ever corporate Planetary Boundaries analysis was created. This kind of assessment makes visible factors that are often omitted. Adding social boundaries to the nine planetary boundaries highlights not only the ecological areas that it is important to consider, but also the conditions that influence the well-being of people working within the value chain. To further increase the transparency of its environmental performance, Houdini publishes a Planetary Boundaries Assessment report every three years, and details about its seasonal sustainability status twice a year. These statistics show the current status of all Houdini products regarding what proportion have been made from recycled, recyclable, biodegradable and renewable, Bluesign®-certified and circular fabrics (Houdini 2018, 2020).

Houdini's achievements have been recognized. In 2010, CNBC named Houdini one of the 25 most creative companies in Europe for its work on sustainability. Svenskt Sportforum awarded the title "Woman of the Year in the Sports Industry" to Houdini's CEO Eva Karlsson in 2011. In 2012, Houdini won the Swedbank Sustainability Prize for their commitment to sustainability and profitable growth. In recent years, several of its products and raw materials have won sustainability awards (the ISPO Eco Achievement 2014, Scandinavian Outdoor Award Sustainability Gold 2015, ISPO Award Gold 2015 and 2020). In 2017, Houdini received one of the most prestigious awards in the fashion industry, the "H&M & Elle Conscious Award," for its commitment to sustainability and its role in renewing the industry. In 2017, Eva Karlsson was winner of the "Pontus Schultz Award for a humane economy" for uniting function and sustainability and adapting research on Planetary Boundaries to make it relevant to Houdini, while also making a profit and inspiring others. In 2018, Karlsson was deemed the winner of the "NMC Award for Sustainable Leadership" for her wholehearted commitment to sustainability and striving to work within Planetary Boundaries (Houdini 2020).

Competition strategy: Houdini has repeatedly confirmed and demonstrated through its operations that its sustainability-related work occurs in close cooperation with other industry players. For example, in a multi-annual research project, the company worked together with raw material suppliers, industry partners (including Bergans, Norrona, and OrganoClick) and a higher education institution (the University of Borås) to develop a revolutionary and innovative material to replace fluorinated carbon-based fabric treatments that are extremely harmful to the environment. It is part of Houdini's strategy to make its discoveries as widely available as possible, so, for example, patents are not used to protect the materials the company has developed in order to make them freely available and usable. Building a sustainable future is possible only through knowledge sharing and innovation via interdisciplinary cooperation. Houdini emphasized this at a sustainable fashion seminar held in the Swedish parliament, for example. Eva Karlsson talked about Houdini's experience with environmental issues and her expectations in relation to decision-makers, including a number of fashion industry players, researchers, and politicians. In the otherwise highly competitive outdoor clothing industry, at the international level (which includes the companies Columbia, Patagonia, and Timberland) and in Sweden (Boomerang, Filippa K), several players are focusing their business models on sustainability and inspiring each other, too (Krarup et al. 2015; Di Ciancia 2017; Houdini 2020).

Interface

Value propositions: The main goal of Interface is to become the most valuable company in the world that offers modular carpets and related services, while leading a world-changing sustainability movement in parallel with eliminating its negative environmental impacts and developing activities that regenerate nature. Even when Interface was founded, there was a focus on making a change, as Ray Anderson began manufacturing modular carpets, which were at that time unknown. In 1994, inspired by Paul Hawken's book *The Ecology of Commerce* (Hawken 1993), he also started on a major transformation of the conventional industrial business model ("take-make-waste") into a renewable, cyclical, more gentle one. Anderson questioned the principles on which petroleum-based carpet manufacturing were based by acknowledging that the traditional business model exploits the Earth and destroys symbiotic ecosystems. He decided that Interface had to become an environmentally sustainable and ultimately nature-regenerating company. The new focus has been on the radical redesign of processes and products and the introduction of new technologies and systems that reduce or even eliminate waste and emissions while increasing the use of renewable materials and energy. Interface's Mission Zero defines the target of eliminating all negative environmental impacts by 2020. Anderson described the process of creating a completely sustainable company as being like climbing a mountain ("Mount Sustainability") that has seven faces, all of which must be scaled. The seven key areas on which Interface focuses its development-related efforts and monitors its performance are the following: (i) eliminating all types of waste related to all activities; (ii) eliminating the use of toxic substances in all products, vehicles, and facilities; (iii) operating its buildings with 100 per cent renewable energy; (iv) redesigning processes and products to close the technological cycle through recovering materials and using bio-based materials; (v) transporting people and products in a more efficient way so that waste and emissions can be eliminated; (vi) creating a culture based on the principles of sustainability to improve the lives of all stakeholders (employees, partners, suppliers, customers, investors, and communities); and (vii) developing a new business model that demonstrates and nurtures the value of sustainability-based business (Gingerich 2016; Interface 2020a).

Market segment: Interface is a global flooring company that specializes in the design, production, and sales of carpet tile and resilient flooring, including luxury vinyl tile and rubber flooring. It is a global leader in the commercial carpet tile industry, accounting for around a third of the approximately 39 billion USD global commercial flooring market. Interface has recently entered into the residential and luxury vinyl flooring segment, and, through acquisition of nora Holding GmbH in 2018 (which also operates in the market segment of rubber flooring for customers mainly in healthcare, life sciences, and education), has more than doubled its customer base in a few years (Interface 2019, 2020b; Grand View Research 2020).

Cost structure and profit potential: The market segments served by Interface are associated with the largest margins for any type of flooring, and the company's products represent a premium category in terms of design, sustainability, and innovation, with the potential for more profit than the industry average. So far, most of the clients of Interface have been architects, designers, and end users, which earns more money for Interface than if it primarily traded with distributors. Since 2017, Interface has been executing a market diversification strategy to increase its presence and market penetration for modular carpet in non-corporate office market segments such as government, education, healthcare, hospitality, and retail space. As a result, in 2019 its net sales increased by almost 14 per cent to over 1.3 billion USD, net income grew by more than 57 per cent to 79 million USD, and earnings per share also significantly increased by nearly 7 per cent. Radical innovations that have steadily and substantially reduced the company's negative environmental impact over the years have also helped cut material and energy costs. The example of Interface indicates that a sustainable business model that supports a livable future has the potential to be financially sustainable (Interface 2019, 2020b; Grand View Research 2020).

Structure of the value chain: During the radical sustainability turn in 1994, Interface did not have the opportunity to rely on past experience, so the company used external knowledge. It set up an expert group of researchers, environmental activists, and entrepreneurs called the Eco Dream Team. These individuals helped Interface to create a business-friendly sustainability framework and keep this under review. The Swedish The Natural Step methodology, developed by Dr. Karl-Henrik Robert to preserve the quality of terrestrial living systems, was also used to create a new business model. According to this concept, in a sustainable society, in order to preserve nature, the concentration of substances derived from the Earth's crust (carbon dioxide, heavy metals, and minerals), substances produced by society (such as antibiotics and endocrine-disrupting chemicals), and the extent of physical destruction (deforestation and ground-water depletion) must not be increased. Additionally, nothing should place limits on human health, potential for impact, knowledge, impartiality, and meaning (The Natural Step 2014, 2020; Interface 2020a).

One of the essential elements of Interface's value chain is radical innovation driven by the principle of sustainability in the field of product and process design and development. Its research and development department is recognized by industry peers, and generated 17.8 million USD in 2019, an increase of more than 27 per cent on 2017. Interface actively collaborates with developers of fabric, yarn, and carpet backing suppliers. Many of its new products and technological innovations were the first in the world to solve industry problems or improve a technological step by making it cleaner, faster, or more environmentally friendly. For example, the environmental footprint of the company's TacTiles glue-free installation system is more than 90 per cent less than that of traditional carpet adhesives. The company's "i2" products, inspired by nature (biomimicry), vary in terms of patterning and coloring so that the tiles can be installed in any order and orientation. In this way, their harmonious appearance is maintained during wear and module replacement over time, and they can be installed faster, repaired more easily, and recycled. Accordingly, they generate approximately 90 per cent less waste than traditional broadloom carpets. As a result of the circularity of the production processes, the carbon footprint of Interface's products has been reduced by more than 50 per cent since 1994, and in 2017 they created the world's first carbon-negative carpet prototype. To offset the remaining carbon footprint and greenhouse gas emissions, the company works throughout the total product life cycle with raw materials that are carbon neutral, as verified by an independent certifier. It also supports emission reduction and offsetting projects worldwide, such as reforestation efforts, anti-deforestation campaigns, wind farms, and domestic heating modernization. Expanding into new market segments, Interface insists on adhering to strict principles. In 2017, the company entered the market for modular resilient flooring (customers can combine fully integrated hard floor and soft carpet coverings on demand). Using this combined system, the materials used for each module can be produced in a controllable way and completely reused at the end of their life through Interface's recycling program, called ReEntry. The company holds numerous United States and foreign patents and multiple trademarks for its products and production processes (Lampikoski 2012; Interface 2019, 2020a).

Interface has eight manufacturing locations in seven countries on four continents (USA, Netherlands, UK, Thailand, China, Australia, and Germany), so it can provide local solutions to its customers in all markets. The raw materials needed for production are also sourced from regional and global external sources to increase security of supply, and recycled fabrics are used for the bottom layer of carpets. An ISO 14001 environmental management system is in place. Four of its product groups have the GreenCircle Certified Environmental Facts label, and 99 per cent of products have an Environmental Product Declaration certified by an independent certifier that authentically describes the related ingredients and environmental impacts, supported by comprehensive life cycle analysis. All Interface products are certified by the United States Green Building Council according to the Leadership in Energy and Environmental Design (LEED) assessment system that transparently demonstrates and certifies the environmental impact of buildings and construction activities. Since 2011, Interface has been actively involved in the development of a new product certification (a "Health Product Declaration") which provides full transparency through disclosing the ingredients in products, and encourages their further development. As a result, the use of halogen- and heavy-metal-based flame retardants, fluorinated stain-repellent coatings, aqueous yarn paints, and liquid carpet adhesives has been eliminated in standard products manufactured in the United States (Lampikoski 2012; Interface 2019, 2020a).

At the end of 2019, Interface was employing more than 4100 employees worldwide, 59 per cent of whom were engaged in manufacturing, with the help of 225 temporary staff. There are also trade unions in several countries, and management maintains a good, active relationship with them. In addition to environmental sustainability, Interface takes the principles of ethical business seriously, with a special focus on respect for human rights and labor standards. Its code of ethics includes anti-corruption and anti-bribery measures, as well as supplier clauses. A hotline and online interface for reporting fraud and ethical misconduct anonymously and confidentially are also available to employees (Interface 2019, 2020a).

The majority of Interface customers (51 per cent of them at the end of 2019) are office building investors or tenants. The company also supplies educational institutions, the hospitality sector, healthcare institutions, retail and financial enterprises, governmental customers, and individual households. The principal geographical markets are the Americas, Europe, and Asia Pacific. In 2019, these areas accounted for approximately 57, 29, and 14 per cent of total net sales, respectively (Interface 2019, 2020b).

Position within the value network: Interface is one of the most innovative companies in the world, with a sincere commitment to sustainability and design. Important recognition of this is the fact that Interface has been the only company to be mentioned, for over 20 years, as being among the best companies for sustainability performance in The GlobeScan-SustainAbility Leaders Survey. Ray Anderson's work and Interface's environmental performance and purpose-driven culture have been recognized many times. From these awards, just a few are highlighted here: in 1996 Ernst & Young's Entrepreneur of the Year award; in 2007 *Time* magazine's Heroes of the Environment award; in 2010 the American Society of Interior Designers' Design for Humanity Award; and in 2016 Ethical Corporation, Responsible Business Awards: Best Company (GlobeScan 2019; Interface 2020a).

Interface sets an example to other market players through its activities and performance in order to help create a viable future. The company demonstrates that it is possible to operate a business model that is carbon neutral or even carbon negative. The results of Interface's Mission Zero so far speak for themselves. Since 1996, the company has reduced its greenhouse gas emissions by 96 per cent, the amount of waste-to-landfill by 92 per cent, water intake by 89 per cent, and the carbon footprint of its products by 69 per cent. Eighty-nine per cent of its energy comes from renewable sources, and 54 per cent of all raw materials used are organic or made of recycled materials. Its sustainability mission, called Climate Take Back, has defined goals that even more ambitious. In addition to having zero negative environmental impacts, Interface wants to change the prejudice against carbon dioxide so it is no longer considered an "enemy" but a resource. The company also strives for revolutionary industrial transformation that can create a better future for humanity, and regenerate the climate control capacity of the biosphere (Interface 2020a, b).

Competitive strategy: Interface operates in a highly competitive global market that includes a large number of players, even after significant consolidation in recent years. The company is a global leader of commercial flooring solutions, but several of its competitors (including Milliken,

Mohawk, Tarkett Tandus Centiva, Shaw, and Bentley) are working in more market segments and may have greater financial capacity. However, Interface is serving an increasing number of segments with hard and soft surface flooring, which generate the largest margins. Its market value has expanded from four to nine billion USD in a couple of years, and it can now successfully compete in terms of design, sustainability, and innovation. The principal competitive factors in Interface's markets are brand recognition, quality, design, service, broad product lines, product performance, marketing strategy, pricing and sustainability. In most of these, Interface has a competitive advantage, so premium prices and industry-leading margins appear to have been secured for the company for the coming years. In addition, Interface's deep commitment and decades-long effort and successful practice of ecological sustainability clearly differentiate and highlight the company, enhance the brand, and represent a strategic advantage (Interface 2019, 2020b).

Natura

Value propositions: Brazil-based Natura's vision is to become a group of global brands with a local and global presence through its responsible operations, caring relationships, and high-quality ethical products and services. The members of this group, as a community committed to building a better world, have better relationships with themselves, others, nature, and with everything as a whole. Natura's mission is to contribute to improving the state of the world by building on its commitment to transparency, sustainability, and well-being. Based on its core values—a close relationship with nature, science and technology—, Natura works to expand and mobilize its network. In addition, the company strives to develop creative solutions for a life inspired by beauty, happiness, and sustainability. Natura recognizes that the source of life is nature, and this provides inspiration and creativity for all the activities of the company. Natura, therefore, turns to nature with great love. The company studies and utilizes both traditional and state-of-the-art knowledge because the products of the best scientific and technological research are essential for making high-quality products. Natura nurtures relationships so that all stakeholders who share similar values can join its mission. The company's basic goal is to have the greatest possible positive impact during its operation and the smallest possible negative one (Natura 2020a, b, c).

Natura established the Natura & Co. global cosmetics group in 2017. At that time the former was a collaboration of three similarly committed brands (Brazilian Natura, Australia-based Aesop, and The Body Shop, headquartered in the United Kingdom) dedicated to creating a positive economic, social, and environmental impact, and co-creating value in all of its businesses, brands, and geographical regions around the world. In line with the spirit of its motto, "Nourish through intelligent interactions," Aesop pays special attention to details, and keeps the importance of maintaining a diversity of needs and sensory pleasure in mind. It strives to display this care in all its products through intellectual commitment, and respect for environmental conditions and the wholeness of life. The Body Shop seeks to fight for a fairer and more beautiful world by offering high quality, naturally inspired products that are produced ethically and sustainably. According to the principles of the founder, Anita Roddick, the company wants to use commerce to change the world and make business a force for good. The Body Shop has been engaged in championing equality, helping women achieve financial independence, sourcing natural ingredients, cooperating with farmers and suppliers in a fair manner, and helping local communities develop. The company has been campaigning against animal testing for cosmetics for more than three decades. In 2019, Natura announced its acquisition of the US-based Avon, which concluded in January 2020. The acquisition expanded the group with an iconic beauty brand that is recognized world over for its strong purpose, business model based on female entrepreneurship and empowerment, powerful relationship-based selling model, and heritage in relation to bringing break-through innovation to the mass market (Natura 2020a, b; Natura & Co. 2020).

Market segment: Natura is present in three major market segments: fragrances, cosmetics (beauty and body care), and gifts. Natura's main market is Brazil but through its group of global brands—Natura & Co.—the company has global reach, and is considered to be the world's fourth-largest multi-channel, direct-selling global beauty giant. The global market value of beauty products increased to more than 150 billion USD in 2019, and its steady rise is expected to continue during the coming years. The biggest global beauty companies include L'Oréal, Unilever, The Estée Lauder Companies, Procter&Gamble, Shiseido, Coty, Louis Vitton Moët Hennessy, Beiersdorf, and Johnson&Johnson. Natura's special market segments are natural and organic cosmetics and premium cosmetics, both of which make up around a quarter of the whole market, with above-average growth expectations of about five to 6 per cent. Geographically, North America and Asia Pacific are the dominant markets, accounting for about 60 per cent of the total market, in which areas Natura has been an increasingly important player in combination with its partner brands. Brazil is the home market for the company, where it has maintained its market-leading position (with market share of 16 per cent in 2019). Natura is followed by the Boticário Group, Unilever, Procter&Gamble and L'Oréal (Melo 2018; Alves 2020; Natura 2020c; Natura & Co. 2020; Shahbandeh 2020, Technavio 2020).

Cost structure and profit potential: Following challenging collaboration among its businesses in diverse regions, in 2019 Natura & Co.'s consolidated net revenue increased by almost 8 per cent to 3.6 billion USD, while EBITDA improved by 7.5 per cent to nearly 500 million USD. As a result, the parent company Natura Cosméticos S.A., headquartered in Brazil, took in more than 2.2 billion USD, thereby growing by almost 7 per cent. The annual sales of subsidiaries also grew significantly (The Body Shop by more than 6 per cent to more than one billion USD, and Aesop by over 22 per cent to 325 million USD, with an improvement of 30–40 per cent in their EBITDA ratios). In Natura's fiftieth anniversary year, collaboration between the three major cosmetics brands thus continued on favorable financial terms.

Nonetheless, the consolidation of the organizations, the revitalization of the business model, and the strengthening of growth potential still required further efforts, so Natura refreshed its strategic pillars. The goals are now to strengthen the power of the brand, pursue greater differentiation and innovation in its product categories, integrate online and offline direct selling, and accelerate digital transformation and entry into international markets to promote scalable growth. With the acquisition of Avon, such synergies permit the group of brands as a whole to invest in the expansion of each brand, particularly in areas such as research and development, digitalization and marketing. Natura & Co. also has greater influence in terms of advocacy of its main causes, including women's empowerment, climate action, the banning of animal testing in cosmetics, and local community development.

As innovation and marketing are key success factors in the beauty cosmetics market, research, and development, and advertising expenses play an important role in driving growth and profitability. In order to keep up with these trends, Natura has continuously been increasing its investment in innovation, which grew from 2.2 to 2.4 per cent of net revenue to over 200 million USD in 2019. The company also initiated an omnichannel strategy to simultaneously implement e-commerce, relationship selling, and traditional retail. In addition, in January 2020 Natura & Co. began trading on the New York Stock Exchange to further improve its financial conditions. Based on these factors and favorable global and domestic market forecasts, Natura's expansion and profit potential seem to be feasible (Alves 2020; Natura 2020c; Natura & Co. 2020; Technavio 2020).

Structure of the value chain: Natura's aim is to create a cyclical production system inspired by nature that is able to reproduce and regenerate itself. Natura wants to become a company that creates a positive impact that not only neutralizes the negative effects of its operations, but also creates value throughout its entire network of relationships. The following results have been achieved in product design, product development, and operations through Natura's "Standing Forest," "More Beauty, Less Waste," and "Every Person Matters" initiatives: Natura has been carbon neutral since 2007, and emission offsetting includes the impact created throughout the value chain until 2018. The company invested over 450 million USD in the Amazon region between 2010 and 2019, thereby helping support the livelihood of more than 5100 supplier families and local communities, non-governmental organizations, and public authorities and conserve rainforests. These efforts have contributed to the conservation of 1.8 million hectares of forest land, which area is equivalent to half the size of Holland. Natura has banned animal testing related to all products and its exclusive raw materials since 2006, and 80 per cent of its portfolio is vegan. In 1984, as the first such initiative in the world, Natura introduced refillable packaging. In Brazil, the company's product packaging now contains 50 per cent recyclable materials, almost 9 per cent post-consumer recycled material, and the half of the volume of waste generated from packaging is recycled in a reverse logistics process. The number of women in management positions grew from 30 per cent in 2017 to over 40 per cent in 2019, and the plan is to reach 50 per cent in 2020. The company allocated more than eight million USD to supplier communities for purchasing inputs and benefit sharing and training, and 7 per cent of its Brazilian work force is disabled. Natura has constantly been improving the income of its consultants and has organized an educational platform for them. The company also aims to increase consumer awareness through its brands, so product information shows the environmental and social impact of products bought via e-commerce (Natura 2020a, c).

Natura only had direct sales until 2012. It then acquired a more balanced sales system, complemented by Aesop and The Body Shop brands and their stores, franchise and wholesale partners, and online platforms. The expansion of the company into a global brand group also had an impact on the geographical area they supply. Natura, which was present in only seven countries in 2012, by 2020 was selling the group's products in more than 100 countries. Among the sales channels, the brick-and-mortar store network has developed dynamically due to the presence of subsidiaries. Starting from a few dozen fully owned Natura shops at the beginning of the decade, by 2019 Natura had 490 stores in ten countries (including franchise stores), while Aesop was operating 240 stores in 23 countries, and The Body Shop about 3000 retail locations in more than 70 countries. In 2019, Natura alone was responsible for 6800 "co-workers," of whom 5000 were working in Brazil, along with 1.8 million personal and online consultants in relationship selling. In the same year, the Natura & Co. group had more than 40,000 employees and associates, and more than 6.3 million consultants and representatives in over 100 countries worldwide. E-commerce is of great importance for the future of the cosmetics industry, and the number of Natura consultants connected to the company's digital platform reached 1.6 million, 680,000 virtual stores having been opened. This increase in the share of e-commerce will be supported by the experience and success of Aesop and The Body Shop in this field and the focus on digitizing Avon's business, as planned by the group (Natura 2020c; Natura & Co. 2020).

Natura & Co.'s corporate governance structure underwent significant changes in 2019 following the Avon acquisition. It led to the creation of the Natura & Co. Holding Company, which took over full control of both Avon and Natura Cosméticos S.A., which in turn controls Aesop and The Body Shop as subsidiaries. The Board of Directors also was complemented with Avon representatives. Via this international group of brands, it should be possible to exploit significant synergies while maintaining the autonomy of the four brands. Accordingly, the four brands will continue to have their own leadership, while the "Group Operating Committee," comprising the chief executive officers of the respective business units, will be responsible for the maintenance of group synergies (Natura & Co. 2020).

Position within the value network: Natura's efforts to operate in an environmentally and socially responsible manner have been recognized in many ways, of which only a few can be mentioned here. Reductions in the company's greenhouse gas emissions and combating climate change are due to the Natura Carbon Neutral Programme that started in 2007 and were recognized by the United Nations Global Climate Action Award in 2019. From the 670 projects that were submitted, Natura was the only Brazilian company featured on the list of 15 winners. The company has been featured on the Canadian media and research company Corporate Knights' Global 100 ranking of the most sustainable companies for 11 years. This list assesses around 7500 companies in 21 countries, and is announced annually during the World Economic Forum at Davos. Since 2004, Natura has been listed on the São Paulo stock exchange Corporate Sustainability Index. The index was developed with the support of the World Bank, the Brazilian Ministry of the Environment, and a number of environmental organizations to demonstrate the sustainability of its member companies based on economic efficiency, environmental sustainability, social justice, and corporate governance. Natura, the only company in the market segment for personal products, has been listed on the Dow Jones Sustainability Index for emerging markets for seven years. Only one-tenth of the 800 largest companies in the Standard & Poor's Global Broad Market Index can be included in the index based on their advanced economic, environmental, and social sustainability practices (S&P Global 2020; Natura 2020c; Natura & Co. 2020).

Natura has been a "B Corp" since 2014, which certifies its commitment to promoting positive economic, social, and environmental impacts. The company was awarded Best for Environment three times, and Best for Changemaker in 2019. Natura was listed in fifth place in the GlobeScan-SustainAbility Leaders Survey, among the best sustainability companies (including Unilever, Patagonia, IKEA, Interface, Danone, Nestlé, Marks & Spencer, Tesla, and BASF). Natura is the most well recognized regional sustainability leader in Latin America, and the only emerging market company to appear in the top ten. For seven years, the Ethisphere Institute, a global organization committed to enhancing business practices, has deemed Natura to be one of the most ethical companies in the world. Chief executive officer João Paulo Ferreira of Natura & Co. Latin America was listed among the ten business leaders with the best reputation in Brazil. Natura does not conduct animal tests, and is thereby certified with the Leaping Bunny certification from the organization Cruelty Free International. The Union for Ethical BioTrade has also attested to the sustainability of the natural ingredient supply chain for the Ekos product line (GlobeScan 2019; B Lab 2020a; Natura 2020c).

Since 2000, Natura has been reporting in accordance with the Global Reporting Initiative standard, the first Latin American company to do so. The company presents, in a transparent manner, the year's results and progress in line with the United Nations' Sustainable Development Goals objectives and its own long-term sustainability plan 2050 Sustainability Vision. Since 2016, Natura has disclosed its negative environmental impacts according to Environmental Profit and Loss methodology. In 2019, its externalities represented 6.8 per cent of its net revenue, a slight decline compared to 2018. Since the first calculation, which was based on 2013 data, the accumulated gain in efficiency in EP&L/Net revenue has been 1.8 percentage points. The valuation of the company's social impacts in the form of Social Profit and Loss started in 2019 among its consultants and supplier communities. Following these achievements, Natura intends to develop its Integrated Profit and Loss assessment to express its environmental, social, and human externalities in an integrated way to support strategic decisions (Natura 2020b, c).

Natura's mission involves serving society and leveraging positive socio-economic impact in its network through promoting education, women's rights, and healthcare. One way of doing this has been investing all the profit from the sale of the Crer Para Ver non-cosmetic product line, launched in 1995, into supporting children's education. This work has been promoted by Instituto Natura (set up in 2010) to further improve the results of earlier initiatives. This includes supporting school projects, education, the training and development of teachers, civil servants, and consultants. In 2019, in Brazil and the Latin American region, almost 100,000 consultants, nearly 10,000 schools, more than 46,000 teachers and coordinators, and over 1.5 million students benefited from this program (Natura 2020c).

Within the group Natura & Co., each brand is committed to social and environmental responsibility. To highlight a few of the recent related achievements, the following may be mentioned: Adding up the brands' contributions so far, over one billion USD has been donated to the most important women's causes, such as women's empowerment, entrepreneurship, and well-being. Since the inception of the Aesop Foundation in 2016, it has donated over three million USD to 30 charitable projects in Australia aimed at promoting literacy and storytelling. Aesop has also transitioned over 70 per cent of its plastic product packaging range into recycled plastic. In 2019, The Body Shop launched a Community Fair Trade Programme in partnership with plastic waste pickers in India, especially women, to incorporate more than 250 tons of recycled plastic into packaging, which amount is expected to reach 900 tons within two years. Last year, The Body Shop joined Natura as a B Corp, and Aesop was also in the process of being certified. To help employees prepare for environmental, social, and business challenges, the brands provide their associates and leaders with sustainability, purpose-driven leadership, and customer care training. They also strive to pay a fair income and living wage to the members of their networks (B Lab 2020d; Natura & Co. 2020).

Competition strategy: In the 1970s and 1980s, the Brazilian cosmetics market was relatively closed to imports. The competitors of Natura were local multinational companies such as Unilever that were not deterred by the country's political instability and hyperinflationary economy. This situation until the early 1990s favored Brazilian companies, which provided local customers with high-quality products. Due to the consolidation of the economy and globalization, consumer markets changed rapidly; new consumer groups with greater purchasing power appeared, and the number of companies that wanted to serve them increased within a short time. In times of intensifying competition, economies of scale, globalized operations, a high level of product development and marketing expenditure were the criteria for success. Natura's main competitor in the direct sales segment was Avon, which is now part of the group. In the retail segment the Boticário Group, Unilever, Johnson & Johnson, Procter & Gamble; and in terms of selective sales channels, L'Oréal, Nivea, Louis Vuitton and Chanel, have been the most important market players. Today, the competition perceived by Natura has intensified in both domestic and international markets, as its competitors sell products with similar functionality. Products of competitors may also appear within Natura's direct sales channels because consultants can sell products from the brands of different companies. Nevertheless, Natura has been the market leader in its largest market in Brazil based on its commitment to environmental and social sustainability and continuous innovation, and respect for and protection of Brazil's values and strong brands. A differentiated strategy of maintaining close relationships via the omnichannel shopping experience of customers and innovative technologies also supports the company's competitive advantage. Becoming a global group of similarly strong and environmentally and socially responsible beauty brands has further increased Natura's potential to be successful both locally and globally (Jones and de Pinho 2006; Melo 2018; Alves 2020; Natura 2020c; Natura & Co. 2020; Shahbandeh 2020, Technavio 2020).

Organic India

Value propositions: Organic India aims at achieving social, environmental and economic sustainability and is committed to the production of genuine organic food and products. It has a unique business model that strives to create benefits for all stakeholders through close-to-nature, sustainable organic farming practices that serve and conserve nature and support the livelihoods and well-being of farmers and tribal "wild-crafters" in rural India. Organic India works with thousands of small family farmers who are engaged in sustainable organic farming on more than 20,000 hectares in seven states of Central and Northern India. Organic India is among the most geographically extensive organizations in the country. Its products symbolize the chain of interconnectedness between Nature, producers, and consumers, which the organization completes by supporting environmental and social sustainability and offering training and a dignified life for farmers, and contributing to the health, happiness, and well-being of consumers (Organic India 2017, 2020a, b).

One important commitment of Organic India is supporting and promoting regenerative agriculture. The goal of the company is not only to produce high-quality organic crops, but also to reverse the degradation of the natural environment. It seeks to improve the condition of land by enriching biodiversity and restoring soil fertility. To this end, Organic India continuously trains its partners, farmers, and tribal wild-crafters about the principles and practices of organic, renewable, biodynamic, and traditional Indian (*vriksha ayurveda*) agriculture, which treats each piece of land as a unique living organism. No synthetic chemicals, genetically modified plants, or radiation are used in its farming activities. However, the following gentle technologies are applied: comprehensive soil fertility management, seed collection, maintaining the diversity of species, composting, water treatment, crop protection and crop rotation, and consideration of lunar and other natural cycles during planting and harvesting. These techniques renew local ecosystems through carbon sequestration and improving water and soil conditions, thus also positively affecting the global ecosystem. By using these methods, more than 4000 hectares of chemically heavily contaminated land had been made fertile again by 2017 (Organic India 2017, 2020a, b).

Market segment: All of the products of Organic India are organically certified. Originally, the company used to make teas and herbal preparations, although its product range now includes a number of other products such as spices, ginger, seeds, beans and cereals, and fresh vegetables. Its main product segments are tulsi teas and tea blends, herbal preparations, and other packaged foods (including ghee, coconut oil, honey, stevia, and apple cider vinegar). The products are sold in 25 of its own shops and 15,000 India-wide partner stores, and online. Organic India also exports to more than 40 countries, accounting for 60 per cent of its annual turnover. Its largest foreign markets are the United States (where a subsidiary has been operating with B Corp certification since 2015), Canada, Europe, the Middle East, and Australia (ASSOCHAM – Ernst & Young 2018; B Lab 2020b; Organic India 2020a, b).

Cost structure and profit potential: From the 1990s onwards, the global market for organic products (food, beverages, textiles, health maintenance, and beauty products) has been growing almost 60 per cent faster than the market for conventional products. This can be mainly attributed to a rise in living standards, growing consumer awareness, and government subsidies. In addition to the countries most engaged in organic production (Australia, Argentina, and China) and consumption (USA, Germany, and France), India is among the world leaders in terms of the number of organic farmers (India is first, with more than 830,000 farmers) and size of area used for growing herbs (after Finland and Zambia, India is third with 4.2 million hectares). The size of the Indian organic market was about 570 million USD in 2018, which may triple by 2020, mainly due to a significant increase in exports, as domestic consumption is still low (1 per cent of total consumption). In addition to this favorable growth potential, the margin on organic products is also above average—ranging from 10 to 40 per cent more than conventional products. Key to Organic India's success, in addition to operating faithfully in line with its core values, is the growing health consciousness of consumers. The company expects further growth in this consciousness and an increase in domestic demand, so annual turnover may rise from 55 million USD in 2017 to 70 million USD by 2020. The successful development of Organic India may also be supported by the planned expansion of the product range, and opening up to new segments (organic skin and body care products) (*The Economic Times* 2017; ASSOCHAM – Ernst & Young 2018; Organic India 2020a, b).

Structure of the value chain: The value chain of Organic India is based on the philosophy of the "whole herb," which is potent and full of life. Herbs are complex, synergistic mixtures of hundreds of compounds (nutrients, minerals, and vitamins) that contain complex "information," as well as natural mechanisms that mitigate harm to health. When plants are used in their entirety, they have a beneficial balancing effect on the human body, unlike when their ingredients are utilized after artificial decomposition into individual ingredients. Four decades of scientific research have proven this fact, which was already known from the ancient Vedic wisdom. Organic India therefore uses whole plants in their purest form for its products. All its ingredients are organic, hand-picked, and locally processed. Owing to advanced processing methods and drying technology, the content of herbs may be efficiently preserved, so that they can be used in the creation of products of the best quality, efficiency, and purity on the market (Organic India 2017, 2020a).

Tulsi tea blends are at the heart of the product offerings. Tulsi, or basil, is one of India's sacred plants and is often referred to as the "queen of herbs" because of its regenerating and healing effects. It is also a key element in Ayurveda, the traditional holistic healing system in use throughout India. Tulsi is considered by science to be an adaptogenic plant that, by exerting complex effects, balances the vital functions of the body, and enhances resilience to environmental, physical, and emotional stress (Organic India 2020a).

Organic India is particularly committed to applying the principles of its holistic quality assurance system (honesty, accountability, and responsibility) throughout the whole production process (seed production, sowing, cultivation, harvesting, processing, marketing, and sales). The quality of its products has been proved through auditing and certification by independent third-party organizations. Accordingly, Organic India is one of the few herbal companies in India to be certified according to the schemes of Good Manufacturing Practices, Hazard Analysis & Critical Control Points, ISO 9001: 2015, and Orthodox Union's kosher quality certification. It also meets the standards of The Indian National Program for Organic Production, the Organic Certification of United States Department of Agriculture, Control Union Certifications World Group, the European Union's Organic Certification, the British Soil Association, GMO Free and Non-GMO Project Verification, and Fairtrade America. Organic India has recently been awarded Cosmos certification by ECOCERT for its core principles of respect for biodiversity, use of clean manufacturing processes, green chemistry, and transparency to consumers (Organic India 2020a).

Organic India values its partners, who include the farming families and tribal wild-crafters that provide the high-quality raw materials, many of whom are women, widows, disadvantaged, elderly, or illiterate people. In addition to organic, biodynamic, regenerative, and traditional Indian agricultural education and practices, it also provides them with training in self-sufficiency and self-development, which also improves the opportunities of future generations. Such relationships are based on mutual respect, dignity, and interconnectedness. The company is committed to improving the lives of its partners in production, which it strives to do on several levels. Organic India buys crops from them on terms more favorable than the normal market ones. It also pays the cost of obtaining the necessary organic certification. Farmers can grow their own crops on their land in rotation, alongside crops for selling to Organic India, giving families a more stable income and source of healthy food. The company provides them with health care, gender equality programs, and improves their infrastructural conditions. As a result of these investments, and through the success of individual families, villages can also start to develop, and the quality of life of local communities can be significantly improved, laying the foundation for more successful, healthier, happier lives for future generations (Organic India 2017, 2020a).

An important element of the value chain is Organic India's close collaboration with Fabindia, India's largest private retail platform for craft-based products. Fabindia acquired a 40 per cent stake in Organic India in 2013 and then increased its stake to 51 per cent in 2015. Fabindia buys home furnishing and decorative items, clothing, and organic food products from rural artisans across the country and exports them around the world (Organic India 2020a).

Position within the value network: For more than 20 years, Organic India has been committed to achieving social and environmental sustainability through a holistic business model. Its successful operation proves that a sustainable and ethical business model can create value not only for the company but also for its entire value network. The quantifiable results of this in 2017 were as follows. Over two decades, more than 2000 marginalized family farmers across India, and hundreds of tribal wild-crafters have been supported. More than 8000 people have been provided with free health care in Organic India's own health center, and almost as many have participated in its health maintenance programs. More than 200 people are employed in the raw material processing plants of the company, 95 per cent of whom are women. Through its restorative agricultural technologies and production partners, the fertility of more than 4000 hectares of chemically ravaged soil has been restored. Organic India's product range includes more than 100 types of herbal teas, dietary supplements, and organic food items, and it sells more than 240 million cups of organic Tulsi tea per year across the world (Organic India 2017, 2020a).

Organic India's products meet a number of quality and organic certification standards, but the company has made further improvements to increase its sustainability performance. The boxes of its flagship product, Tulsi tea, are made from 100 per cent recycled paperboard, at least a third of which is post-consumer content. Additionally, Tulsi tea bags are made entirely of unbleached, chemical-free, biodegradable, and therefore safely compostable paper, and will soon be produced without metal clips. In the near future, Organic India will open a new factory on the outskirts of Lucknow, housed in a building that will be certified in line with the world's strictest environmental certification scheme, LEED (Leadership in Energy and Environment Design). In order to preserve natural ecosystems, tribal wild-crafters will be supported through training about environmentally sustainable, regenerative, and hygienic methods of procuring herbs and fruits on more than 1.4 million hectares of certified organic forests, and will then be supervised and certified (Organic India 2017, 2020a, b).

Organic India is also seeking to add value to its entire value network by establishing the "Earth Seer" standard, which certifies ethical, trustworthy, holistic, socially, environmentally, and economically responsible sustainable business behavior that is committed to promoting well-being and respect for all beings and Mother Nature. The certification also encourages its holders to go beyond expected requirements by improving their business processes and supports the emergence of conscious consumption through collaboration with all stakeholders (Organic India 2017, 2020a).

To support its stakeholders, the company created the Organic India Dharti Mitr Award, which is given to organic farmers whose work significantly contributes to improving the agricultural, ecological, and social conditions of India. In addition to moral recognition, the award, which also includes a cash prize, also serves as a platform for strengthening relationships and sharing knowledge between committed actors in the field (IFOAM 2017).

In recognition of its accomplishments in relation to the common good, social and environmental sustainability, local communities, and ethical behavior, Organic India's US subsidiary received the "Best for Changemakers" award in 2018 and the "Best for the World Overall" award in 2018 and 2019. With this result, Organic India is now recognized as being among the best ten per cent of 2500 B Corporation certified companies in 150 industries in 50 countries around the world (B Lab 2020b).

Competitive strategy: The production and consumption of organic products has been increasing more than that of conventional products, mainly due to rising living standards, the spread of health-conscious lifestyles, the expansion of the urban population, and government subsidies (for example, in 2016, the entire agricultural area of the Indian state of Sikkim was converted to organic production). In India, a number of successful organic enterprises share a market that is growing 50 per cent faster than the world average. In addition to the market leader Organic India (with 25 per cent market share), the largest companies are Sresta Natural Bioproducts (14 per cent), Morarka Down to Earth (9 per cent), Conscious Foods (8.5 per cent), and Ecofarm (6 per cent). As a world market leader, Organic India is also very popular among consumers, the loyalty of its regular customers is strong, and brand recognition is significant. However, organic products are still considered a luxury in India, so Organic India intends to change this to "reintroduce" natural, healthy, and sustainable organic products to the market and make them available to all members of society. The company seeks to make producers, consumers, and decision-makers realize that organic production is the only long-term sustainable solution. Thus, Organic India encourages a growing range of stakeholders to raise their awareness and change their lifestyles. It uses digital and print media marketing tools based on the potential of the positive impact of experts and opinion leaders, influencers, and community building to increase brand awareness. Organic India is further developing its unique product offerings and is strengthening its presence in different geographical regions (for example, in China) and is also targeting new market segments (skin and body care products). The goal is to make its holistic health-conscious conception available in most product segments related to consumer welfare (ASSOCHAM – Ernst & Young 2018; Srivastav 2018; Organic India 2020a, b).

Patagonia

Value propositions: Patagonia's mission is to produce the best products without causing unnecessary harm, and to contribute to resolving the environmental crisis through the company's activities and continuously developing new ways to do things. These goals are rooted in a love of nature and a feeling of close connection with it, which constantly encourage Patagonia to act in order to preserve nature, improve its condition, and protect and restore the stability, integrity and beauty of the web of life. The culture and practice of mountaineering based on simplicity and utility can also be seen in Patagonia's value provisions, and is most evident in their product design and process organization (Patagonia 2020).

Market segment: Patagonia manufactures and distributes sustainable, ecological outdoor women's, men's and children's clothing, accessories (caps, scarves, and gloves), and apparel (sleeping bags, backpacks, blankets, and mugs) worldwide in its own stores as well as via resellers and the internet. According to industry analyses, the global outdoor clothing market accounted for almost 13 billion USD and is expected to reach more than 19 billion USD by 2026. The total value of the global outdoor apparel market was about 43 billion USD in 2016 and has been forecast to increase steadily by around 5 per cent per year until 2024. North America is one of the leading regions, with continuous demand for innovative products, but the Asia Pacific market is expected to expand significantly in the coming years owing to emerging demand by fitness inclined young generations. So far, offline stores have been dominant in sales, but the convenience of buying and online discounts could drive the e-commerce segment (Gaille 2018; GlobeNewswire 2019; Statista 2019; Business Wire 2020; Patagonia 2020).

Cost structure and profit potential: Committed to environmental and social responsibility, Patagonia's production costs are higher than the market average. For example, the company pays higher wages than its competitors in order to respect fair trade principles and to support the livelihood of local communities. However, Patagonia's brand, which is well known for its genuine commitment to sustainability, is strong, and, thanks to customer loyalty, the company has sufficient stable profit for its long-term success. This is true despite the fact that the company's advertising is minimal, and Patagonia sometimes uses anti-consumption campaigns such as its famous New York Times advertisement "Don't Buy This Jacket" in response to a Black Friday campaign in November 2011. The latter drew attention to the catastrophically harmful environmental effects of overconsumption and ways of mitigating these. Patagonia has since encouraged its consumers to consciously change their shopping habits. Despite these unusual methods, Patagonia's profits are enough to finance its functioning and long-term value commitments. For instance, as co-founders of the One Percent for the Planet initiative, the company donates the equivalent of 1 per cent of its annual gross sales in the form of a combination of monetary, in-kind, and promotional support directly to local environmental nonprofit organizations that are working to protect air, land, and water around the globe (since 1985, over 89 million USD has been donated to support environmental protection). Market expectations are favorable for Patagonia, as they expect further expansion by more than 5 per cent in the outdoor clothing and apparel segment during the coming years. North American and Asia Pacific markets may well develop significantly due to increasing consciousness among people about fitness and healthy lifestyles. After having stagnated in the middle of the decade, the European outdoor market has recovered and is expected to grow at a moderate speed. Most of the market segments for Patagonia products (clothing, backpacks, hiking, and climbing apparel) have also strengthened due to an increase in marketing activity aimed at encouraging women and families to engage in more outdoor activity. The Latin American outdoor clothing and apparel market may also grow moderately over the next five to six years, but because of the low per capita income in the region, interest in Patagonia's products may be subdued (Jans 2017; Chen 2018; GlobeNewswire 2019; Statista 2019; Business Wire 2020; Patagonia 2020).

Structure of the value chain: Patagonia strives to promote environmental and social sustainability throughout its entire value chain. Its fundamental goal is to mitigate and neutralize the negative effects of climate change, and to develop and apply practices that can improve the state of nature. In this endeavor, product design based on simplicity, usability, durability, and repairability is the starting point. In addition, Patagonia focuses on the following goals: (i) mitigating the environmental impact of its supply chain through monitoring its carbon footprint, and increasingly using renewable energy, brownfield investment, and green offices. Patagonia supports "car-free to work" initiatives. Its high-quality raw materials come from traceable sources. (ii) The company helps local grassroots environmental movements worldwide by making donations funded by a self-imposed "Earth tax" that is equivalent to 1 per cent of its total annual revenue. Patagonia also provides employees with an extra 320 hours for voluntary work, in lieu of working hours. (iii) Patagonia promotes systemic change by leveraging the company's popularity and brand strength. This includes embracing various environmental campaigns and environmental issues. (iv) The company provides consumers with high-quality, repairable products and also ensures the reuse of second-hand products via its Worn Wear project. (v) Patagonia is engaged in regenerative organic agriculture, and operates its own business called Patagonia Provisions to trade in sustainable and organic food. (vi) The company has been helping develop new ways of doing business. In 2013, a venture capital fund (Tin Shed Ventures) was launched with share capital of 20 million USD to fund start-up companies that address environmental issues, including the development of renewable energy infrastructure and regenerative organic agriculture, protecting water resources, neutralizing waste, and developing sustainable materials (Patagonia 2020; Patagonia Provisions 2020; Tin Shed Ventures 2020).

Patagonia has been operating as a certified B Corp oration since 2012. In line with this certification (by nonprofit organization B Lab), in 2019 Patagonia achieved over 151 points out of a maximum of 200 points, maintaining the level of a year earlier. This independent external evaluation appraises company performance in terms of impact on the environment, employees, customers, community, and transparency. Patagonia prepared an Annual Benefit Corporation Report in 2018, which was organized around its six main commitments. In relation to these, the following primary results were achieved in terms of increasing environmental and social sustainability.

One per cent for the planet: between 1985 and 2017, 89 million USD was donated to environmental organizations. In 2017, nearly half a million dollars was donated to 1000 organizations. Five hundred Patagonia employees participated in the distribution of grants.

Best product without no unnecessary harm: in 2017, according to the company's internal product evaluation system related to sustainability impact, Patagonia products achieved an average of 8.77 out of a maximum of 10 points. Almost a third of the materials they used were recycled, while a fifth were renewable, and the proportion of newly produced polyester and nylon-based materials was reduced from 58 to 51 per cent in two years. More than 70,000 of Patagonia products were repaired at 54 repair centers worldwide. Due to Patagonia's Fair Trade program, 26,000 of its workers received higher-than-market wages in 2017. The number of workers supported by the program has recently grown to 66,000. Patagonia now uses 100 per cent traceable down for its products, 72 per cent of its product line is made from recycled materials, 76 per cent is Fair Trade Certified sewn, and more than 150 farmers are taking part in its Regenerative Organic Cotton Pilot Program.

Operations causing no unnecessary harm: an environmental impact assessment focusing on the topics of energy, water, and waste was conducted in 32 Patagonia stores. The company reduced car commuting by almost 900,000 km under its "Drive-Less Program" (a decrease of 6 per cent). Although in Europe and South America Patagonia managed to reduce both electricity use and pollutant emissions (by 13 and 31 per cent, respectively) in 2017, both indicators increased worldwide by about 10 per cent due to growth in the largest markets (United States, 6–7 per cent, and Japan 24–37 per cent). Since then, Patagonia has met 100 per cent of its electricity needs in the United States with renewable electricity, and 76 per cent of its global needs. The company's goal is to switch to 100 per cent renewable electricity and become a zero-waste-to-landfill company by 2020, and to become carbon neutral by 2025.

Sharing best practices with other companies: five factories held chemical treatment training for 130 workers that they shared at meetings of the Outdoor Industry Association. Patagonia made over 400 public appearances at industrial and public forums, and organized more than 2000 free community events in its North American stores. Over 30 Patagonia employees held management or consulting positions in more than 40 industry, academic, and environmental organizations.

Transparency: Patagonia has published a list of its factories for 14 years, placing the company at the forefront of the industry. More than 30 studies and posts have been published about various sustainability topics. In 2017, the Fair Labor Association, an independent third party, accredited Patagonia's social responsibility program with regard to improving the position of employees.

Supportive work environment: all employees are provided with health insurance premiums free of charge. Half of Patagonia's employees are women. Almost half of the company's executives are women, while the US average is around one third (Rose Marcario was formerly Patagonia's CEO, while fewer than 5 per cent of US companies are led by a woman). Nearly every other weekend included a three-day break for Patagonia workers. Around 80 employees took part in internships at environmental organizations using fully paid leave of up to two months. Eighty workers received non-violent civil disobedience training about how to participate in civil movements in a dignified manner. There were also on-site child-care and development programs for 130 children of mothers employed by Patagonia (Patagonia 2018, 2020; B Lab 2020c).

Position within the value network: For almost five decades, Patagonia has been a committed advocate of sustainability and fair trade, and is playing a pioneering role in solving environmental and social problems and supporting the social movements involved in those efforts. As an "activist company," Patagonia has embraced a number of initiatives and used its brand and the organization's resources to serve the common good. In 2018, the company launched an activism platform ("Action Works") to connect committed individuals to organizations working on environmental issues in the same community. Via this initiative, it has supported initiatives for improving and regenerating agricultural soil, the health of terrestrial and marine ecosystems, climate and atmosphere, species diversity, and fostering environmental justice, civic democracy, sustainable communities, and indigenous people. Patagonia's recent environmental campaigns include attempts to protect Tasmania's rainforests, dismantle offshore oil rigs, prevent the construction of river hydropower plants in the Balkans and the expansion of the Keystone oil pipeline in North America, and the extraction of oil and gas in protected areas in Alaska. To increase consumer consciousness, Patagonia has actively participated in the creation, publishing, and distribution of several books and films on relevant topics, including *Unbroken Ground,* and *DamNation.*

Over the years, Patagonia has increasingly engaged in political activism in order to live up its values and foster the specific goals related to these. An example of Patagonia's peremptory attitude is the several cases in which it has stood up for its value commitments against the US government. In 2017, for example, Patagonia organized a boycott against a prestigious outdoor fair in Utah in protest of the state's intention to reverse a previous presidential decision to establish the Bears Ears National Monument National Park. Also in 2017, in line with several environmental and local community organizations, Patagonia sued the US government and president in relation to a decision to significantly reduce the area of two national parks. In 2018, the company urged over 400 companies to provide employees with a day off to vote on election day, as Patagonia itself did, also becoming the first brand to endorse specific political candidates (Friedman 2017; Gelles 2018; Fast Company 2020; Patagonia 2020).

The performance of Patagonia has continuously been recognized. Just a few of the most recent awards are highlighted here: in 2018, *Fast Company* magazine ranked Patagonia the sixth best among "The World's 50 Most Innovative Companies" for leading the way in promoting long-term solutions to environmental problems and growing its business in line with amplifying its social mission. The list included more than 350 companies, including Apple, Netflix, and Amazon. In addition, in the category of "Social Good" Patagonia was first that year. In the GlobeScan-SustainAbility Leaders Survey, Patagonia has been mentioned among those companies with the best sustainability performance since 2013. The company was ranked second behind Unilever, followed by companies such as IKEA, Interface, Natura, Danone, Nestlé, Marks & Spencer, and Tesla. Patagonia has been certified as a B Corporation since 2012. It has been awarded the "Best for the World" prize in the categories of Environment, Community, Changemaker, and Overall since 2017. With these results, Patagonia has continuously been among the best 10 per cent of the 2500 B Corporation certified companies in 150 industries in 50 countries around the world for years (Fast Company 2018, 2020; GlobeScan 2019; B Lab 2020c; Patagonia 2020).

Competitive strategy: Trends in the North American outdoor clothing and apparel market are dominant factors in the industry. Although the market is fragmented and several brands are responsible for the majority of the market, the most significant US-based players (such as Nike, VF Corporation, Hanesbrands, Under Armour, Canada Goose, North Face, Columbia Sportswear, G-III Apparel Group, Newell Brands, and Outdoor Research) also have global influence. Other major non-US brands include Adidas, Puma, Hugo Boss, ASICS, and the Mizuno Corporation. However, due to some new market trends, even smaller or recently established players have been able to rapidly gain significant market share and brand awareness. One notable phenomenon is that traditional sports retail chains have been losing ground against direct-to-consumer selling manufacturers. Direct sales, associated with transparent pricing, have been gaining popularity among consumers and allow manufacturers to benefit from higher margins by omitting interaction with merchants. The other trend is the proliferation of lifestyle marketing, which, along with direct sales, can result in deeper and more credible customer relationships. Younger generations increasingly prefer wellness and outdoor lifestyle and lifestyle brands. The owners and managers of these brands, partly through their own approaches to life, are sincerely committed to representing brand values. In addition to Patagonia, Fjällräven, HippyTree, Kühl, and Blackyak are also popular and increasingly successful emerging brands. It seems that the values faithfully represented by Patagonia for almost half a century are becoming more favorably received in the market owing to the increase in consumer consciousness and emphasis on genuine value commitment. Patagonia can thus continue to rely on its genuine core values, work closely with its stakeholders, and focus on customers committed to sustainability and a minimalist style. It can also maintain its success in financial terms, creating a sufficient financial base on which to advocate and support the sustainability issues it considers important (Chen 2018; GlobeNewswire 2019; Statista 2019; Business Wire 2020; Patagonia 2020).

Scott Bader

Value propositions: As set out in its long-term plan, "2036 Vision," based on the three pillars of Business, Ecology, and Humanity, Scott Bader seeks to become a sustainable chemical company that operates in line with the company's ethos and the founder's vision. Scott Bader imagines a world in which humanity develops without endangering the natural systems that support its living conditions. Scott Bader promotes this development by responding to the challenges of a changing society by using the chemical industry to promote the common good. The company defines its values in terms of seven strategic goals. (i) Through leading the way in the creation of a circular economy, Scott Bader works with its partners to create a closed-loop value chain free of waste and negative impacts. (ii) Acting beyond compliance: As a reliable, proactive market leader of a safe, transparent, and ethical chemical industry, it produces products and processes that protect people and the planet while setting an example to others. (iii) Scott Bader promotes sustainability by sharing its vision, knowledge and profit and supplying products and services that create real value for society. (iv) While striving for excellence, as a purpose-driven company Scott Bader continually improves all of its processes and products to deliver world-class business performance and meet all the expectations of its customers. (v) Through initiatives and innovations at all levels of the organization, the diversity and creativity of empowered and motivated employees gives the company its strength. (vi) In order to encourage working together for success, Scott Bader is an integral part of a global network of mutually beneficial supplier, consumer, and alliance-based partnerships that share its vision and mission. (vii) Through the ethical use of natural resources and as a proponent of clean energy, freshwater conservation, and zero emissions, Scott Bader is a strong advocate of environmental protection and a shift toward sustainability (Scott Bader 2019, 2020).

Market segment: Scott Bader's main activities are the production of adhesives and composites such as resins and gelcoats and specialty polymers in the geographical markets of Europe, North America, South Africa, the Middle East, the Far East, and Australia. Its products and services are mainly used in the automotive-, construction-, chemical-, transportation-, marine-, and wind energy industries, and, to a lesser extent, in households and the sector of graphic arts in more than 100 countries worldwide. In 2018, 58 per cent of total annual sales came from Europe (131 million GBP), about a third of which came from the UK, where Brexit has not yet had any tangible effects, but the uncertainty of potential scenarios has been a constant subject of discussion and the company has taken various measures to prepare for all possible outcomes. European sales of composites increased by 1.3 per cent due to strong demand for specialty products, resulting in an increase of over 5 per cent on the margin per ton. The adhesives sector continued to grow significantly (by 13 per cent) in all product types and geographic regions, but the margin per ton declined by 3 per cent because price increases could not be passed on to customers. Sales volumes and margins of the specialty polymer business remained at the levels of the prior year, although the polidene business was divested as it no longer fit with strategic plans (Scott Bader 2019, 2020).

Cost structure and profit potential: Due to the uncertainty of Brexit scenarios, threats of further consolidation of the industry, and a continuously tightening regulatory framework, market conditions were fairly difficult for Scott Bader in 2018. However, the company was able to grow in terms of volume, sales, turnover, and profit, and considered the year to be a successful one, with progress on several fronts. Sales increased by almost 12 per cent to 226 million GBP, while profit before tax and exceptional items grew by 7 per cent to six million GBP. Net cash at year end amounted to more than 13 million GBP, including profit on the sale of the polidene business. Mainly due to the impact of the product mix, which was due to significant growth in resin sales, margin per ton decreased compared to 2017. Operating costs increased, while trading cost per ton declined by about 5 per cent. In the Middle East, 2018 was an outstanding year with record sales that increased by 54 per cent based on strong local sales and exports. However, major construction projects have been hit by the regional geopolitical situation. South Africa has yet been suffering from recession. Challenges in relation to ensuring safe access to key raw materials continued, as in previous years, along with a significant number of force majeure incidents along the supply chain, and the adverse effects of climate change (for instance, the low level of the Rhine River) notably impacted raw material supply. Scott Bader expects this trend to continue and is preparing to mitigate such effects. As a part of efforts to spread the risk and enter potential markets, in 2018 Scott Bader established a Japanese subsidiary to better serve the growing demand in the region, and planned to upgrade its Indian production facility during 2019. In June 2020, after ten years of cooperation, the company also acquired the assets of Summit Composites Pty, thereby establishing an Australian subsidiary to distribute resins, gelcoats, and adhesives, all built on Summit's reputation for product quality, performance, and reliability (Scott Bader 2019, 2020).

Structure of the value chain: A special feature of Scott Bader's value chain is the corporate governance structure derived from community ownership and leadership. The company has no external shareholders, so it can take into account different, longer-term aspects in its operation compared to the average company. The three main governing bodies are the Group Board, the Commonwealth Board, and the Members'

Assembly. The nine-member Group Board acts as a board of directors and is responsible for business affairs (finance, legal, and strategy), but also ensures that the business is run in accordance with defined principles and helps promote charitable objectives. An Audit Subcommittee also assists its work. The Group Chief Executive Officer reports to the Group Board on a monthly basis. The nine-member Commonwealth Board operates as a holding company and is the company's registered charity. It is responsible for discussing strategic matters, acquisitions, profit distribution, decisions determining long-term development, and monitoring compliance with community principles and legislation related to charity law. The Members' Assembly is the group's democratic forum to which the Group Board is accountable. The Assembly monitors adherence to the principles defined by The Commonwealth. The five largest sites (UK, South Africa, Dubai, Croatia, and France) each have Local Councils that represent members when discussing local issues (Scott Bader 2019, 2020).

Scott Bader puts meeting special customer needs at the heart of its strategy, so innovation is the soul of the value chain. Every year, the company comes up with new solutions and new products so that it can maintain its technological leadership in the markets for its key products. Scott Bader does not intend to compete on price and volume, but on the highest possible technological and service solutions that will enable the enterprise to firmly differentiate itself from its competitors (Jacob 2012; Scott Bader 2020).

Based on its core values and its deep commitment to sustainability, Scott Bader embarked on a long-term strategic sustainability project across the whole group in 2017. In collaboration with advisory groups The Natural Step Germany, and Sustainable Growth Associates, it redefined its vision and strategic goals and the steps required to achieve these. The three main pillars of the plan "Vision 2036" are Business, Ecology, and Humanity (Sustainable Growth Associates 2018, Scott Bader 2020; The Natural Step 2020).

Within the pillar "Ecology," the reduction of the company's environmental impact is captured in the following achievements from 2018. The first 100 tons of bio-based resin were sold, serving as a cost-effective, "greener" solution for the market. All but one site made significant energy savings through process optimization and a higher rate of asset utilization. Over the last two years, energy consumption per ton has been reduced by almost 14 per cent. A third of all production sites now run on 100 per cent renewable electrical energy. The Croatian site has signed the "Responsible Care" global charter, showing Scott Bader's commitment to dealing with chemicals safely over the total life cycle. Also in Croatia, 50 per cent of the factory's cooling water consumption came from rainwater. Scott Bader played a very active role in an industry initiative led by its UK subsidiary involving defining what a circular composite economy could look like, which project also continued during 2019 (Scott Bader 2019).

In the pillar "Humanity," Scott Bader focused on the strategic goal of "Striving for Excellence." Project management, sustainability, and coaching training were provided to more than 200 employees across the group. A "Company Employee Ownership Day" was organized at all manufacturing sites to improve teamwork and strengthen community spirit. Scott Bader's largest site in the UK won the "Investors in People Gold Award" for its outstanding performance at effectively managing and supporting the development of its people. The company's "Values in Action" program was extended globally for a second year to recognize the teams or individuals who exemplify commitment to organizational values, fairness, teamwork, and responsibility through their activities. In 2017, the general manager of the Amiens site was given new management responsibilities at a group level, and his position was reorganized into a collegial management structure, thereby strengthening Scott Bader's progress toward increasing self-management. In the same year, a major project was launched to review the reward and appraisal system to make Scott Bader more competitive at finding and retaining talented and skilled people, and also to create the right human resource background for underpinning the company's progress toward sustainability (Scott Bader 2018, 2019).

In 2018, progress was made in the field of occupational health and safety performance, but more work is needed to improve this as planned. "Lost Time Accidents" related to chemical processes and engineering declined from three to one, but the "Lost Time Incident Frequency Ratio" (the number of accidents at work that resulted in an injury, per one million working hours) disappointingly rose above the industry benchmark. In 2018, the first global Health and Safety survey was conducted, while local improvement action plans have been implemented at all sites during 2019, and this area has remained a key focus, with behavioral safety and change management targeted for improvement (Scott Bader 2019).

Position within the value network: As a leader in innovation and technology, Scott Bader prioritizes getting its internal processes and products audited and certified by third-party organizations in all markets and geographical regions. The company has been certified according to ISO 9001:2015, ISO 14001:2015, ISO 45001:2018, and BS OHSAS 18001:2007. In addition to these qualifications, Scott Bader's commitment to social and environmental sustainability has already been recognized by several awards. In 2017, the UK site received the "Gold Investors In People Award" (IIP); an important form of recognition for organizations that effectively ensure the development and leadership of their employees. This represented a major step forwards after Scott Bader joined the IIP evaluation scheme in 1997, and took home the IIP Silver Award for several years. A significant milestone in the company's environmental sustainability development was the acquisition of a Silver EcoVadis rating for its sustainable procurement practices in 2018. EcoVadis evaluates the global supply chain of more than 40,000 companies according to four main themes (Environment, Ethics, Labour/Social, Supply Chain) in line with 21 indicators based on several sustainability schemes (such as the UN Global Compact, and ISO 26000). Through an online platform, EcoVadis provides ratings, based on which Scott Bader was in the top 17 per cent of companies in its industry in 2019. This represents a slight decline compared to 2018 but reassessment has been under way, with several performance indicators now addressed in Scott Bader's "2036 Vision" (Scott Bader 2020).

Scott Bader plays an exemplary role through its active charitable activities. It donates to charities with similar goals at local, national, and international levels. Each year, grants go to beneficiaries through the "Global Charity Fund," which is financed by a specified percentage of the company's yearly income. In 2018, donations worth 222,100 GBP were distributed to more than 200 charities in 23 countries in relation to initiatives involving education, training (41 per cent), health and disabilities (23 per cent), community (22 per cent), and sustainable livelihoods (14 per cent). Through the Central Fund, four large African projects were also supported with 25,000 GBP each, based on the votes of Commonwealth Members at the annual general meeting. In addition, group companies paid 30,000 GBP of donations directly to numerous charities. Scott Bader also helps charities by giving employees an extra day off each year to contribute to the work of an NGO of their choice through volunteering. In 2018, there was record uptake in the number of volunteering hours, which reached almost 1000 hours (Scott Bader 2019).

Competitive strategy: The market segments targeted by Scott Bader are characterized by high volatility and an extremely strong competitive situation involving numerous players of different sizes. The key success criteria are low production costs and sales prices, and increasingly personalized customer service through adequate innovation and customer care. Scott Bader is strong mainly in terms of innovation capability and meeting specific customer needs, as well as in some niche markets. There is also strong emphasis on working closely with partners because, being true to its core values, Scott Bader does not want to engage in fierce industry struggles (involving cost and price competition). Accordingly, it is also a member of 12 trade associations on four continents. By concentrating on these strengths, Scott Bader can hold its own against its competitors, which are usually larger. In addition, the special ownership and management structure can give the company a competitive advantage because short-term market effects have less influence on its operations. Scott Bader can be successful based on its long-term strategic approach and committed, mutually beneficial partnerships. During 2017–2018, the company had to cope with significant challenges due to several factors (rising raw material prices, issues with security of supply, economic and geopolitical uncertainty in certain geographical regions, and tightening of the regulatory environment). It is generally expected that competition in its key markets, cost, and price pressure will remain significant, raw material supplies will be uncertain, regulation of the chemical industry in line with safety and environmental standards will increase, and the impacts of Brexit will be unpredictable for the time being. As a strategic response, Scott Bader plans to continue to focus on its core values and leverage its strengths in product development and technical service to intensify its presence in niche markets where it can maintain its long-term success through serving specific customer needs. In addition, expansion and quality improvement have been planned in geographical regions (e.g., Brazil, India) that have not received much emphasis so far (Jacob 2012; Scott Bader 2019).

TerraCycle

Value propositions: The essence of TerraCycle's business model is that, based on the triple principle of "People, Planet, Profit," the company creates profitable business solutions for the selective collection, recycling, and reuse of waste traditionally classified as non-recyclable for its individual, corporate, institutional, and civil partners. Such waste includes, for example, cigarette butts, empty coffee capsules, drink pouches, yoghurt cups, confectionery packaging, worn-out baby seats that usually end up in landfills or incinerators, and even plastic ocean waste. For wastes that are difficult to recycle, TerraCycle develops national selective collection platforms (such as voluntary waste collection teams, and Zero Waste Boxes) whose operation is financed by the producers of the related brands, retailers, municipalities, small businesses, NGOs, or individual consumers. Waste that is collected can be used in its original form (reuse), with a changed, even upgraded function (upcycled), or as processed raw material for use in the manufacture of new products (recycled). When possible, TerraCycle seeks to create new products with the least possible additional energy input, thus moving closer to the implementation of the circular economy concept (by which the by-products or end products of material flows serve as inputs for another process cycle at the end of their lives). Costly selective collection can be still profitable for TerraCycle's partners because, in addition to generating measurable benefits in financial terms (working in an ecologically conscious way can generate more sales, greater market share or margins, and lead to more donations), the company can provide its partners with numerous additional services (e.g., media, PR consulting, social media, communication and marketing services) and added value. Such value may include the auditing and innovative redesign of products and production processes, creating more sustainable operations, higher customer turnover, brand enhancement, stronger consumer loyalty, and cleaner city streets. From this point of view, therefore, waste is positively evaluated rather than negatively. With the help of innovative solutions, the amount of waste and raw materials that need to be extracted can also be reduced. Another important value proposition of TerraCycle is that by focusing on the principles of "better, greener, cheaper" of eco-capitalism, it can promote solutions that are environmentally, socially, and economically sustainable in the long term (Vaccaro 2009; Eliances 2017; TerraCycle 2020a).

Market segment: TerraCycle is a global market leader in the collection and recycling of waste. It deals with all kinds of waste, in addition to the traditionally selectively collected paper, glass, metal, and rigid plastic waste. The company has four principal business divisions: (i) National recycling operations, whose shipping, storage, and recycling costs are sponsored by the manufacturers and traders of over 80 consumer brands. Major brand partners include Procter & Gamble, Garnier, Bausch & Lomb, Santa Fe Natural Tobacco, Colgate, Tom's of Maine, and Kiehl's. (ii) Zero Waste Boxes, which can be used to collect a specific waste stream (such as coffee capsules) or a defined category of waste (such as non-compostable kitchen waste). Including pre-paid shipment, these boxes are paid for by waste-collecting customers, and, once filled, can be delivered by UPS to a TerraCycle recycling facility. (iii) The sale of raw materials refers to the sale of pellets made from recycled plastics to plastic product manufacturers such as Green Mantra, Nexus Fuels, Akshar Plastics, and Ultra Poly Corporation. (iv) Regulated waste recycling helps businesses to manage regulated, universal and hazardous waste such as fluorescent lamps, bulbs, batteries, e-waste, organic waste, medical waste, and other types of waste potentially harmful to the environment (TerraCycle 2020a, b).

TerraCycle US Inc., the largest subsidiary of TerraCycle that accounts for the majority of the company's total annual sales, generated annual sales of more than seven million USD in 2018 from about 80 brand-sponsored national recycling programs. The subsidiary generated revenue of 3.6 million USD from more than 100 categories of (available-to-the-public) Zero Waste Boxes, 2.3 million USD from material sales, and over seven million USD in the regulated waste division. Within the material sales division, since 2017 TerraCycle has developed a "storied plastic" segment to sell recycled materials at premium prices. The name refers to the fact that the origin of these materials is known (for example, beach/ocean plastic), so new products (such as cosmetics packaging or shoes) made from these plastics can be associated with special meaning and unique marketing-related value. Main partners in this segment are Procter & Gamble, REN Skincare, AXE, Microsoft, and a network of organizations that is cleaning up beaches in more than ten countries. This initiative is now one of the world's largest beach/ocean plastic collection programs (TerraCycle 2020a, b).

Cost structure and profit potential: TerraCycle's business model is based on zero or negative cost waste, because producers, retailers, and consumers sponsor their waste collection and recycling programs. The company can thus potentially operate at below-market cost and generate greater profits even while providing additional services to differentiate itself from its competitors. The company's financial results, based on 2018 and 2019 data available for the US subsidiary, were as follows. In 2019, in the United States, TerraCycle saw a significant increase in total annual sales in the largest market by 35 per cent to 27 million USD. The reason for this was primarily the significant growth of the two largest business units, as new Sponsored Waste programs were launched and the Zero Waste Boxes segment was expanded to individual and corporate customers. In addition, large new deals in the Material Sales business were concluded. Costs were mainly due to logistics expenses, the redemption of charity points, the cost of employees engaged in delivery services, sales-related expenses, and general and administrative expenses. Total expenses increased, but by less than net sales, so net income grew by more than 160 per cent to a record 3.2 million USD. With regard to general market trends, TerraCycle expects that the awareness-raising impact of the World Economic Forum and Ellen MacArthur Foundation will strengthen global efforts to achieve a circular economy, and more and more companies will incorporate a sustainability element into their operations. In the United States, the government's denial of climate change may also stimulate individual initiatives to reduce personal environmental impacts; as a result, demand for TerraCycle products and services will continue to increase in its market segments. Based on its strong and positive reputation, TerraCycle plans to launch new business platforms and improve its product mix to increase premium-priced offerings that will allow the company to expand and grow profitably (PR Newswire 2020; TerraCycle 2020b).

Structure of the value chain: the core elements of TerraCycle's value chain are innovation and applied scientific research, through which the composition of different types of waste and the conditions of their generation are studied. Through such analyses, TerraCycle explores the recyclability of these wastes and investigates how to reduce the content of waste streams. Innovation activity also includes the development of new business models and marketing and communication plans, with which the collection, processing, and sale of specific types of waste can be made profitable in a strictly financial sense. Most waste management companies only deal with the collection and shipment of waste to landfills, so TerraCycle's concept of innovation and research and development activity is among the few exceptions (TerraCycle 2020a, b).

Important actors in the value chain include waste collectors and civil, corporate, and official communities. Over 202 million people have participated in related waste collection programs since the founding of TerraCycle. These people represent all areas of society regardless of age, gender, religion, financial status, and political sympathy. The role of employees is also an important human factor in the value chain. At TerraCycle, a strong work ethic prevails, and the atmosphere is cheerful and informal. Office furnishings are made from waste, and walls are covered in graffiti and are repainted every few weeks. Dress style is relaxed, but expectations about employee and managerial conduct are very high. However, TerraCycle's mission, value background, and meaningful jobs are very attractive to committed people, so dozens of students spend their internships with them in the summer, for example, and the most experienced employees have been working at TerraCycle for almost two decades (Bhasin 2011; *Parvati Magazine* 2017; TerraCycle 2020a, b).

TerraCycle does not have its own processing capacity because of capital requirements and technological inflexibility, as the processing of each type of waste can require different equipment, requiring new investment and financing. On the other hand, TerraCycle's processing partners are happy to cooperate with the company because TerraCycle now has two decades of experience and much know-how in market segments ignored by other market players, offering promising opportunities for expansion for those who work with them. The company also relies on third-party services for the shipment of collected waste (principally UPS in the United States), and warehouses for storage (TerraCycle 2020a, b).

Significant contributors to the performance of the value chain are the companies that produce and distribute the global brands that are cooperating with TerraCycle within the framework of national recycling programs and the brand-sponsored waste collection programs that have been developed for each type of waste. Some of the more than 100 corporate partners and brands include 3 M, AstraZeneca, Barilla, Bausch & Lomb, Coca-Cola, Colgate, ESPN, Garnier, GlaxoSmithKline, Gillette, Hasbro, Home Depot, L'Oréal, Nespresso, Nestlé, PepsiCo, Procter & Gamble, Royal Canin, Rubbermaid, Schwarzkopf, Subaru, Tom's of Maine, and Walmart (TerraCycle 2018, 2020a, b).

An important goal of TerraCycle is to increase consumer awareness, so activities directly related to consumers have been awarded priority in the value chain. To support these, in addition to major trading partners, TerraCycle opened its own retail outlets in Princeton in 2009. In 2010, the company set up so-called "Hot Spots" eco-displays in Walmart stores, which pair original products such as Capri Sun soft drinks pouches with upcycled products made from waste (e.g., bags, caps, balls, and penholders). Individual consumers were targeted by the 2009 reality show series *Garbage Moguls* on The National Geographic Channel, in 2014 in the multi-season TV series *Human Resources*, and by *Trash Tycoon*, the first online game about repurposing waste. Recently, TerraCycle has launched Loop, a circular shopping system for delivering popular brands to households and recollecting them in durable, refillable packaging to eliminate single-use plastics (TerraCycle 2018, 2020a).

Position within the value network: TerraCycle is a global market leader at recycling "unrecyclable" waste, and a committed and tireless advocate of waste elimination initiatives, the promotion of the circular economy, and eco-capitalism. The founder, CEO Tom Szaky, is the author of three books on the topic, and a frequent author and interviewee on professional online platforms, websites about social and public life, and notable newspapers such as Inc., Entrepreneur, Chief Executive, and Smart CEO. Szaky and his company have been recognized more than 200 times as leading sustainable social and environmental actors for their pioneering work. A few of these awards include *Inc.* magazine awarding Szaky the "# 1 CEO in America under 30" award in 2006, and the United States Environmental Protection Agency recognizing the company's business model as consistent with the agency's solid waste strategy. In 2011, the founder was included in *Forbes* magazine's "Impact 30" list. In 2012, Szaky was awarded the United Nations' "Leader of Social Change Award" and TerraCycle was awarded the *Entrepreneur* magazine's "Brilliant 100 Companies" award. In 2013, Szaky won the Schwab Foundation's "Social Entrepreneur of the Year" award. In 2017, TerraCycle was awarded the United Nations' "Momentum for Change Lighthouse Activity" award for the first fully recyclable shampoo bottle made from ocean plastic waste, developed in collaboration with Procter & Gamble and the French water, energy and waste management company SUEZ. In 2018, TerraCycle received the "BIG Innovation Award" and Szaky was awarded the Ernst & Young "Entrepreneur of the Year Regional Winner." In 2019, the company was number ten on *Fortune* magazine's "List of 52 Companies Changing the World," as well as on "Inc's 5000 List," and Tom Szaky was named one of the "Top Most Influential Doers and Shakers in Purposeful Retail" (TerraCycle 2020a, b).

TerraCycle's significant impact within its value chain is clearly illustrated by the fact that it is working on solving a global social and economic problem that no other company has taken on so far. Coping with more than 23,000 billion pieces of consumer waste (including five million tons of plastic waste per year) is an extreme challenge for the planet and for human health. The recycling rate of waste is very low, and environmental education and awareness raising have not been awarded high priority so far. TerraCycle's mission is to change people's perceptions of waste, mobilize individuals to engage in selective waste collection, encourage companies to support recycling programs, and support educational, community, religious, and charitable NGOs from its income. The company also takes responsibility for its local environment. It has chosen economically disadvantaged New Jersey as its headquarters to create jobs for locals through its operations, and to provide education and training for young people who can participate in service of a good cause. TerraCycle also provides urban artists with a creative, exhibition space (TerraCycle 2018, 2020a).

TerraCycle's Sponsored Waste Programs involve offering individuals and organizations that collect waste a reward in the form of "TerraCycle charity points," which can be converted into donations to charities. Points are awarded based on the net weight of waste items that are collected, and can usually be converted into a 0.02 USD payment per waste item to a charity or school selected by the collector. So far, almost eight billion waste units have been collected and more than 44 million USD has been donated.

In 2018, TerraCycle formed TerraCycle Global Foundation to address the complex challenges of collecting and recycling waste in emerging countries such as Thailand, China, Indonesia, Vietnam, Sri Lanka, and the Philippines. Through this organization, the company engages with local governments, communities, and nonprofit and corporate partners to offer waste collection and recycling platforms, with special emphasis on preventing plastic pollution from entering aquatic systems. It creates charitable organizations that cooperate with social, environmental, and educational organizations to enhance recycling efforts and improve the conditions of the waste pickers who work to clean up waterways and are engaged in land-based plastic collection (TerraCycle 2020a).

Competition strategy: No other company is collecting and recycling most of the waste categories treated by TerraCycle, so there is no direct competition. In its main market segments, the following factors characterize the development of competition. National recycling programs and Zero Waste Boxes are unique in the market. Working closely with its partners, TerraCycle helps them with not only sustainability, logistics, marketing, and PR services, but also with innovative research and development activities to become more sustainable and operate more effectively. There is significant competition in the regulated waste recycling segment, with larger, stronger companies than TerraCycle dominating the market. Nevertheless, the company can successfully differentiate itself from traditional players and gain market share in the medium term by building on the company's brand, reputation, special services, and competencies that contribute to improving the awareness and popularity of partners' brands. In the material sales business, TerraCycle does not deal with conventionally collected and recycled HDPE and PET-type plastics, but with categories not treated by others. In this special segment, the company has a market-leading position, which gives TerraCycle significant room to determine supply and pricing. Competition has been growing in the "Storied Plastic" segment, as other companies are increasingly inclined to deal with ocean plastics due to their increasing market value. However, TerraCycle's competitive advantage as a market leader may last over the coming years. An important phenomenon in terms of competition is that, in addition to working closely with its stakeholders in more than 20 countries, TerraCycle has also partnered with the largest conventional waste management companies (Progressive Waste Solutions in Canada, Landbell GmbH. in Germany, and SUEZ in France) since 2013. The latter companies are becoming closer partners and now own 20–30 per cent of all shares in TerraCycle's subsidiaries in Canada and nine European countries. Thus, in addition to the strategy of focusing on special market segments, innovation, and differentiation, the current direction of development also involves more industry collaboration (TerraCycle 2020a, b).

The main characteristics of the ecologically conscious business models in the international sample are presented in Table 6.1.

Table 6.1 Main characteristics of ecologically conscious business models in the international sample

	Value propositions	Market segment	Cost structure and profit potential	Structure of the value chain	Position within the value network	Competitive strategy
Green Monday	Construction and operation of a plant-based food ecosystem to counter planetary devastation, food insecurity, food-related illness, and animal suffering, and promoting local and global community building, implementing triple-bottom-line approach	Social: awareness-raising campaigns for promoting a sustainable lifestyle; Business: grocery stores for plant-based food, sustainability, CSR consulting, impact investing	Rapid turnover growth since foundation, stable profitability, good prospects, geographical expansion because of successful business based on increasing consumer consciousness and lifestyle changes	R&D, sustainable, ethical suppliers, knowledge sharing, community platform, own store network, program for sustainable event management, seed investment fund for sustainable start-ups, prestigious advocates and partner organizations	Good reputation, popularity, real impact on the culture of consumption and people's lifestyle in Hong Kong and the region, numerous entrepreneurial awards immediately after its foundation, market leader with intention to further develop the whole market segment and promote feasible lifestyle change around the world	Shaping consumer consciousness, inspiring other businesses, increasing network of partners involving knowledge sharing and value transfer, strengthening sustainable food industry cooperation

	Value propositions	Market segment	Cost structure and profit potential	Structure of the value chain	Position within the value network	Competitive strategy
Houdini Sportsware	Nature is of intrinsic value and is blueprint for creating circular systems through creativity, high quality, timeless style, sustainability-based business model for highly innovative, ethical and sustainable functional clothing	Ethical and sustainable production and retail sale of functional outdoor clothing for ecologically conscious consumers living an active lifestyle who prefer timeless minimalist style	Steady organic development of revenue with fluctuating profitability due to significant increase in assets and number of staff, continuous innovation, stable demand due to popular brand	Highly innovative product design based on sustainability, recycled, biodegradable raw materials, open-source innovation, local R&D, production in Europe, few suppliers, short supply chain, redefining consumption and consumer care	Leader in sustainability, transforming consumption patterns, first ever Planetary Boundaries Assessment, social boundaries added to measure well-being of value chain, interdisciplinary collaboration, transparency, prestigious business and industry awards	Striving for sustainability in close cooperation with other industry players, even competitors to inspire each other, joint research projects, making company-developed materials freely available

(continued)

Table 6.1 (continued)

	Value propositions	Market segment	Cost structure and profit potential	Structure of the value chain	Position within the value network	Competitive strategy
Interface	Production of modular carpets and services, elimination of negative environmental impacts, restoration of nature, renewable, cyclical, gentle business model, radical redesign of processes, products, improvement of life of all stakeholders	Manufacture and sales of mainly commercial modular carpets for office and retail sector, expanding to household and luxury vinyl floors and rubber flooring mainly for healthcare and education sectors	Largest margin segment, premium products, profit potential greater than industry average, radical innovation to lessen environmental impacts and reduce cost, increase turnover and profitability	Company culture based on sustainability, own sustainability advisory group, "The Natural Step" methodology, radical product and process innovation, close cooperation with value chain, fair and ethical relationship with stakeholders	Pioneering innovator committed to sustainability, design that mimics nature, recognized among the best performers, numerous entrepreneurial awards, role-model for others approaching zero negative impact, striving for revolutionary transformation of industry	Highly competitive global market, global leader in its segment, successfully competes on sustainability and innovation, excellent brand due to ecological commitment and performance

	Value propositions	Market segment	Cost structure and profit potential	Structure of the value chain	Position within the value network	Competitive strategy
Natura	Production and sales of natural, ethical, sustainable, innovative beauty care products through four iconic brands, nature-inspired cyclical system to protect nature, fair, caring relationships and value creation throughout entire partner network	Fragrances, beauty and body care cosmetics, gifts mostly positioned in the natural, organic and premium market segment, originally in Brazil, now worldwide through its group of brands Natura & Co.	Significantly increasing group size, revenue, profitability, synergistic cooperation of global brands Natura, Aesop, The Body Shop, Avon, plans to harmonize and strengthen brands, R&D, digital transformation, innovation	Striving for cyclical production system neutralizing negative impact on nature and creating value for stakeholders in rainforests and partner network, product development, fair relationships, multi-channel sales, consumer awareness raising	Exemplary transparency, continuous monitoring, reporting of sustainability performance, improving environmental and social impacts, widespread recognition of leading practices, support for local communities, reforestation, public education, social justice, women's empowerment and well-being, B Corp	Strong global competition, significant product development, marketing expenditure is key, Brazilian market leader, globally fourth based on environmental, social performance, innovation, strong brands, reputation

(continued)

Table 6.1 (continued)

	Value propositions	Market segment	Cost structure and profit potential	Structure of the value chain	Position within the value network	Competitive strategy
Organic India	Production of genuine organic teas, herbal products and food, holistic business model for the benefit of all stakeholders, promoting restorative agriculture that serves and protects Nature, support for farmers, wild-crafters in rural India	Production and sales of organic teas, herbal preparations, spices, ginger, seeds, fresh vegetables and packaged foods (ghee, coconut oil, and honey), main product Tulsi (basil), certified organic, organic, GMO free	Growth rate of global and Indian organic product market, margins significantly higher than for conventional products, increasing consumer health consciousness and product range can further improve profit potential	Based on "whole herb," traditional holistic healing system Ayurveda, cutting-edge technology, holistic quality assurance system and certification, fair trade, development of farmer communities, regenerative agriculture, close cooperation with market-leading distributor	Business model restores health of soil and plants, continually adds value to marginalized rural farmers, tribal wild-crafter families, quality and organic certification, initiatives to improve environmental, social sustainability performance, recognized by awards, created own ethical standard, award for organic farmers, B Corp	Global market leader in a growing market segment, popular brand, strong customer loyalty, expansion of product range and geographic reach, wants to make organic products available to all

	Value propositions	Market segment	Cost structure and profit potential	Structure of the value chain	Position within the value network	Competitive strategy
Patagonia	Provision of the best quality, simple, useful outdoor products without causing unnecessary harm, using business to solve the environmental crisis via leading society to protect and restore nature, supporting activism and regenerative agriculture	Manufacture and retail of sustainable, ecological outdoor clothing, accessories, apparel designed for utility in simple, minimalist style via own and partner stores worldwide and internet	Higher costs than market average due to application of natural, innovative materials, fair trade in the supply chain, steady growth, sufficient profit expected based on strong brand, customer loyalty, increasing awareness	Design of durable, repairable, reusable products using recycled, renewable, fair trade materials, transparency of factories, footprint of operation monitored and reported, supportive working environment, women's empowerment, regenerative agriculture	Pioneer in industry, promoting environmental, social, political activism, raising consumer consciousness, publishing books, films, 1% of annual sales to grassroots environmental NGOs worldwide, venture capital to environmental start-ups, sharing of best practice, recognized by many awards, B Corp	Fragmented market with global players, emerging brands, growing consumer consciousness, direct sales, transparency, focusing on genuine core values and customers committed to sustainability and minimalist style

(continued)

Table 6.1 (continued)

	Value propositions	Market segment	Cost structure and profit potential	Structure of the value chain	Position within the value network	Competitive strategy
Scott Bader	Three pillars: Business, Ecology, Humanity, avoid jeopardizing natural systems, safe, ethical and transparent chemical industry for common good, knowledge, profit sharing, circular economy, workforce development, commonwealth, charity	Production and distribution of composites such as resins, gel coatings, adhesives specialty polymers, and fireproof, smoke- and toxic- resistant materials, mainly in Europe, North America for various industries	Growing sales, profitable, but margins fluctuating and uncertain conditions due to tightening regulations, Brexit, and issues regarding safe access to raw materials, expansion in Japan, India, Australia to mitigate negative effects	Trusteeship as unique form of community ownership, corporate governance system, continuous innovation, using "The Natural Step" sustainability methodology to reduce environmental impact, develop employees, improve health and safety conditions	Leading role in its segments in innovation, technology, certified processes and products, professional awards for social and environmental sustainability performance, exemplary role in donating to local, national, and international charities in education, health care, community development, sustainable livelihoods	Strong competition with variety of competitors, innovation, customer care, long-term strategic commitment, enhancing niche market presence, close partnerships, supply and geographic expansion

	Value propositions	Market segment	Cost structure and profit potential	Structure of the value chain	Position within the value network	Competitive strategy
TerraCycle	Eliminating the concept of waste, recycling the non-recyclable, implementing eco-capitalism, circular economy, awareness raising, enhancing partners' brands and consumer loyalty, additional consultancy services (PR, social media)	Collection, recycling, reusing waste considered non-recyclable, national brand- and customer-funded waste collection programs, material sales including "Storied Plastic," regulated waste recycling	Zero or negative cost raw material due to sponsored waste collection programs, all products made of waste, increasing awareness, growing sales and profitability, improving product mix to achieve higher margins	Scientific research, product, process, business model, marketing innovations, 200 + million individual waste pickers, global brands as engaged partners, fair partnerships in logistics, plastic processing, consumer awareness raising (TV shows, online game)	Global market leader, opinion leader, pioneer in eliminating the concept of waste, solving global social and ecological problems, mobilizing brands, individuals, and communities, donating to charities selected by waste pickers, funding programs in emerging countries, over 200 business, sustainability awards	Unique service, no direct competition, strong brand, effective differentiation, close and intensifying cooperation with stakeholders, even with large waste management companies abroad

Source Author's construction

Ecologically Conscious Business Models in the Hungarian Sample

This section briefly introduces the business models of 14 well-known Hungary-based ecologically conscious business organizations along the same six dimensions as the international cases. The descriptions are based on the interviews with the businesses and information available on the companies' websites and in company documents such as mission statements, annual reports, ethical codes of conduct, sustainability reports, compliance reports, programs, and product catalogues.

Skanska Hungary

Skanska Hungary is a leading commercial property developer with a focus on project development, leasing, and facility management. The main goal of the company, as part of the Sweden-based Skanska group, is to help build a better society and generate shareholder value alongside promoting social responsibility. Through innovative and sustainable solutions, the company is involved in creating a sustainable future for all stakeholders, including employees, customers, and communities. It is committed to safety, ethical behavior, green, and community investment, as well as diversity and inclusion. Skanska's core values are caring for the lives of people, the environment, and future generations by supporting health and well-being and promoting green solutions, doing business in an ethical and transparent way, development, innovation, learning and sharing expertise, and cooperation and commitment to its customers, as summarized by András Schmidt, sustainability manager of Skanska Hungary.

The company has developed more than 200,000 square meters of commercial space for offices and retail in Hungary, and more projects are under development. It is located in Budapest, and since 2015 has also involved a construction business. Skanska Hungary is focusing on green real estate development, and promotes close cooperation through tailor-made solutions primarily for business clients.

The company has steadily grown into a medium-sized company and had annual turnover of over five million USD and around 70 employees in 2019. It has been profitable since its inception in 1987. The market outlook seems to be favorable, but depends on economic growth trends. The sensitivity of its clients to the sustainability features of buildings is still evolving, so it considers it important to keep its prices at around the market average, despite the higher costs of its social and environmental responsibility practices. However, in many cases improvements in environmental performance can lead to financial savings due to decreases in material and energy consumption.

Skanska Hungary's value chain includes property development, architectural design, construction, and project management. The Hungarian construction department is smaller than in the Polish, Czech, and Romanian sister companies, although innovation performance (such as creative interior design) is of particular importance. The company strives to minimize the negative environmental impact of the construction process and facility management through industry-leading practices like reducing energy, water, material consumption, the amount of waste, and carbon emissions.

Within its value network, in addition to social responsibility (with a focus on supporting employee well-being through occupational safety and clients via applying the "WELL" standard, helping local communities through corporate community investment, and considering future generations), the protection of the environment is of particular importance to Skanska Hungary. It is the only developer in Hungary that has been audited and certified according to the standards of ISO 14001 environmental management schemes. Its buildings are also certified in line with the LEED standard. The company also pays attention to adhering to its core values when choosing potential partners. Skanska Hungary has developed and applied a code of conduct to its own employees and suppliers. Over the years, the professional and sustainability performance of the company has been recognized by awards such as "The City of the Future Award" in the For a Greener City category in 2016, Leading Green Building Development of the Year in the "Construction and Investment Journal Award" in 2016, and the "Office of the Year" in 2016 and 2018.

Skanska Hungary is a medium-sized player in the Hungarian construction industry and faces strong competition, while the Swedish group is one of the world's leading real estate developers. The industry has recently been a key economic sector in Hungary. The company can maintain its competitive advantages by continuously developing its industry-leading environmental and work and health safety practices that go beyond strict regulatory compliance.

Nekedterem Hungary

The mission of owner-manager Eszter Molnár and her company is to provide its customers with food products from local producers in Budapest and its surroundings via a webshop, home delivery, food-box delivery, and subscription services. Its product range includes high quality, pure, local, small-scale, artisan Hungarian products. Nekedterem's goal is to minimize the length of the supply chain and transportation, use recyclable, biologically degradable packaging materials, and selectively collect the waste produced during its operation.

Nekedterem serves consumers who are searching for healthy, high-quality fresh or processed food products from Hungarian small-scale, artisan producers who prefer to make purchases online, with home delivery. Products in many cases are organically certified or guaranteed additive-free and chemical-free. Its customers are mainly individuals and families, and Nekedterem supplies corporate partners who want to provide their employees with fresh seasonal fruit.

The company offers premium products at higher than average prices. Due to its stable customer base, Nekedterem is still a micro enterprise, but it has experienced an increase in annual turnover since its inception. Its profitability is stable, and the owner is optimistic about the future due to strong customer loyalty and the growing popularity of health-conscious food consumption.

The structure of Nekedterem's value chain is simple. It involves purchasing goods from local Hungarian farmers and suppliers around Budapest. The processing of online orders, home delivery of orders, and serving consumers at the company's site (which also functions as a pick-up point) are key elements of the business. The company is continuously searching for new products and partners, monitoring existing ones, and keeping contact with its stakeholders.

Within its value network, Nekedterem maintains close contact with its stakeholders and monitors the operations and conditions of suppliers. Through this, the company has a strong desire to support the local economy and local communities. Nekedterem strives to promote environmental consciousness and minimize environmental harm during its operations. It also promotes the health of future generations by providing fresh, healthy, clean food to families. The company is strengthening its connections and relations with civil society, occasionally serving them and people in need by fundraising and donating food.

Nekedterem perceives the competition to be only moderately strong because of its unique niche in the market. Compared to the home delivery services of large supermarket chains such as Auchan, Tesco, or specialty stores, the company's service is special because it offers fresh organic, healthy products. Compared to food-box delivery schemes based on community-supported agriculture (such as that of Szatyorbolt), Nekedterem offers home delivery as a convenience service.

Réthy Fashion

Réthy Fashion's value proposition is recycling and reusing denim jeans fabric by rethinking the connection between environmental protection and fashion. The company designs and creates high quality, innovative fashion products made of denim fabric. At fashion shows and environmental events, it also engages its stakeholders with environmental awareness-raising activities, and promotes selective waste collection. Réthy Fashion is also engaged in social value creation through employing disadvantaged people. The idea is to protect the climate and fight global warming by striving toward a circular economy.

The company's market segment is designing, producing, and selling unique fashion products to individual and business customers who are interested in environmentally conscious and custom-designed clothing.

Réthy Fashion sells its products in its own showroom, via a webstore, in a sample shop, and at fairs.

Réthy Fashion was established in 2012 as a family business, but started to develop after moving to a new site with its own recycling center in 2016. It still counts as a micro enterprise, and its profitability is around the break-even point. The company's high-quality design products can cost more and reach higher prices than the market average. Based on a satisfied and growing customer base, Réthy Fashion expects to undergo steady but limited development due to the relatively low level of demand and average level of ecological consciousness of consumers in Hungary.

The structure of Réthy Fashion's value chain consists of collecting, cleaning, processing, color sorting, tailoring, and sewing denim fabric. Raw materials can include lace, empty coffee capsules, buttons, and zippers, of which uniquely designed products, uniforms, and accessories are made. The company also operates a jeans repair service, does tailoring work, and runs an upholsterers. High-quality product development, awareness-raising activities at fashion shows, and educational events are key parts of its value chain, too.

Within its value network, Réthy Fashion promotes social engagement in environmental protection, recycling, environmental education, and employment of disadvantaged people. After its inception, the company participated in the fourth NESsT-Citibank Social Enterprise Development Program because of its commitment to environmental and social responsibility. Réthy Fashion's goal is to set an example, for example, by having been a 100 per cent carbon-neutral enterprise since 2015. The company has continuously been working on raising consumer awareness. It advocates eco-fashion and recycling at various environmental events such as eco-fashion shows, eco-days, jeans waste collections, and exhibitions. In 2013, Réthy Fashion was awarded the special Ozone Green Award, which was founded by a Hungarian television channel (*Ozone Network*) for environmentally conscious viewers.

The company has faced very strong competition, even if it has only one main competitor in Hungary. This competitor competes on price, while Réthy Fashion's main competitive advantages are high quality, unique design, increasingly efficient operations, and an expanding brand

and customer base. The company expects that the ecological consciousness of consumers will increase, and more and more committed partners will find Réthy Fashion due to its strong value commitments.

Bio-Drog Berta

This small-scale, family-based company, founded by Zoltán Berta, has been engaged in organic farming for almost three decades and is committed to value creation for its local and national community. Its goal is to use the talent and skills of its owners and employees in a form of agriculture that is still not common in Hungary. Bio-Drog Berta strives to produce as many innovative products as possible, to utilize through processing the greatest proportion of plants grown on excellent agricultural land as possible, and to supply the needs of primarily Hungarian consumers.

Bio-Drog-Berta has the potential to produce, process, and trade approximately 160 kinds of organic, health-preserving, health-improving, processed plant-based products made of fruit, vegetables, cereals, seeds, herbs, spices, and flowers. Its product range includes fruit and vegetable products, flours, creams, syrups, pastes, tea, vinegar, chocolate, and aperitifs. The company's primary focus is serving Hungarian consumers of organic products through resellers or direct sales, although its products are well known in 14 European Union countries, in Switzerland, and in the United States.

The company has been investing its profits into asset and product development since its foundation in early 1990s. The goal has been to achieve the highest level of processing in order to develop the broadest possible product range. Company products have been well received, and are successful on the market. Production costs are higher than for conventional production, and the main limiting factors have been a growing labor shortage and the potential weakness of an environmentally conscious consumer base. During recent years, the annual sales of Bio-Drog Berta have significantly decreased, along with the number of employees. Profitability is sustainable at a much smaller scale, in the form of a micro-sized enterprise, and due to unfavorable industry trends, significant growth cannot be expected.

The organization's value chain originally covered a full range of organic farming activities such as production, processing, storage, and sales of products cultivated on high-quality agricultural land in Central Hungary. Production involved family farmers from the local community. Bio-Drog Berta's continuous research and development was aimed at the waste-free processing of crops. Its processing capacity has been continuously expanded during the years with increased drying capacity since 1995, a canning industry since 2000, an artisanal chocolate program since 2012, and cold-press oil production since 2014. However, due to significant labor shortages in the industry and the region, the company has faced capacity constraints for years; also, its sales activity has been limited. As a result, a considerable part of the potential product range has recently become unavailable.

Bio-Drog Berta has obtained Hungarian, American, and Swiss certification for its products. Because of the need to make up for the operational limitations of its value chain related to labor-market difficulties, different types of collaboration have been developed over the years. For instance, the company has employed prisoners and people with disabilities recruited by an NGO for several years. For its outstanding social responsibility performance, it was awarded the "KÖVET" award by the KÖVET Association for a Sustainable Economy. The company has also been an active partner of "Bátor Tábor," a foundation that supports seriously ill children and their families, for more than five years. In serving its local community, Bio-Drog Berta also established a community park called "Musical Instrument Trees" to boost tourism in the region based on the Danube river.

Regarding its competitive strategy, the company has faced very strong competition, although the complexity of its value chain, capacity for innovation, and excellent health-promoting features of its products may represent competitive advantages in the domestic market. In multiple cases, Bio-Drog Berta has suffered from unfair competition such as patent infringements by former partners, but due to its weak human resources and financial capacity, it was not able to engage in lengthy litigation processes.

Ecoizm/Zöldbolt

The mission of the founder, Adrián Zelenák, and his enterprise is to distribute sustainable and ecological household products in its brick-and-mortar stores and via e-commerce. Its products serve consumers' everyday needs, and in all cases have some significant environmental beneficial characteristics in relation to substitute products. Ecoizm is continuously monitoring product innovations and promoting environmental awareness.

In its stores and via its webshop Ecoizm sells a wide range of environmentally friendly household products, as substitutes for conventional products, to domestic and foreign individuals, retailers, and wholesale partners. Its mix includes products in the segments of beauty and health care, cleaning and washing, food staples, eco paper and stationery, home and garden, eco clothing, toys and baby products, books and DVDs, and energy-efficient products for water-, power- and heat-saving.

Ecoizm saw sharp growth after its start in 2008. It then witnessed a moderate but steady increase in sales and profitable operation as a small enterprise. Its profit margins are significantly lower in the wholesale than in the retail segment, but the potential for profit has been favorable for years.

Its value chain consists of search- and purchasing-related activities, in some cases the retail packaging of larger items, own brand packaging, serving retail and wholesale partners, a webshop operation, and after-sales services. Ecoizm also has an online magazine with fresh environmental news and provides awareness-raising services for its partners. The enterprise is continuously striving to make its office operation greener through the recycling of paper and packaging materials. Its office and stores have been powered by 100 per cent renewable energy (from a nearby wind power plant) since 2017.

Ecoizm is always searching for household products with the best characteristics in terms of environmental load, and raising its consumers' awareness. Its aims are to be a flagship player in the market, and a reference point for purchasing eco-products. The enterprise has developed its own rating system for products in different categories for much more convenient and transparent searching. It is continuously monitoring the latest innovation and product developments, and cooperates with the authorities when introducing innovations such as new packaging-free solutions in the retail sector. It has close relationships with civil society organizations such as the Association of Conscious Consumers, the Hungarian Cyclists' Club, the Hungarian Ornithological and Nature Conservation Association.

Ecoizm perceives its competitive situation to be moderately strong because its wide range product offerings, value commitments, and customer base do not fully overlap with those of its competitors and some other major Hungarian eco shops such as Zöldpolc and Ökokuckó. Consumers can thus differentiate, so these shops are able to coexist. Over the years, some of its products have appeared on the shelves of large international supermarket chains including Tesco and Drogerie Markt, and also in specialized stores (for example, washable diapers in shops for baby products). In these cases, its competitive advantage based on specific and unique products does not exist, but one of Ecoizm's main goals, to disseminate eco-products, has been achieved, which the enterprise considers much more important than improving sales figures at all costs.

ALTEO Energy

ALTEO's main value proposition is related to sustainable energy production, energy trade, and bespoke energy services that will help the company achieve its climate protection goals at an appropriate cost level and security of supply. The company is committed to promoting energy democracy, customized services, and sustainable business benefits for its customers. It strives to create solutions that are easy-to-adapt, flexible, innovative, and efficient. ALTEO aims to be an industry leader by example, and to generate long-term economic, environmental, and social value. The company believes that everyone has right to a healthy environment, to which ALTEO can contribute by adhering to the principles of operating in an ethical and transparent way, responsible corporate governance, the protection of human life and security, environmental responsibility throughout the supply chain, ensuring equal treatment, and fostering lasting, mutually beneficial partnerships.

Company market segments include the construction, operation and maintenance of power plants based on natural gas and renewable energy sources such as solar, wind, water, landfill gas, biomass, biogas power. ALTEO mainly serves large business partners by producing electricity, and retail customers primarily through natural gas trading.

General market conditions have been challenging for years due to labor-market-related supply issues and a stringent regulatory environment. Taking advantage of its favorable market segment potential, ALTEO has rapidly grown since its foundation in 2008. In 2019, annual sales reached nearly 90 million USD (increasing by almost 40 per cent at the group level). This positive trend has been supported by the successful extension of several long-term key contracts and the acquisition of guaranteed profit contracts. The number of employees has significantly increased to more than 250. ALTEO's energy trading segment is price sensitive, but in other segments the company's customers are strongly loyal. Return and profit potential is outstanding, and stock prices have been increasing year by year, while credit ratings are favorable and improving.

ALTEO's value chain includes the construction, operation, and maintenance of power plants. The company has a leading role in research and development and customer-oriented product development. Through its energy provision and trading services, ALTEO is a qualified supplier for numerous partners. In line with strong industry trends, the company is striving to develop and implement high-quality information and communication technologies specific to the energy sector. Its support functions (such as human resources, legal affairs, finance, accounting, and administration) all are based on respect for and the financial and moral recognition of employees, good working conditions, and responsible relationships. The company culture is strong and optimistic; employee turnover is well below the average for the industry and national economy. In 2017, ALTEO introduced the ALTEO Academy to organize sustainability training for employees to enhance their holistic vision.

ALTEO plays an exemplary role in its value network, striving for the position of opinion leader and benchmark actor in the industry. The company's intention is to set an example by being a highly innovative, environmentally and socially responsible player in its market. ALTEO's energy production has been "certified green" by the Hungarian Energy Office. The company has applied environmental management standards, and published award-winning sustainability reports. In 2019, it published an integrated report to present its economic, sustainability, and social impact. For years, its chief executive officer, Attila Chikán, Jr., has been an active member and president of the Business Council of Sustainable Development Hungary (BCSDH), and a member of the Hungarian Business Leaders' Forum. ALTEO has trust-based, mutually beneficial relationships with its stakeholders through providing tailor-made solutions to clients, closely cooperating with nonprofit organizations (WWF, the National Council for Sustainable Development), and following the activities of flagship organizations such as Greenpeace. The company promotes sustainability along its value chain and ecological orientation via its participation in BCSDH. ALTEO also aims to educate and motivate people to consider and live out the principles of sustainability.

ALTEO's competitive strategy is to maintain its leading position, and continue setting an example. The company has some competitive advantages and a moderately strong competitive position, for example, in relation to the largest market player E.On, due to its genuine value-based commitment and ecological consciousness. These factors are increasingly important to its customers, so ALTEO can efficiently use its market potential. The company collaborates with other market players based on the principle of reciprocity. It is also actively developing its competencies, adaptability, and learning abilities because it expects significant changes in the world.

Hajtás Pajtás

Hajtás Pajtás is a bicycle courier company that delivers mail and parcels to its customers in Budapest. Its service is based on high level of trust and reliability. Unlike the typical companies in this industry, Hajtás Pajtás operates in a transparent and law-abiding way. Its value proposition includes sustainability, environmental consciousness, striving to avoid unnecessary pollution, social responsibility, and, no less importantly, promoting bicycling activism for a more livable environment in Budapest.

Its market segment is a bicycle courier service mainly for business customers, primarily for those who are seeking a reliable and confidential service. Hajtás Pajtás has recently complemented its service portfolio with an electric truck service for packages not suitable for bicycle courier delivery.

The company's services are the highest priced ones on the market due to significant labor costs and its fully transparent, law-abiding mode of operation. The company dynamically grew after its inception in 1999 until the 2008 financial crisis. During the big downturn, Hajtás Pajtás reorganized its operations and decided to totally move away from the "grey" practices typical to the industry. Accordingly, headcount was reduced, prices increased, and all profits were reinvested into becoming economically sustainable again. The company restarted developing in new directions, including increasing package sizes, webshop deliveries, cooperating with traditional courier companies such as GLS, and enhancing trust by ensuring a high level of confidentiality. Hajtás Pajtás has increased over the years into a medium-sized company, and in 2019 had annual revenue of about 1.5 million USD, and nearly 60 employees.

Hajtás Pajtás's value chain is based on its couriers. The company strives to provide them with salaries higher than the industry average. In the case of price increases, it also pays a higher wage to couriers. Couriers are also supported through in-kind benefits such as the provision of a benefits system, bicycles, repairs and parts. Hajtás Pajtás maintains a good company atmosphere and a special culture, thus for the employees who work there the job represents a lifestyle. Based on these factors, the company has played a major role in promoting bicycling in Budapest and throughout Hungary. Beyond the core business, due to its environmental consciousness Hajtás Pajtás has always been committed to pioneering other activities, including those related to office administration, such as the selective collection of waste and empty batteries, even years before this became popular in Hungary.

The goal of the owners, László Géresi and Zoltán Tóth, and their company is to set an example by promoting ecological and social responsibility, social equality, and cycling activism. For two decades, Hajtás Pajtás has been the greatest advocate of expanding the cycling movement in Budapest and has undertaken negotiations with local governments and authorities and cooperated with them to improve conditions for bicycling in the capital. The company started and organized the biggest Hungarian cycling movement (Critical Mass), and it is the founder and committed supporter of the Hungarian Bicyclist's Club.

Hajtás Pajtás has been a pioneer of bicycle courier services in the Hungarian market, and is still the market leader. Its brand is well known and respected, despite very strong competition. Due to its social impact, all courier companies now have a bicycle courier business. The area served by bicycle couriers in Budapest is expanding and the company has started collaborating with traditional players to meet demand for all those orders that can be managed by bicycle couriers as well. Hajtás Pajtás's competitive advantage is based on its strong commitment, reliability, credibility, and a brand that has gained strength over two decades.

Pannonhalmi Abbey Winery

The company's value proposition is based on the Benedictine monks' values, derived from the Regula of Saint Benedict, marketing and sales manager Tamás Illés says. These include balance between spiritual life and physical activity, love of work, self-sustenance, education, and the dissemination of culture. Its goals are to revive the culture of viticulture and wine-making, to make the nonprofit activities—including operating social, educational, cultural institutions—of the abbey economically sustainable, and to produce their own sacramental wine. The company's core values are responsibility, sustainability, and the preservation and nurturing of values and resources for future generations.

Pannonhalmi Abbey Winery's market segment is the production of high-quality premium wine for domestic sales and exporting to commercial companies and direct consumers.

Since its foundation in 2002 and the full renewal of its grapevine plantations, the company has dynamically developed in terms of quantity and quality. Its annual turnover fluctuated at first along with seasonal changes, but then increased for years to about 1.8 million USD in 2019. The company's position has been stable in the quality wine segment, and customer loyalty is strong. Profits from operations are favorable, and after the repayment of a large portion of the initial investment loan, it became financially profitable two years ago. Regarding the state of ecological consciousness, Pannonhalmi Abbey Winery expects significant improvements and changes in consumers' attitudes in the future that will positively influence the sale of its quality products.

In its value chain, the company cultivates its vineyards and engages in viticulture, grape processing, and wine-making with the least possible environmental harm. Due to expanding sales, the number of employees has increased, and during seasonal work it typically doubles. The organizational development project of the abbey is complete, and the operation of the company group has become more transparent, due to which the company expects better synergy.

Pannonhalmi Abbey Winery evaluates its position within its value network in relation to the application of a holistic approach. Its stakeholders include the beneficiaries of the social, educational, and cultural activities of the abbey. It strives to strengthen environmental consciousness through its operations and supporting functions such as administration and transportation, thereby minimizing the environmental impact of the whole company. It has actively supported the civil society organizations established by the abbey, through which the company would like to fulfill its broader mission of integrating social and environmental responsibility.

The company faces strong competition, because in the premium market segment for wines there are many competitors in the limited Hungarian market. The export orientation is thus given special emphasis. In the premium market segment, other players are also increasingly ecologically conscious, so Pannonhalmi Abbey Winery can operate successfully by relying on its traditional values and the strong loyalty of its consumers.

NES Shoes

The most important value proposition of NES Shoes is the individual and customized design, production and distribution of handmade, tasteful, and environmentally friendly shoes in a sustainable manner. Its products are made exclusively of natural, biodegradable, vegan, recycled and/or recyclable raw materials. Since 2015, its increasingly sustainable production facility has been located in Eastern Hungary, where the factory plant is heated by biomass, gets its electricity from solar panels, and strives to use local raw materials. NES Shoes also promotes the self-expression of self-aware women and men, the spread of ecological consciousness, social and individual responsibility, and the importance of aesthetics and joy.

Its market segment is specifically the production of unique shoes and making direct sales primarily to female customers who are self-aware, wish to care for the environment, and want to express their personality through their shoes. Since 2020, alongside ongoing individual orders, a limited series of sophisticated yet affordable shoes have been available in brick-and-mortar stores in Budapest as part of an expanded product range.

In the years after the inception of this micro-sized enterprise in 2008, annual sales fluctuated and profitability remained around the break-even point as performance was based solely on the individual effort and engagement of the owner-manager, Andrea Szilágyi. Costs have been higher than the market average due to the tailor-made products with strong ecological commitment. NES Shoes has recently joined forces with a manufacturer through which changes have been made to the offerings, and the product range has been extended to a limited series of shoes. The company predicts that domestic purchasing power is now sufficient to support growth, but consumer awareness is still developing. The aim of the company is to strengthen serial production while maintaining a personalized service and improving profitability.

NES Shoes' value chain consists of customized and serial design, production and sales. It has had no inventory until recently, only making new collections every three months in advance of exhibitions that could be purchased at its show room. In 2020, this model was completed with a limited serious of sophisticated shoes to be sold in brick-and-mortar stores in Budapest. The value chain also includes high-quality personal customer management, a lifetime repair warranty for every product, and continuous research and development for monitoring and testing ecological raw materials such as "mushroom leather." NES Shoes also nurtures personal, fair, and caring relationship with employees at the manufacturing plant.

The company typically sells directly to customers and resellers, with whom it has personal relationships. The ecological aspects of its operation are particularly important, so the sources of raw materials and circumstances of production are carefully considered and selected. NES Shoes also strives to minimize the need for transport and to enhance the proportion of pre- and post-consumer recycling in its operations. Its civil society relations and collaboration are of significant importance, and the company promotes and supports artistic, educational activities. The designer-owner and her brand are well known in the special market segment for their value commitments, which reputation the designer has also used to increase customer awareness.

Due to the special product range of customized environmentally friendly shoes and the global geographic fragmentation of its competitors, NES Shoes has not really perceived strong competition so far. In general, Hungarian customers are still less environmentally conscious than in richer countries such as those of Northern Europe. Ecological consciousness is slowly evolving in Hungary, and customers typically seek out NES Shoes because of the products' aesthetic features, not due to their minimal environmental impact.

Béres Pharmaceuticals

Béres's most important value proposition is natural health preservation by using a unique trace element supplement, "Béres Drops," and other natural products that strengthen the immune system. Its operations are based on the principles of human-centeredness, supporting people in need, and acknowledging and promoting education, culture, and science. Béres also has a subsidiary that is engaged in environmentally friendly viticulture and wine-making. Its core values include righteousness, reliability, predictability, social-, national-, and environmental responsibility, perseverance, diligence, compassion and commitment, most of the latter based on the founder family's deeply felt Christian values. Béres aims to produce high-quality products and operate a successful, transparent, and efficient organization imbued with team spirit.

The company's products are made of natural ingredients and are designed for health-conscious consumers who are actively seeking to prevent illness or are living with special needs. Its main markets are Hungary, Russia, and Eastern Europe. Béres is one of the best known Hungarian brands, and the company is a market leader in several product segments such as immune-system strengthening products, complex multivitamin, mineral and trace element products, and vitamin C and complex vitamin D supplements.

The predecessor of the company was established in the late 1980s based on an invention of the late József Béres Sr. known as "Béres Drops," and has organically grown from a small family business into a large company with annual revenue of about 45 million USD, and is still owned and managed by the Béres family. It predominantly makes domestic sales, but exports are dynamically expanding. Indirect costs are a large part of its cost structure, but their proportion is decreasing. Béres has succeeded in making production more efficient, so its profitability has increased. Its profit margins are below the average of international competitors, but for the company this is not the focus, but rather behaving ethically with all of its stakeholders and promoting their well-being, while remaining sufficiently profitable.

Key factors in Béres's value chain are intense, successful product innovation, the licensing of new products, and the protection of industrial property rights. During recent years, the company has successful expanded its factory and warehouse in Szolnok, Eastern Hungary, thereby creating new work places. Its logistical capacity has also been developed, and strategic alliances built with its main sales channel, pharmacies. The company's contract manufacturing business has significantly increased and several successful products made in collaboration with competitors are now sold outside Hungary. Béres's outstanding marketing and communication performance has been recognized with numerous awards. Its human-centered approach to human resource management is manifest in an attractive and supportive working environment.

Considering Béres's position within its value network, the following facts can be highlighted. The company is committed to using a code of ethics and behaving fairly to a wide range of stakeholders, from employees to the natural environment. The Béres Foundation was established to support people in need, education, culture, and science through financial and in-kind donations. Béres Winery supports the development of the Tokaj wine region, its traditions, and wine culture. Béres strives to retain employees and build respectful, positive relationships with its consumers, partners, and the authorities; it has even cooperated with several of its competitors. Through its environmental policy, Béres aims to create a safe working environment, to reduce pollution and material consumption, to provide training, and to motivate its employees to develop an environmental consciousness.

In the industry in which it operates, regulation is very strict, and competition with big international manufacturers is strong but fair. However, Béres also has to compete with dubious Far Eastern producers in terms of cost and price competition. Continuous product innovation, a high-quality management system rolled out along the whole value chain, and strict compliance with the law are key success factors, besides Béres's strong value commitment that differentiates the company and helps maintain a loyal consumer base and the organization's organic development.

Kör Architect Studio

The company's value proposition includes designing ecological and sustainable buildings, providing education, developing the wider industry, and continuously striving to implement an ecological approach. Kör Architect Studio is committed to Rudolf Steiner's anthroposophy and the idea of a self-governing society, which represents an alternative approach to the prevailing socio-economic system. The principles of this concept are freedom of spiritual life, direct democracy, and a solidarity economy; moreover, the healthier functioning of the economy based on an alternative economics that is not built on the principle of profit and the exploitation of natural and human resources. Core values are cooperation, preferring human-scale solutions, and harmony between humanity and the natural environment.

Kör Architect Studio's market segment is the architectural design of ecological and sustainable buildings like autonomous houses and resilient houses. This is complemented by educational and industrial development services and awareness-raising activities. The company has no permanent, long-term partners, unlike the situation in some segments of the Hungarian construction industry.

The company is a micro-sized enterprise whose annual revenue and economic situation fluctuates depending on the number of orders. It has successfully maintained its business operations so far, but it appears that the future is impossible to forecast. In general, market conditions in the construction industry now favor big players with integrated operations and significant lobbying capacity. Although the company does not have permanent partners, conscious customers who have similar value commitments may find it, and other designers recommend the company when they are faced with special demands concerning ecological design. Satisfied and loyal customers also recommend it to others. Overall, ecological consciousness is still rewarded because some customers specifically choose the company because of its commitment and attitude.

Kör Architect Studio's value chain is short. It is a micro enterprise that works with trusted partners in architectural design in accordance with the organic social concept of anthroposophy. It has its own product development process, the results of which are the concepts of the autonomous house, the resilient house, and bio toilets. It strives to make its office operations more environmentally friendly, for instance, through the use of solar cells. However, the success of this goal will depend on future resources. The owner-manager's activities in higher education, professional training, and talks and writing about awareness raising also play an important role in the company's value chain.

Within its value network, Kör Architect Studio pays particular attention to the stakeholders of its operations in accordance with an ecological approach. It closely cooperates with its partners, and potential customers and partners can find the company even in the absence of a company website. Its goal is to set an example and to maintain committed relationships with social organizations. The owner-manager, Attila Ertsey, participated in the establishment of a green party, and has been the leader of the Nádasdy Academy, which holds annual ecological and artistic symposiums about facilitating a paradigm shift in the socio-economic system.

The company has faced very strong competition in the architectural design market in general, but the ten to twenty actors in the eco-design segment tend to coexist and cooperate. Through strong commitment to ecological and sustainable building design, Kör Architect Studio strives to create a network of self-governing, collaborative, and human-scale companies through which the healthier, organic operation of companies can prevail, instead of a situation of competition and "overgrowth."

Pipacs Bakery

Pipacs Bakery produces organically certified bread. Its main value propositions are ecological values, connection to nature, the protection of life and health, human-centeredness, and the material and moral appreciation of employees. The company respects all members and communities within its value chain. It strives to avoid night shifts, in sharp contrast to other bakeries. Its relationships with partners are based on trust and mutual responsibility over the long term, owner-manager Ádám Fülöp expresses.

The company's market segment is producing organic bread in an artisan bakery, primarily for health-conscious and environmentally aware individual consumers. Pipacs Bakery has only a few distributor partners with limited sales volumes. It operates no night shifts, so the product range is narrow and focuses on organic bread and a few kinds of pastries. In its shop, Pipacs sells raw materials for bread making and some staple food products made by its partners such as cheese, butter, jam, honey, and sausages, which are mainly organically certified.

Pipacs Bakery has grown rapidly since its foundation in 2014, both in terms of production volumes and sales. The company has been reinvesting its profits into machinery and human resource development for years. It pays better salaries than the industry average, and its organically certified raw materials are much more expensive than non-certified ones, so its products are more costly than conventional ones. Customer loyalty to Pipacs Bakery has been strong since the beginning. In many cases, it is not environmental consciousness but health issues that drive its clients to buy organic products.

With organic bread baking, the value chain starts with the purchase of high quality, certified raw materials, and the quality of flour is the key factor. Only organic raw materials from organic farms are used by Pipacs Bakery, which uses flours from Hungarian mills plus one Hungarian farmer, while only some additional raw materials are sourced from overseas for which there is no local alternative. Flour is ground by selected, trustworthy Hungarian mills with which Pipacs Bakery has long-term relationships. The company sells its products at an organic market, in its own shop, and via a few resellers. Pipacs Bakery has been maintaining salaries at a more favorable level than the industry average and does not use night shifts to avoid violating its own strict value commitment to human-centeredness.

Pipacs Bakery was the first bakery in Hungary not to use night shifts, and only a few have followed its example so far. However, owing to this practice the company is not as affected by labor shortages because employees are more engaged and loyal. Its focus is on direct consumers, because Pipacs Bakery has not been able to supply larger quantities of products to reseller partners due to the unconventional work shifts. The company wants to set an example, promote artisanal organic bread, and prioritize direct human relationships with partners. An important element of its value network is promoting the products of its key partners.

The company perceives competition as moderate, and even strives to cooperate with industry competitors on the basis of shared values. It consciously pursues a strategy of non-competition, mutual support, and learning from other players. Its commitment to organic production is its strongest value, but customer loyalty may be due to the high quality, healthy products it offers rather than the organic certification. Nevertheless, these circumstances do not change the depth of Pipacs Bakery's commitments, as the company believes in making long-term changes in a positive direction, because there is no other way to create a livable future.

Pendits Vineyards and Winery

Pendits' main value proposition is following the principles and values of biodynamic farming and winemaking as a philosophy of life. Through this spiritual approach based on anthroposophy, the company seeks to live and work in harmony with nature, acknowledging the organic unity of man, plants, animals, and terroir. Pendits takes into account the impact of natural cycles in its operation and believes that the economy is an organism that should have the ability to be self-sustaining and autonomous. By building on regional values, products, and relationships, Pendits's goals are to produce a clean and healthy product of excellent quality that is associated with social responsibility and the conservation of the integrity of terroir, to develop sustainably, and to strengthen and utilize environmental, cultural, and local community values, as well as engage in innovative research, and support family farms.

The company's market segment is ecological, biodynamic farming, viticulture, and winemaking, and producing biodynamic and organically certified products for some major partners such as merchants and restaurants, as well as direct consumers.

Pendits is a micro-sized family enterprise that was established in 1991. It employs one permanent employee and seasonal workers, besides the two owners. Its economic situation fluctuates depending on the demands of the main partners. Various trends and difficulties associated with the wine market cause significant challenges from year to year. Pendits has also been suffering from labor shortages. The company believes that organic products do not necessarily have to be more expensive than conventional ones, but the consumption of organic products in Hungary is not yet widespread. However, due to improving living standards and purchasing power and a gradual increase in consumer awareness, this market segment has significant development potential in Hungary. Pendits can also build its development on its loyal customer base.

The company's value chain includes viticulture, winemaking, and trade. Innovation is also a key element. For instance, Pendits has been researching how to use covering vegetation among rows of vines to improve soil characteristics and increase biodiversity. It also has studied how to manage manual labor shortages in the region, and thereby increase economic viability. Most of its employees do not share Pendits's ecological commitment, but respect the owners' genuine commitment, fair attitude, and persistent activity.

Pendits was the first certified biodynamic wine estate in the traditional Hungarian Tokaj-Hegyalja wine region, being certified since 2011. It operates according to the strict regulations defined in European Union wine law, and the international trademark of Demeter International. The company is the only Hungarian member of the international biodynamic association established by Nicolas Joly, a leading organization in the biodynamic market segment. Pendits has excellent personal and professional relationships with the Bioculture Association, Biokontroll Hungary certification organizations, and the Hungarian Research Institute of Organic Agriculture. It continuously supports regional and local communities and intends to represent an example. It also participates in youth education, and awareness-raising activity. The owner-manager, Márta Wille-Baumkauff, is president of the Hungarian Association of Independent Winegrowers (Vindependent), and strives to help shape favorable regulation for the industry through long-term cooperation with the Hungarian authorities.

In Hungary, there are very few biodynamic wineries, and those that exist are typically small enterprises. The size of the area on which organic farming is practiced is growing, but there is not yet significant domestic demand for the associated products. Pendits faces strong price competition because the company competes with conventional wine producers. However, through its special technology and products, Pendits can differentiate itself from the market average and build its development on its value commitments that represent a competitive advantage.

MagNet Bank

MagNet Hungarian Community Bank is a commercial bank that was founded as a savings cooperative in 1995. It became a triple-bottom-line focused, values-based community bank that strives to promote sustainability and environmental and social responsibility a decade ago. This kind of organization is also known as an ethical or sustainable bank. Respect and trust, consciousness, innovation, transparency, and human-centeredness are key elements of MagNet Bank's value proposition, deputy chief executive officer Gábor Pozsonyi explained.

MagNet Bank works within the market segment of commercial banks. It does not provide some services, such as securities account management or credit cards, to its customers for historical, business, and moral reasons. Its main customers are individuals and business enterprises, and MagNet Bank is one of the most important banking partners of nonprofit organizations in Hungary.

MagNet Bank has developed organically through some minor acquisitions into a company with about 20 million USD annual turnover. It has lower than industry average profitability, but income is more stable, and the bank has never made a loss in its entire history, including during the 2008 financial crisis. Its values-based practices have proved to be resilient due to a more favorable credit portfolio grounded on a customer base with a higher than average willingness to cooperate and repay loans.

In its value chain, MagNet Bank focuses on banking services such as account management, loans, deposits, receivables management, and factoring for retail, corporate, and nonprofit customers. Supporting Hungarian civil organizations and communities is of particular importance to the company.

MagNet Bank is a member of the international organization Global Alliance for Banking on Values due to its values-based approach related to ethical and sustainability principles. The company has been a supporter of the long-term social, environmental and economic development and cultural life of its stakeholder communities. It provides community banking products such as a Community Donation Program through which MagNet Bank donates 10 per cent of its annual profit to nonprofit organizations selected by customers. The company also has community deposits and loans and a Supportive Bank Card program, Pay as You Like account management, and an Individual Donation Program. In the course of credit scoring, MagNet Bank applies positive screening measures to projects that have a beneficial impact on the environment and society. The company supports civic events and community-building activities through providing space in its MagNet Community House.

Competition has been very strong in the retail and small business banking sector in Hungary, but MagNet Bank's values-based functioning successfully differentiates the company from conventional banks. Social and environmental responsibility, credibility, transparency based on its value commitments, and a loyal clientele are MagNet Bank's main competitive advantages.

Table 6.2 describes the main characteristics of the Hungary-based ecologically conscious business models.

To conclude, the business models of the businesses in the Hungarian sample are based on the various value commitments of the representatives of the enterprises (founders, owners, and managers). The market segments of their operations and the structure of their value chains are characterized by great variety. Most of them operate at a higher cost level and/or lower but more stable level of profitability (due to their more committed customer base) compared to the market average of their competitors who have no ecological consciousness, or only a lower level. Micro-sized enterprises face a more challenging economic situation as they struggle with financing, human-resource-related issues including labor shortages, and deficiencies in general business knowledge such as in marketing, sales, and finance. In terms of their position within the value network, most of them have a smaller environmental load than the market average and a strong social responsibility orientation. They pay special attention to their stakeholders (as defined in a broad sense, usually starting with human relationships and expanding to local communities and future generations). More than two-thirds of the respondents stated that setting an example, having a progressive approach, or pioneering or fulfilling the role of opinion-leader are central elements of their operation. The development and promotion of value-oriented, ethical business practices are among the basic goals of most

Table 6.2 Main characteristics of ecologically conscious business models in the Hungarian sample

	Value propositions	Market segment	Cost structure and profit potential	Structure of the value chain	Position within the value network	Competitive strategy
Skanska Hungary	Building for a better society with social responsibility and generating shareholder value	Commercial property development, project management, leasing, facility management, construction	Steady growth to medium size, positive profitability, influenced by economic trends, cost savings	Design, construction, project management, innovation, minimizing negative impacts	Strong social responsibility focus and environmental protection, certification, awards	Very strong competition in a key sector, continuously developing environmental, work, safety practices
Nekedterem Hungary	Sale of healthy, high-quality local food via webshop, home delivery	Fresh/processed food from local, small-scale, artisanal producers, home delivery	Higher costs and prices, micro-sized but improving trend, sales and profits	Purchasing food, products, processing orders, delivery, monitoring, updating	Close contact with stakeholders, supporting local economy, communities	Moderate competition, unique position between supermarkets and box deliveries

(continued)

Table 6.2 (continued)

	Value propositions	Market segment	Cost structure and profit potential	Structure of the value chain	Position within the value network	Competitive strategy
Réthy Fashion	Recycling, reusing jeans in new fashion products, awareness raising	Design, unique fashion products for ecologically conscious customers	Micro enterprise at break-even, with steady but limited growth potential	Collecting raw materials, unique product design, repair, awareness raising	Carbon neutral, promotes social engagement, environmental education, awards	Very strong competition on price, advantages of quality, design, brand
Bio-Drog Berta	Organic farming, value creation for local, national community, innovative products and processes	Organic health-improving processed plant-based products, primarily for Hungarian organic-consumers	Micro-sized, significant decrease in sales, so far sustainable profit, negative industry trends	Potential for full range of organic farming, significant processing capacity, labor shortage in production	Domestic and international organic certification, supporting NGOs, local community, awards	Very strong competition, patent infringements, innovative, healthy products as competitive advantage

	Value propositions	Market segment	Cost structure and profit potential	Structure of the value chain	Position within the value network	Competitive strategy
Ecoizm /Zöldbolt	Distribution of innovative, eco-household products, awareness raising	Wide range of eco-household products for individuals and resellers	Sharp then moderate and steady increase in sales, positive profit potential	Purchasing, packaging, selling, awareness raising, using renewable electricity	Monitoring latest innovation, cooperating with authorities, NGOs	Medium strength competition, differentiation, spread of eco-products
ALTEO Energy	Sustainable energy production, trade and customized service	Construction and operation of power plants using natural gas and renewable energy	Continuous growth in annual revenues, profits, improving stock prices	Leading role in R&D, digitization, loyal stakeholders, strong co.culture	Exemplary role, opinion leader, active cooperation with NGOs	Maintain leading position based on genuine value commitments

(continued)

Table 6.2 (continued)

	Value propositions	Market segment	Cost structure and profit potential	Structure of the value chain	Position within the value network	Competitive strategy
Hajtás Pajtás	Bicycle courier, trust and reliability, transparency, sustainability, activism for a livable capital	Reliable, confidential, bicycle courier service mainly for businesses, electric truck delivery possible	High labor costs, medium-sized company, increasing sales for years depends on general market trends	Couriers are key, providing support to them, special culture, cycling activism in Budapest and the country	Setting an example, promoting ecological, social responsibility, bicycling activism, commitment	Pioneer, market leader in environment of very strong competition, cooperation, reliability and brand reputation
Pannonhalmi Abbey Winery	Reviving winemaking, sustainability based on Benedictine values	Production of high-quality white and red wine for commercial and direct customers	Small company with dynamic development, net profit, growing demand	Viticulture, winemaking, growing workforce, synergies in the company group	Holistic, broad stakeholder definition, environmental, social consciousness	Strong competitors are also eco-conscious, values, consumer loyalty is key

	Value propositions	Market segment	Cost structure and profit potential	Structure of the value chain	Position within the value network	Competitive strategy
NES Shoes	Customized design, increasingly sustainable production, self- and eco-consciousness	Individual orders of unique shoes, limited series, direct sales, stores in Budapest	Fluctuating performance in sales, break-even profit, micro-sized, expansion into serial production	Customized and serial design, production, sales, customer care, lifetime warranty, innovation	Ecological, social commitments, personal relationships, good reputation, awareness raising	Less strong competition due to special product, fragmented global competitors, aesthetics is key
Béres Pharmaceuticals	Health preservation, natural products, social, environmental responsibility, Christian values	Natural immune system, vitamin, mineral, trace element products in Hungary and Eastern Europe	Organic growth into a large family enterprise, margins below those of competitors, focus on sufficiency	Product innovation, licensing, patents are key, factory expansion, strategic alliances, many awards	Fair, long-term relationships with wide range of stakeholders, donations, local development	Strong competition, product innovation, quality management, compliance key, value-based differentiation

(continued)

Table 6.2 (continued)

	Value propositions	Market segment	Cost structure and profit potential	Structure of the value chain	Position within the value network	Competitive strategy
Kör Architect Studio	Ecological architectural design based on anthroposophy, education	Architectural design of autonomous, resilient houses, development	Micro-sized, fluctuating revenues, profitability, based on specific demands	Product development, working with partner designers, education, writing	Close cooperation with stakeholders, setting example, promoting paradigm shift	Fierce competition in gen eral, collaboration in eco-design, creating a network
Pipacs Bakery	Production of organic bread, ecological values, trust, human-centeredness	Organic, artisanal bread, a few pastry and food products, for individuals with specific needs	Rapid growing but still micro enterprise, higher costs, stable customer base	High quality, organic raw materials, long-term cooperation, no night shifts	Setting an example, promoting artisanal products, direct relationships	Moderate competition, striving for collaboration, strong value commitments
Pendits Vineyards and Winery	Biodynamic winemaking, holistic approach, social responsibility	Ecological, biodynamic, organic-certified farming, viticulture, winemaking	Micro-sized family company, labor shortages, loyal customers	Viticulture, winemaking, trade and innovation, labor shortages	Leader, advocate of biodynamic practices and regional values, awareness raising	Strong competition with conventional wine producers, differentiation

	Value propositions	Market segment	Cost structure and profit potential	Structure of the value chain	Position within the value network	Competitive strategy
MagNet Bank	Community and impact bank based on sustainability, transparency	Commercial bank mainly for individuals, NGOs, no securities, credit cards	Organic development, lower but more stable profitability than average	Core banking services for retail, corporate, NGOs, support for communities	Member of GABV, supporting development of stakeholder communities	Very strong competition, successful differentiation, loyal customers

Source Author's construction

of them. In general, these organizations are in a strong or very strong competitive position. If they consider the competition to be less challenging, this is clearly attributable to their competitive advantage due to their special attributes that originate in their ecological consciousness or other value commitments. Such competitive advantages include having a strong brand, a loyal customer base, superior quality products, or being innovative, adaptive, and resilient. Almost two-thirds of the sample companies emphasized that they tend to seek out mutually beneficial forms of cooperation and knowledge sharing, rather than look to compete. Their flexibility and self-sustaining capabilities are strong and they rarely receive subsidies and nor do they usually require them, as they are willing and able to prosper on their own.

Summary

This chapter has presented the business models of ecologically conscious business organizations in an international and Hungarian sample. Their common attributes, compared to the market average, are less environmental impact, greater social responsibility, and the employment of a wider definition of and special attention to a wide range of stakeholders. Additionally, the companies under analysis seek to set an example and lead the markets in which they operate. They may also be required to operate with higher costs and/or lower but more stable levels of profitability, and often face strong competition. However, in this fierce competition, the competitive advantages derived from the material benefits obtained through their genuine value commitments (as confirmed by Cornell economist Robert Frank [2004]) support their long-term survival, pursuit of cooperation, knowledge sharing, and resiliency.

The business models of Hungary-based organizations are similar to those of the leading ecologically conscious international businesses in terms of value propositions and cost structure in most cases. The market segments and the structure of value chains may nevertheless vary greatly, representing the multifarious external and internal operating environments of the individual companies. There are also remarkable differences regarding the size of the international enterprises, which are significantly bigger than the Hungarian ones. The potential for profit of the companies in the international sample is greater, as they operate in a global competitive environment. The smallest Hungarian organizations typically face a more challenging economic situation because they struggle with finance- and human-resource-related challenges and deficiencies in general business knowledge, such as marketing, sales, finance, or human resource management.

Naturally, some of the differences identified above arise due to sampling bias, because in the international sample only world-renowned companies were studied, but—because of our local knowledge—even very small businesses were examined in the Hungarian sample.

The results of the empirical research were also compared to those found in the literature. The main findings of Parrish (2010) in relation to the differences between sustainable enterprises and conventional businesses were confirmed. The objective and raison d'être of successful sustainable enterprises is to preserve and strengthen the quality of human and natural resources over the longest possible time. Sustainable enterprises do not solely pursue a strategy of cost-effectiveness, but rather strive to increase the benefits and synergies of their activities. They harmonize their objectives, seeking satisfactory results from different goals simultaneously, and do not just aim to maximize the value of a single set of goals.

The characteristics of the ecologically conscious business organizations in the sample are also similar to those identified in the most sustainable business models, or "truly sustainable businesses," as Dyllick and Muff (2015) describe them. These businesses go beyond making an effort to minimize negative impacts and seek to understand how to maximize positive impact in relation to critical social and ecological issues. Such enterprises forge solutions to the most pressing sustainability challenges into business opportunities, and try to exert an influence throughout their entire sector or beyond it. This is done through increasing their transparency, sharing examples of good practice, building cooperation with non-economic actors, and striving to change the rules of the game.

In several cases, the businesses described above can be considered "entrepreneurs inspired by nature," based on Hofstra (2015). These enterprises have a different worldview and innovate products and services that help regenerate natural services, re-connecting people with nature. They recognize that humankind is a part of nature, and very much depends on it. Resources are limited, and sustainable businesses are required to solve real ecological problems in parallel with meeting human needs and generating economic value. The main principles of successful nature-inspired businesses are very similar to those that the ecologically conscious businesses in the sample adhere to. They take into account the different interests of business, society, and nature. They achieve long-term economic efficiency combined with taking an ethical perspective and being ecologically effective. They discover but do not exploit nature, and they generate business value in a broad sense, including economic, ecological, moral, social, and spiritual value.

Gast and her co-authors (2017) define ecological entrepreneurship as the process of identifying, evaluating, and implementing business opportunities that minimize the environmental impact of an enterprise, and operating for the benefit of local communities and society as a whole. As the authors argue, intrinsic motivation, personal values, and volunteering is of paramount importance in terms of commitment to sustainability. Ecological enterprises do not primarily pursue economic goals, but seek to change people's mindsets and society as a whole, and to create value for the environment and society. Operating in an ecologically sustainable way involves redesigning business practices and networking. The role of networks and cooperating with stakeholders is extremely important, as individual enterprises alone have less chance of influencing public policies.

Most of these characteristics are revealed in all of the businesses under analysis.

In conclusion, future research directions are defined that partially address the limitations of the findings, and support the further study, identification, and promotion of ecologically sustainable economic activities. The relevance of the present results can be tested by repeating the research on the same samples, or on extended samples. A variety of significant issues related to ecological consciousness in a business context could not be focused on in this book, but such issues—as the following questions indicate—may be studied in the future. What is the context of ecologically conscious business?—with particular attention to the differences between inclusive and extractive institutional systems and social systems (Acemoglu and Robinson 2012). What environmentally beneficial impact may ecologically conscious business functioning potentially have? Does drawing on the best practices of ecologically conscious businesses allow specific recommendations to be made in relation to the ecological turn of the present economic system? How can a spiritual value orientation in a business organization foster this ecological turn, and how can the evolution of consciousness and spirituality in business be supported? We have examined ecologically conscious businesses, but the extensive literature about sustainable and ecologically conscious consumption has not been addressed. It would be worthwhile studying how an ecological consciousness and value orientation may appear and take effect in both supply-side and demand-side economies.

References

Acemoglu, D., & Robinson, J. A. (2012). *Why Nations Fail: The Origins of Power, Prosperity, and Poverty*. New York: Crown Business.

Alves, B. (2020). *Cosmetics Industry in Brazil—Statistics & Facts*. https://www.statistacom/topics/5424/cosmetics-industry-in-brazil/#dossierSummary. Accessed 15 July 2020.

ASSOCHAM – Ernst & Young. (2018). *The Indian Organic Market: A New Paradigm in Agriculture*. https://www.ey.com/Publication/vwLUAssets/ey-the-indian-organic-market-report-online-version-21-march-2018/$File/ey-the-indian-organic-market-report-online-version-21-march-2018.pdf. Accessed 30 June 2018.

AVPN – Asian Venture Philanthropy Network. (2016). *Green Monday Movement—One Mission, Many Ventures*. https://avpn.asia/case-study/portfolio-management-green-monday-movement-one-mission-many-ventures/. Accessed 14 June 2020.

Bhasin, K. (2011, Aug 25). *PHOTOS: Inside TerraCycle's Incredible 99% Recycled Office*. https://www.businessinsider.com/terracycle-office-tour-2011-8. Accessed 10 June 2020.

B Lab. (2020a). *B Impact Report: Natura Cosméticos SA*. https://bcorporation.net/directory/natura-cosmeticos-sa. Accessed 8 June 2020.

B Lab. (2020b). *B Impact Report: Organic India USA*. https://bcorporation.net/directory/organic-india-usa. Accessed 8 June 2020.

B Lab. (2020c). *B Impact Report: Patagonia, Inc*. https://bcorporation.net/directory/patagonia-inc. Accessed 9 June 2020.

B Lab. (2020d). *B Impact Report: The Body Shop International*. https://bcorporation.net/directory/the-body-shop-international. Accessed 8 June 2020.

Business Wire. (2020). *Analysis on New Product Launches in Covid-19 Related Markets. Outdoor Apparel Market 2020–2024: Technavio*. https://www.businesswire.com/news/home/20200501005038/en/Analysis-New-Product-Launches-Covid-19-Related-Markets-Outdoor. Accessed 9 June 2020.

Chen, C. (2018, January 26). *Retail May Be Sinking, but Outdoor Apparel Is Killing It*. TheStreet. https://www.thestreet.com/story/14463219/1/retail-may-be-sinking-but-outdoor-apparel-is-killing-it.html. Accessed 9 June 2020.

Chesbrough, H., & Rosenbloom, R. S. (2002). The Role of the Business Model in Capturing Value from Innovation: Evidence from Xerox Corporation's Technology Spin-off Companies. *Industrial and Corporate Change, 11*(3), 529–555. https://doi.org/10.1093/icc/11.3.529.

Di Ciancia, C. (2017, December 29). *Houdini's Trick: Adventure Without Impact. A Sportsware Brand of Clothes to Wear with a Clear Conscience*. Welum. https://www.welum.com/article/houdini-sportswear-trick-maximum-experience-zero-impact-2/. Accessed 16 June 2020.

Dyllick, T., & Muff, K. (2015). Clarifying the Meaning of Sustainable Business: Introducing a Typology from Business-as-Usual to True Business Sustainability. *Organization & Environment, 29*(2), 156–174. https://doi.org/10.1177/1086026615575176.

Eliances. (2017, October 17). *Tom Szaky Founder Terracycle Eco-Capitalism King Over 20 Countries*. http://eliancer.com/tom-szaky-founder-terracycle-eco-capitalism-king-over-20-countries. Accessed 10 June 2020.

Fast Company. (2018). *The World's Most Innovative Companies 2018*. https://wwwfastcompany.com/most-innovative-companies/2018. Accessed 9 June 2020.

Fast Company. (2020). *Most Innovative Companies: Patagonia*. https://wwwfastcompany.com/company/patagonia. Accessed 9 June 2020.

Frank, R. H. (2004). *What Price the Moral High Ground? Ethical Dilemmas in Competitive Environments*. Princeton: Princeton University Press.

Friedman, V. (2017, July 27). A Call to Activism for Outdoor Apparel Makers. *The New York Times*. https://www.nytimes.com/2017/07/27/fashion/bears-ears-patagonia-rei-adidas-activism.html. Accessed 7 May 2018.

Gaille, B. (2018, August 4). *21 Outdoor Apparel Industry Statistics, Trends & Analysis*. https://brandongaille.com/21-outdoor-apparel-industry-statistics-trends-analysis/. Accessed 7 June 2020.

Gast, J., Gundolf, K., & Cesinger, B. (2017). Doing Business in a Green Way: A Systemic Review of the Ecological Sustainability Entrepreneurship Literature and Future Research Directions. *Journal of Cleaner Production, 147*, 44–56. https://doi.org/10.1016/j.jclepro.2017.01.065.

Gelles, D. (2018, May 5). Patagonia v. Trump. *The New York Times*. https://www.nytimes.com/2018/05/05/business/patagonia-trump-bears-ears.html. Accessed 7 May 2018.

Gingerich, E. F. R. (2016). Case Studies: Spirituality and Sustainability. Interface Global (Ray C. Anderson, Atlanta, GA, USA). In S. Dhirman, & J. Marques, J. (Eds.), *Spirituality and Sustainability: New Horizons and Exemplary Approaches* (pp. 169–171). Switzerland: Springer International Publishing. https://doi.org/10.1007/978-3-319-34235-1_11.

GlobeNewswire. (2019). *Global Outdoor Clothing Market Will Reach to USD 19.45 Billion By 2026*. Zion Market Research. https://www.globenewswire.com/news-release/2019/07/11/1881331/0/en/Global-Outdoor-Clothing-Market-Will-Reach-to-USD-19-45-Billion-By-2026-Zion-Market-Research.html. Accessed 9 June 2020.

GlobeScan. (2019). *The 2019 GlobeScan-Sustainability Leaders Survey*. Report. https://globescan.com/wp-content/uploads/2019/07/GlobeScan-SustainAbility-Leaders-Survey-2019-Report.pdf. Accessed 7 June 2020.

Grand View Research. (2020). *Flooring Market Size, Share & Trends Analysis Report By Product (Soft Coverings, Resilient, Non-Resilient), By Application (Residential, Non-residential), By Region, And Segment Forecast, 2020–2027*. https://www.grandviewresearch.com/industry-analysis/flooring-market-analysis/methodology. Accessed 7 June 2020.

Green Monday. (2020). https://greenmonday.org/en/. Accessed 6 June 2020.

Hackett, J. (2019, August 14). *Flexitarian (Semi-Vegetarian) Diet Definition*. The Spruce Eats. https://www.thespruceeats.com/what-is-a-flexitarian-3376803. Accessed 6 June 2020.

Hawken, P. (1993). *The Ecology of Commerce: A Declaration of Sustainability*. New York: HarperBusiness.

Hofstra, N. (2015). Entrepreneurship Inspired by Nature. In L. Zsolnai (Ed.), *The Spiritual Dimension of Business Ethics and Sustainability Management* (pp. 159–166). New York: Springer. https://doi.org/10.1007/978-3-319-11677-8_13.

Houdini. (2018). *Planetary Boundaries Assessment 2018. This Is Houdini.* https://api.houdinisportswear.com/storage/2A69199BFCBA925CC9260D61F41301EA566C760FB9A727B5DABB2C330C13D1BC/08df8496f36f49f0bb821fdeafdd775e/pdf/media/e5eec5e201b242e9a2aa14aba9c3b696/Houdini_Planetary_Boundaries_Assessment_2018.pdf. Accessed 6 June 2020.

Houdini. (2020). https://www.houdinisportswear.com/en/. Accessed 6 June 2020.

IFOAM – International Federation of Organic Agriculture Movements. (2017). *19th Organic World Congress—Awards*. https://owc.ifoam.bio/2017/awards. Accessed 8 June 2020.

Interface. (2019). *Annual Report 2019*. https://s22.q4cdn.com/139673446/files/doc_financials/2019/ar/tile20191229_10k_web.pdf. Accessed 7 June 2020.

Interface. (2020a). https://www.interface.com/US/en-US/homepage. Accessed 7 June 2020.

Interface. (2020b, May 8). *Investor Presentation*. https://s22.q4cdn.com/139673446/files/doc_presentations/2020/05/Investor-Presentation-05.08.2020-FINAL.pdf. Accessed 7 June 2020.

Jacob, A. (2012, August 28). Scott Bader—A Long Term Commitment to Composites. *Reinforced Plastics Magazine*. https://www.materialstoday.com/composite-industry/features/scott-bader-a-long-term-commitment-to-composites/. Accessed 9 June 2020.

Jans, G. (2017, June 19). Outdoor Sector Shows Good Performance: ISPO, "There is High Growth". https://www.ispo.com/en/markets/id_79708040/good-figures-for-the-outdoor-sector-high-growth-.html. Accessed 9 June 2020.

Jones, G., & de Pinho, R. R. (2006, August). *Natura: Global Beauty Made in Brazil*. Harvard Business School Case 807-029 (Revised October 2012). https://edisciplinas.usp.br/pluginfile.php/…/807029-PDF-ENG.pdf. Accessed 25 June 2018.

Krarup, M., Kiørobe, N., & Sramkova, H. (2015). *Moving Towards a Circular Economy: Successful Nordic Business Models*. Copenhagen: Nordic Council of Ministers.

Lampikoski, T. (2012). Green, Innovative, and Profitable: A Case Study of Managerial Capabilities at Interface Inc. *Technology Innovation Management Review, 2*(11), 4–12. https://timreview.ca/article/624. Accessed 11 June 2020.

Largest Companies. (2020). http://www.largestcompanies.com/company/Houdini-Sportswear-AB-227636. Accessed 7 June 2020.

Mannion, L. (2018, May 24). *Saving the Planet…One Dumpling at a Time.* Thomson Reuters Foundation. https://www.reuters.com/article/us-hongkong-food-environment/saving-the-planet-one-dumpling-at-a-time-idUSKCN1IP2VN. Accessed 14 June 2020.

Melo, A. (2018, Abril 16). Natura ultrapassa Unilever e lidera mercado de beleza no Brasil. Valor. http://www.valor.com.br/empresas/5456961/natura-ultrapassa-unilever-e-lidera-mercado-de-beleza-no-brasil. Accessed 25 June 2018.

Natura. (2020a). https://www.naturabrasil.fr/en-us/. Accessed 8 June 2020.

Natura. (2020b). https://www.naturabrasil.com/. Accessed 8 June 2020.

Natura. (2020c). *Natura 2019* (Annual Report). https://static.rede.natura.net/html/home//br_07/relatorio-anual/natura_annual_report_2019.pdf. Accessed 8 June 2020.

Natura & Co. (2020). *Creating the Best Beauty Group for the World* (2019 Report). https://naturaeco.com/report_2019_en.pdf. Accessed 8 June 2020.

Organic India. (2017). *Sustainability Report 2016/2017*. https://organicindia.com.au/wp-content/uploads/2017/08/Ebook-Sustainability-Report-2017-Final-Digital.pdf. Accessed 7 June 2020.

Organic India. (2020a). https://www.organicindia.com/. Accessed 8 June 2020.

Organic India. (2020b). https://organicindiausa.com/. Accessed 8 June 2020.

Parrish, B. D. (2010). Sustainability-Driven Entrepreneurship: Principles of Organization Design. *Journal of Business Venturing, 25*(5), 510–523. https://doi.org/10.1016/j.jbusvent.2009.05.005.

Parvati Magazine. (2017, April 17). Community: TerraCycle, with Tom Szaky. *Parvati Magazine.* http://parvatimagazine.com/2017/04/community-terracycle-tom-szaky/. Accessed 10 June 2020.

Patagonia. (2018). *Annual Benefit Corporation Report. Fiscal Year 2017*. https://www.patagonia.com/static/on/demandware.static/-/Library-Sites-PatagoniaShared/default/dw824fac0f/PDF-US/2017-BCORP-pages_022218.pdf. Accessed 9 June 2020.

Patagonia. (2020). http://eu.patagonia.com/hu/en/home/. Accessed 9 June 2020.

Patagonia Provisions. (2020). https://www.patagoniaprovisions.com/pages/our-ambassadors. Accessed 9 June 2020.
PR Newswire. (2020, May 6). *TerraCycle US Reports 2019 Year-End Financial and Operational Results*. https://www.prnewswire.com/news-releases/terracycle-us-reports-2019-year-end-financial-and-operational-results-301053497.html. Accessed 10 June 2020.
Scott Bader. (2018). *Annual Report 2017*. http://www.scottbader.com/wp-content/uploads/Scott-Bader-Annual-Report-2017.pdf. Accessed 27 June 2018.
Scott Bader. (2019). *Annual Report 2018*. https://www.scottbader.com/wp-content/uploads/Annual-Report-2018.pdf. Accessed 9 June 2020.
Scott Bader. (2020). https://www.scottbader.com/. Accessed 9 June 2020.
Shahbandeh, M. (2020). *Breakdown of the Cosmetic Market Worldwide 2011–2019, by Product Category*. https://www.statista.com/statistics/243967/breakdown-of-the-cosmetic-market-worldwide-by-product-category/. Accessed 10 June 2020.
Srivastav, T. (2018, May 2). *Organic India's CEO on Organic Product Popularity and Why They Don't Advertise on TV*. The Drum+. https://www.thedrum.com/news/2018/05/02/organic-indias-founder-organic-product-popularity-and-why-they-dont-advertise-tv. Accessed 8 June 2020.
Statista. (2019). *Sports & Outdoor Segment Relevant Market Data. Highlights Worldwide*. https://www.statista.com/statistics/979475/outdoor-clothing-market-value-forecast-worldwide/. Accessed 9 June 2020.
Sustainable Growth Associates. (2018). *References*. https://www.sg-associates.eu/references/. Accessed 29 June 2018.
S&P Global. (2020). *The Sustainability Yearbook 2020*. https://www.spglobal.com/esg/csa/yearbook/. Accessed 8 June 2020.
Technavio. (2020). *Top 10 Cosmetics Companies in the World 2020*. https://blogtechnavio.com/blog/top-10-cosmetics-companies. Accessed 8 June 2020.
TerraCycle. (2018). https://www.terracycle.com/en-US/. Accessed 5 July 2018.
TerraCycle. (2020a). https://www.terracycle.com/en-US/. Accessed 10 June 2020.
TerraCycle. (2020b). *Offering Circular Dated January 27, 2020*. TerraCycle US Inc. https://www.sec.gov/Archives/edgar/data/1714781/000110465920007210/tm205941d1_253g2.htm. Accessed 10 June 2020.
The Economic Times. (2017). Organic India Eyes Rs 500 Crore Turnover by FY20 End. https://economictimes.indiatimes.com/industry/cons-products/fmcg/organic-india-eyes-rs-500-crore-turnover-by-fy20-end/articleshow/62240529.cms. Accessed 8 June 2020.

The Natural Step. (2014). *Case Study: Interface*. https://thenaturalstep.org/project/interface/. Accessed 9 June 2020.

The Natural Step. (2020). *Approach*. https://thenaturalstep.org/approach/. Accessed 9 June 2020.

Tin Shed Ventures. (2020). http://www.tinshedventures.com/. Accessed 9 June 2020.

Vaccaro, A. (2009, March 24). *10 Tips from an Eco-Capitalist: Tom Szaky Shares TerraCycles' Secrets to Success*. GreenBiz. https://www.greenbiz.com/blog//03/24/10-tips-eco-capitalist-tom-szaky-shares-terracycles-secrets-success. Accessed 9 June 2020.

Yusof, H. (2020, May 29). David Yeung, CEO & Co-Founder of Green Monday. *The Business Times* 2020, https://www.sgsme.sg/news/david-yeung-ceo-co-founder-green-Monday. Accessed 6 June 2020.

Zimberoff, L. (2017, June 12). *This Guy Is Trying to Break Hong Kong's Meat Addiction*. Bloomberg. https://www.bloomberg.com/news/articles/2017-06-12/this-guy-is-trying-to-cure-hong-kong-s-meat-addiction. Accessed 7 June 2020.

Part IV

Conclusions

This book has addressed the problematic of ecological consciousness in business organizations through studying the theoretical concepts, value orientation, and business models of ecologically conscious businesses. The role of an ecological consciousness has become increasingly important as the ecological degradation of the Earth and the role of business in causing ecological harm has become more obvious. It is also well known that the values of economic actors influence their business decisions. A large body of scientific literature has thus investigated the value orientations of businesses. However, the topic of ecological consciousness has been much less deeply studied in this context, and there are hardly any examples of research about the relationship between value orientations and ecological consciousness in business.

To fill this gap, the present book has gone beyond previous studies to investigate the value orientations of ecologically conscious business organizations through the application of qualitative methodology, and analyzed the value commitments and business models of exemplary ecologically conscious businesses from all over the world.

7

The Future of Ecologically Conscious Business

This chapter explores the current global context of ecologically conscious business and makes some conclusions about the future of business.

In a seminal book, *The Entropy Law and the Economic Process*, published in 1971, Nicholas Georgescu-Roegen argues that physical reality influences economic reality, and natural resources are irreversibly degraded when the economy exploits them. The Earth has a carrying capacity that limits the sustainability of human populations and material consumption (Georgescu-Roegen 1971). Recent concepts such as the Great Acceleration, Planetary Boundaries, and the Anthropocene reinforce the claim that Earth system processes have environmental boundaries, and human activities since the Industrial Revolution have been responsible for the global environmental changes of our times. Defining and respecting a "safe operating space for humanity" at all levels of Earth system functioning is a prerequisite for achieving sustainability (Steffen and Broadgate et al. 2015; Steffen and Richardson et al. 2015).

At present, humanity does not fulfill the requirements of sustainability. Humankind has been in a state of ecological overshoot since the early 1970s, and now uses as many ecological resources as could be sustainably produced by 1.6 Earths. The ecological footprint is a measure that

A. Ócsai, *Ecologically Conscious Organizations*, Palgrave Studies in Sustainable Business In Association with Future Earth,
https://doi.org/10.1007/978-3-030-60918-4_7

compares the resource demand of individuals, populations, governments or businesses, and activities (such as consumption, production, exporting and importing) to the resources the planet can sustainably supply. It quantifies, usually in units of global hectares, what area of biologically productive land and water would be required to produce all the resources the object of measurement consumes, and to absorb the waste it generates. In 2020, due to coronavirus-induced lockdowns around the world, Earth Overshoot Day (the day when humanity's demand for ecological resources and services exceeds the amount Earth can regenerate in a year), was August 22, more than three weeks later than a year before. This reflects a reduction in humanity's footprint of almost ten percent, mostly owing to the decrease in wood harvesting and CO2 emissions from fossil fuel combustion. Efforts to respond to the global health crisis have demonstrated that acting swiftly and improving ecological resource consumption trends is possible when people put health above all else, collaborate in the pursuit of shared goals, and consider the consequences of their decisions at all levels (Global Footprint Network 2020). Can economies and business organizations also meet these criteria?

Mainstream economics is in conflict with the potential solutions that are required to reach a state of "one-planet compatibility." It does not recognize the biophysical limits of ecosystems on which the economy depends, and it considers technology to be the ultimate solution to resource constraint problems. The mainstream economic approach tends to focus only on markets, and to ignore externalities (Costanza 2010). Modern economic activities are based on individualistic and materialistic values, have a focus on maximizing short-term profit or utility, and promote infinite economic growth. They do not acknowledge the intrinsic value of nature, but use living beings as mere natural resources, threaten the integrity and diversity of natural ecosystems, destroy the autonomy and culture of local communities, and decrease future generations' prospects for a decent life (Zsolnai 2015; Shrivastava et al. 2019).

The late London Business School professor Sumantra Ghoshal criticized mainstream management ideology and its orientation toward competitiveness. Companies strive to increase their market share and power to maximize profits, so their ultimate goal is to make markets as imperfect as possible. According to this understanding, social value creation is not part of business organizations' set of core values, but only an add-on element at most (Birkinshaw and Piramal 2005; Tencati et al. 2009). Rushing after productivity in order to increase profitability and shareholder value is often counterproductive for businesses themselves. An analysis of trends in the business sector in the United States in the early 2000s showed that what appeared as productivity gains actually ruined many companies. Tendencies such as extreme power centralization, cooking the books, cashing in goodwill that had been cultivated for years, trashing brands, drastic downsizing and cost cutting, exploiting employees, and neglecting environmental harm ultimately undermined the long-term health of the business enterprises concerned (Mintzberg 2007). Regrettably, short-term shareholder value maximization and competitiveness, self-interest and aggression, and an exclusive focus on financial performance negatively affect stakeholders and ultimately business itself (Tencati et al. 2009).

In contrast to mainstream economic thinking, ecological economics, a transdisciplinary field that emerged from the work of Georgescu-Roegen and other scholars in the 1980s, strives to understand how the economy is embedded in the natural ecosystems that support human existence. It recognizes the carrying capacity of the Earth and the limits and constraints that the natural environment creates, and shows that the economy cannot grow indefinitely in a finite environment. Nature, society, and the economy are interconnected in dynamic unity. Humans should act as responsible and caring members of the community of life and respect the intrinsic value of natural beings. Taking a holistic and organic perspective and considering intergenerational equity and the issue of sustainable development play an important role in this (Daly and Farley 2004; Costanza 2010; Brown and Timmerman 2015; Capra and Jakobsen 2017). Ecological economics also claims that if the economy can reach and be maintained at a sustainable scale, with fair distribution and efficient allocation, well-being can be improved without economic growth. Just as in nature when a field fills up with early successional plants that coexist and cooperate, so the economy should also go through a transition from a phase of growth to a steady state, in which, instead of cut-throat competition, more collaborative, alliance-building, stable kinds of relationships prevail (Costanza, 2010). Nevertheless, the propositions of ecological economics have mostly remained at the academic level and are not reflected in business practices (Shrivastava et al. 2019).

Recently, the concept of the circular economy has emerged. This is based on an organic worldview based on the unity of economy, nature, and society, solidarity, sustainability, respect for pluralistic values, and transparent cooperation. It strives for change at structural and individual levels in terms of values, consumption, and lifestyle habits. Circularity means that production, distribution, consumption, and recycling take place in a cyclical way, not through linear processes. Instead of one-off consumption events, the circular economy promotes products with extended lifetimes, and supports waste reduction through the recycling, upcycling, and reuse of materials, commodity efficiency, and local-level solutions (Ims and Jakobsen 2006; European Commission 2015, 2018; Jakobsen 2017). The principles of the circular economy have started to spread throughout the world economy, and have been applied in several ecologically conscious business models that are described in the empirical research contained in this book, such as in the operations of Houdini, Interface, Natura, Scott Bader, and TerraCycle. However, the future development of the concept and its success at contributing to a sustainable future depend on how much it is applied to boost competitiveness and sustainable growth (for example, the explicit goals of many business organizations, and the target of the European Union's development agenda). Transitioning to a steady state economy with the primary goal of respecting planetary boundaries seems to require a different approach.

Business leaders are familiar with the meaning of the term VUCA (volatility, uncertainty, complexity, and ambiguity) nowadays. The US Army War College introduced the notion in the late 1980s, and the acronym itself gained general recognition after the September 11, 2001 attacks (Lawrence 2013; US Army Heritage and Education Center 2019). Strategic business management has also adopted the term and has been using it to describe the way that a volatile, uncertain, complex, and ambiguous business environment is becoming the new normal due to consecutive financial, economic, and geopolitical crises. Risks have been continuously increasing, and not only in a business context. In the epoch of the Anthropocene, human activities are dramatically influencing the trajectory of the Earth System toward a "Hothouse Earth" pathway that is likely to cause serious disruption to ecosystems, societies, and economies. To steer the Earth System onto a "Stabilized Earth" trajectory, deep transformation and a fundamental reorientation of human values, equity, behavior, institutions, culture, science, technologies, and economies are vital (Steffen et al. 2018; Shrivastava et al. 2019, 2020). Some transformations have been initiated but much more needs to be done to preserve humanity's chance of a livable future.

Mainstream economics and the market economy in their current form have not been able to overcome their shortcomings; for instance, by acknowledging biophysical limits, or internalizing market externalities. Ecological economists suggest that governments, quasi-government institutions, and community institutions at local, regional, and global levels, such as watershed trusts, global atmospheric trusts, or ocean trusts, may be able to fix markets (Costanza 2010)—at least theoretically. The majority of states seem to be unwilling to move beyond their short-term economic interests and collaborate to create comprehensive and coercive international environmental agreements due to their weak political, economic, and social commitments to sustainability. While the successful and effective implementation and enforcement of such international regulations and a real start to the green transformation of the global economy have been awaited for decades, the voluntary environmental assessments, reporting frameworks, and environmental and social labelling schemes applied by business, civic, and professional organizations have proliferated. Such self-regulatory initiatives include, inter alia, the Forest Stewardship Council, Fair Trade, the Marine Stewardship Council, Rainforest Alliance, and ISO 14001. However, the outcome of the voluntary environmental programs of business organizations can be shallow, contradictory, and serve as tools for greenwashing, without leading to measurable improvements in environmental performance. Thus, private and voluntary environmental regulations and standard setting by business organizations cannot be an alternative to comprehensive and legally binding international environmental regulation (Fekete 2015).

The number of socially and environmentally oriented businesses such as B-corporations and social enterprises has been steadily increasing, but the majority of business organizations, corporations, and unincorporated businesses still use the natural environment and humans in an instrumental way to achieve their own business goals, thereby furthering ecological and social degradation (Shrivastava et al. 2019).

In terms of translating ethics, social, and environmental responsibility to the organizational level, the discipline of business ethics has more than three decades of history. It has developed concepts, normative models, and organizational solutions, provided evidence, and created a deeper understanding of the social role of business. However, business ethics has failed to contribute effectively to solving the social, ethical, and environmental problems caused by business organizations. The reason for this is that it has been unable to realize that some of the features that are inherent to mainstream business prevent business enterprises, and primarily global corporations, from becoming genuinely ethical organizations. The instrumental and opportunistic use of ethics to make business gains, the moral disengagement strategies of business executives, a legal and financial institutional setting that dilutes responsibility, the superficial and contingent results of voluntary compliance, a lack of market mechanisms for rewarding responsible behavior, and weak stakeholder activism have hindered business ethics from meeting its stated goal of fostering the ethical behavior of businesses. Business ethics should aspire to generate an open and critical approach toward the basic assumptions of current management practices and their underlying theoretical concepts. The transformation of business is much needed. To this end, moving away from the corporate model via fostering a collaborative logic, more efficient management accountability, more strongly empowered non-governmental organizations, political commitment and leadership, and the international regulation of business that protects and promotes basic human rights, the well-being of people, and sustaining natural ecosystems are potential solutions (Boda and Zsolnai 2016; Boda 2019).

Business practitioners should consider and face up to the great challenge of humanity's survival and the future of the planet. Sustainable change is crucially needed through changing existing business systems and enterprises, and creating new ones in order to reduce the negative externalities caused by companies and contribute to solving social and environmental problems. Although some sustainable changes and much incremental change have been achieved, the vast majority of sustainable change initiatives fail. The conceptual and practical challenges of engaging with unfamiliar and complex problems, evaluating these problems via previously unused methods, developing and applying brand new solutions, motivating internal and external stakeholders to change their mindsets and habits and improving their skills can make implementing changes difficult. In general, five major obstacles can be identified that may impede sustainable change: (i) an ambiguity of objectives, (ii) a narrow frame of reference based on a focus on shareholder value maximization and a lack of systemic thinking about sustainability problems, (iii) the adversarial external perspective and lack of understanding and cooperation of external stakeholders who are also directly affected by changes, (iv) the functional isolation of the issue and change management teams and organizational inertia, (v) performance standards that are unsuited to assessing the complex creation of value related to sustainable change. Behind these obstacles to sustainability change is the mainstream paradigm of business that is disembedded from the social, planetary, and ethical context. To overcome the obstacles to sustainable change, a new paradigm is required which promotes sustainability at all levels and brings into being a resilient and humane economy that serves the common good (Wallis 2020).

A new economic paradigm is needed which promotes sustainability at all levels and creates a resilient and humane economy that serves the common good. Ecological consciousness in its broadest sense embraces the well-being of all living beings, and can be the cornerstone of this new paradigm. What kinds of economic and business initiatives and practices can we see now that may be promising in relation to the new and emerging business paradigm?

Current economic systems use Gross Domestic Product (GDP) as an indicator of development, thus encouraging the pursuit of infinite material growth. Acknowledging that human well-being depends on the well-being of the living Earth, and that a focus on one-dimensional GDP growth has led to escalating social and ecological ills and imperils the human future, in 2017 the governments of Scotland, Costa Rica, Slovenia, and New Zealand created an organization to promote well-being as the goal of development. Since then, numerous organizations, networks, individuals, academics, and businesses have joined the initiative (Costanza et al. 2018; Korten 2020b). The vision of a "wellbeing economy" is based on principles of prioritizing long-term human well-being, ecological stability, and environmental justice, protecting environmental standards, developing green infrastructure, satisfying basic needs, and ensuring a decent living for everyone, fairly distributing resources, and engaging in transparent and inclusive democratic processes at all levels, as well as prioritizing social and ecological goals in all business ventures, as well as cooperation and the public control of money. The city of Amsterdam and countries like New Zealand, Iceland, and Scotland are already implementing some of the principles of wellbeing economics and many more experiments such as renewable energy developments, ecovillages, Transition Towns, and regenerative agriculture projects are happening around the world (Costanza et al. 2018; Wellbeing Economy Alliance 2020).

The Club of Rome, through its Reframing Economics project, is also working to support the shift away from mainstream economics and policy making toward a philosophy of "beyond GDP," which considers the well-being of people and the planet and strives to reverse ecosystem, climate- and social tipping points of vital importance. Through the Emerging New Civilization Initiative, the Club of Rome's fundamental goal is to develop a new paradigm for the greater Earth Community. Based on ancient wisdom, African Ubuntu philosophy, Buddhist and other indigenous knowledge systems, its core values are promoting human dignity, promoting the understanding that human beings are inherently relational, respect for nature, and preserving the commons for future generations. The awareness of the need for a paradigm shift is a manifestation of developing human consciousness and the understanding that people have to reflect on who they are as humans and what values should determine their relationships with each other and nature (Club of Rome 2020).

The concept of "Ecological Civilization" acknowledges that the future of humanity depends on a successful transition to a state where humans can live in a balanced and harmonious relationship with Earth's living systems, and every person's needs for health and happiness are met. The current structure of civilization is designed to dominate nature. Capitalism inherently exploits and destroys nature, the biosphere, and human civilization. To reconcile humanity with nature, the system should be based on a long-term approach, cooperation, and production aimed at satisfying needs, not generating profit. A successful transition should include the transition of governance from transnational corporations to national governments, from competition to cooperation, and from GDP growth to an economy in which everyone's spiritual and material needs are fulfilled while respecting planetary limits. Economic performance should be measured according to the health of the biosphere, local ecosystems, communities, and individuals (Korten 2020a).

"Diverse economies" scholarship draws attention to the fact that, while thinking of how to fundamentally transform society and build more resilient economies based on the need for a paradigm change away from the mainstream capitalist economy, one should consider the "iceberg" of diverse community economies. Capitalist economic practices such as waged labor, the production of commodities for exchange in markets, and capitalist businesses are "above the waterline," thus make up only a smaller part of the whole iceberg. Below the waterline are many other practices, including economic ones (such as cooperatives, informal loans, and in-kind payments), social ones (such as housework, family, and childcare), and ecological (photosynthesis, composting, and metabolic processes) ones. The systemic framing of such diverse economies has led to the development of an inventory of economic activities arranged into five types, including enterprise, labor, transactions, property, and finance, containing dozens of items in each type. Representing this economic heterogeneity can help to undermine the power of a capital-centric worldview and realize the diversity of economic practices. The approach thus supports the identification of a variety of economic pathways for transitioning toward sustainable community economies (Gibson-Graham and Dombroski 2020).

According to the concept of the "collaborative enterprise," a change of paradigm is also needed. Business should consider ecological and social boundaries and embed themselves in natural and social systems through creating long-term, mutually beneficial relationships with every stakeholder. The basis of the sustainable future of the collaborative enterprise is value creation that is shared throughout the entire value network (Tencati and Zsolnai 2010). Such a radical renewal of business models is key to the future of business organizations. The most important goal is to maintain consistency with the flourishing of life on Earth, including human, nonhuman, and future life. To this end, progressive business models should be able to embrace sustainable, circular, frugal approaches and practices, pursue different and sometimes even opposing goals simultaneously, apply systemic thinking and coordinate activities harmoniously, and efficiently balance environmental, social, and economic dimensions. They should be stakeholder oriented and be led from the top by visionary, committed leaders and teams. Progressive businesses are exceptional compared to mainstream businesses. However, they can be considered as only the first step in a long and difficult journey to transform business enterprises into organizations that fit the reality of the Anthropocene (O'Higgins and Zsolnai 2018).

From this analysis of several concepts and initiatives, we see that in order to meet the requirements of the ecological and social reality of the Anthropocene, business organizations should develop radically new solutions, extraordinary innovations, and unusual business models. Important inspiration for this may be ecological consciousness. Driven by an ecological consciousness, businesses can reconsider and change their underlying assumptions, core values, raison d'être, and definitions of success. From where can businesses gain such inspiration and creativity, and the persistence to cope with such an enormous challenge?

Ian I. Mitroff has developed a model that analyzes four dimensions of any problem, defined as scientific/technical, interpersonal/social, systemic/ecological, and spiritual/existential. In the mainstream business world, the scientific/technical perspective is the dominant viewpoint, and leads to a preference for technical solutions to every problem. The interpersonal/social perspective considers problems from a social, group, or community point of view. The systemic/ecological perspective analyzes the long-term consequences and local, regional, and global impacts of problem solutions. The spiritual/existential perspective serves to promote respect for the dignity, beliefs, and feelings of individuals, and the consideration of existential issues such as the meaning of life, or the purpose of one's life (Mitroff 1998; Ims and Zsolnai 2009). Through using this model, businesses can obtain deeper insight and understanding of their own situations. The process also supports critical thinking, the re-evaluation of basic assumptions, out-of-the-box thinking, and creativity.

Spirituality is the search for the meaning of life—in various sociocultural forms—that connects people with every living being and with God or Ultimate Reality (European SPES Institute 2020). It can contribute to a deeper understanding of different decision-making situations and systems, the creation of new categories, and the overcoming of existing paradigms (Zsolnai 2014). Spirituality can be the background and intrinsic motivational basis for economic actors to behave ethically, serve the common good, and measure their success multidimensionally (Zsolnai 2015). Through spiritual-based leadership, spirituality as the internal source of wisdom can help direct the actions of leaders, and this consciousness will be reflected in their decisions in the external business world (Pruzan 2011, 2015). Spirituality can lead to a qualitative change in leaders, and through them also in organizations, ultimately affecting every stakeholder of an organization (Lozano and Ribera 2004).

What can we say about the ecologically conscious business organizations studied in the empirical research described in this book? Are they exemplars of future businesses in the face of the new ecological reality?

All of the internationally well-known global corporations, medium-sized companies with a regional reach, and small-scale local lifestyle businesses in the samples are genuinely committed to ecological values and ecologically conscious functioning. They all have their own initiatives for reducing or eliminating environmental harm and improving their social performance, even considering the needs of future generations in several cases. They are taking important steps in the long journey of transforming economic activities into sustainable ones, but they cannot magically make themselves carbon neutral or carbon negative in an instant. Furthermore, their endeavors are increasingly challenging as company size increases. However, they are also very far from following a mainstream business approach, and are moving in a good direction. Ecologically conscious business models, and the exemplary people behind them, can play a significant role with regard to sustaining the future of humankind and the Earth. Their noble goals and innovative practices give us hope, inspire people, and help and invite others to become more conscious and to work creatively for the betterment of the world.

Not only do ecological impacts and "real world" consequences really matter, but it is also critical that people do authentic things without craving the fruits of their action. In other words, to create a livable future, and to achieve our ecological and ethical goals, we should act with genuine commitment, and "with non-attachment and for God's sake only" (Huxley 1947, p. 312). As psychologist Stanislav Grof argues, the obstacles that stand in the way of the transformation of the future of the planet for the better are located deeply inside the human personality, and reflect the level of evolution of the consciousness of humanity. Untamed forces and absurd, violent, and greedy elements of human nature prevent the reorientation of our concern for ecological priorities. The global crisis is thus of a psychospiritual nature, and a radical inner transformation of humanity that increases emotional maturity and spiritual awareness appears to be a prerequisite for reinforcing ethical values, displaying sensitivity toward others, accepting voluntary simplicity, and developing an ecological consciousness. A hopeful solution is—as Grof (1998, p. 220) suggests—"systemic and deep inner work with inspired action in the external world [...], if it were practiced on a sufficiently large scale."

References

Birkinshaw, J., & Piramal, G. (eds.) (2005). *Sumantry Ghoshal on Management: A Force for Good*. Harlow: Prentice Hall.

Boda, Z. (2019). Business Ethics as Critical Thinking: Moral Motivations and the Limits of Ethics. In J. B. Ciulla & T. K. Scharding (Eds.), *Ethical Business Leadership in Troubling Times* (pp. 41–59). Cheltenham – Northampton: Edward Elgar. https://doi.org/10.4337/9781789903058.00011.

Boda, Z., & Zsolnai, L. (2016). The Failure of Business Ethics. *Society and Business Review, 11*(1), 93–104. https://doi.org/10.1108/SBR-11-2015-0066.

Brown, P. G., & Timmermann, P. (eds.) (2015*). Ecological Economics for the Anthropocene: An Emerging Paradigm.* New York: Columbia University Press. https://doi.org/10.7312/brow17342.

Capra, F., & Jakobsen, O. D. (2017). A Conceptual Framework for Ecological Economics Based on Systemic Principles of Life. *International Journal of Social Economics, 44*(6), 831–844. https://doi.org/10.1108/IJSE-05-2016-0136.

Club of Rome. (2020). http://www.clubofrome.org/. Accessed 20 July 2020.

Costanza, R. (2010). What is Ecological Economics? Interview by Ted O'Callahan on May 11, 2010, Yale Insights. Ideas from the Yale School of Management. https://insights.som.yale.edu/insights/what-is-ecological-economics. Accessed 20 July 2020.

Costanza, R., Fioramonti, L., Kubiszewski, I., Henry, L., Lovins, H., McGlade, J., Mortensen, L. F., Philipsen, D., Pickett, K., Ragnarsdóttir, K. V., Roberts, D., Sutton, P., Trebeck, K., Wallis, S., Ward, J., Weatherhead, M., & Wilkinson, R. (2018). Toward a Sustainable Wellbeing Economy. The Solutions Journal, Vol. 9, No. 2, April 2018, https://www.thesolutionsjournal.com/article/toward-sustainable-wellbeing-economy/. Accessed 20 July 2020.

Daly, H. E., & Farley, J. (2004). *Ecological Economics: Principles and Applications.* London: Island Press.

European Commission. (2015). Closing the Loop: Commission Adopts Ambitious New Circular Economy Package to Boost Competitiveness, Create Jobs and Generate Sustainable Growth. Press release, Brussels, 2 December 2015, http://europa.eu/rapid/press-release_IP-15–6203_en.htm. Accessed 27 March 2020.

European Commission. (2018). Circular Economy. Implementation of the Circular Economy Action Plan. 2018 Circular Economy Package. http://ec.europa.eu/environment/circular-economy/index_en.htm. Accessed 27 March 2020.

European SPES Institute. (2020). Definition of spirituality. http://www.eurospes.org/content/our-mission-spiritual-based-humanism. Accessed 20 July 2020.

Fekete, L. (2015). Regulatory Roadblocks to Environmental Sustainability. In B. Fryzel (Ed.), *The True Value of CSR* (pp. 211–230). London: Palgrave Macmillan. https://doi.org/10.1057/9781137433206_13.

Georgescu-Roegen, N. (1971). *The Entropy Law and the Economic Process*. Cambridge: Harvard University Press.

Gibson-Graham, J. K., & Dombroski, K. (2020). Introduction to The Handbook of Diverse Economies: Inventory as Ethical Intervention. In J. K. Gibson-Graham & K. Dombroski (Eds.), *The Handbook of Diverse Economies* (pp. 1–24). Cheltenham – Northampton: Edward Elgar. https://doi.org/10.4337/9781788119962.00008.

Global Footprint Network. (2020). https://www.footprintnetwork.org/. Accessed 20 July 2020.

Grof, S. (1998). *The Cosmic Game: Explorations of the Frontiers of Human Consciousness*. Albany: State University of New York Press.

Huxley, A. (1947). *The Perennial Philosophy*. London: Chatto & Windus. https://archive.org/download/perennialphilosp035505mbp/perennialphilosp035505mbp.pdf. Accessed 20 July 2020.

Ims, K. J., & Jakobsen, O. D. (2006). Cooperation and Competition in the Context of Organic and Mechanic Worldviews—A Theoretical and Case Based Discussion. *Journal of Business Ethics, 66*(1), 19–32. https://doi.org/10.1007/s10551-006-9055-z.

Ims, K. J., & Zsolnai, L. (2009). Holistic Problem Solving. In L. Zsolnai & A. Tencati (Eds.), *The Future International Manager* (pp. 116–129). London: Palgrave Macmillan. https://doi.org/10.1057/9780230274068_7.

Jakobsen, O. (2017). *Transformative Ecological Economics: Process Philosophy, Ideology and Utopia*. London – New York: Routledge. https://doi.org/10.4324/9781315205434.

Korten, D. C. (2020a). Ecological Civilization: The Vision. https://davidkorten.org/home/ecological-civilization/. Accessed 20 July 2020.

Korten, D. C. (2020b). Wellbeing versus GDP: The Challenge and Opportunity of Human Development in the 21st Century. International Science Council, May 16, 2020. https://council.science/human-development/latest-contributions/wellbeing-versus-gdp-the-challenge-and-opportunity-of-human-development-in-the-21st-century/. Accessed 20 July 2020.

Lawrence, K. (2013). Developing Leaders in a VUCA Environment. UNC Executive Development, UNC Kenan-Flagler Business School. https://www.emergingrnleader.com/wp-content/uploads/2013/02/developing-leaders-in-a-vuca-environment.pdf. Accessed 20 July 2020.

Lozano, J. M., & Ribera, R. (2004). A New Chance for Management, A New Challenge for Spirituality. In L. Zsolnai (Ed.), *Spirituality and Ethics in Management* (pp. 175–185). Boston, Dordrecht, London: Kluwer Academic Publishers. https://doi.org/10.1007/1-4020-2365-0_15.

Mintzberg, H. (2007). *How Productivity Killed American Enterprise*. https://mintzberg.org/sites/default/files/article/download/productivity2008.pdf. Accessed 20 July 2020.

Mitroff, I. I. (1998). *Smart Thinking for Crazy Times. The Art of Solving the Right Problems*. San Francisco: Berrett-Koehler.

O'Higgins, E., & Zsolnai, L. (2018). Future of Business. In E. O'Higgins & L. Zsolnai (Eds.), *Progressive Business Models: Creating Sustainable and Pro-social Enterprise* (pp. 289–303). London: Palgrave Macmillan. https://doi.org/10.1007/978-3-319-58804-9_13.

Pruzan, P. (2011). Spirituality as the Basis for Leadership. In L. Bouckaert & L. Zsolnai (Eds.), *The Palgrave Handbook of Spirituality and Business* (pp. 287–295). London: Palgrave. https://doi.org/10.1057/9780230321458_35.

Pruzan, P. (2015). Further Reflections on Spirituality and Spiritual-based Leadership. In L. Zsolnai (Ed.), *The Spiritual Dimension of Business Ethics and Sustainability Management* (pp. 191–198). New York: Springer. https://doi.org/10.1007/978-3-319-11677-8_19.

Shrivastava, P., Stafford Smith, M., O'Brien, K., & Zsolnai, L. (2020). Perspective: Transforming Sustainability Science to Generative Positive Social and Environmental Change Globally. *One Earth*, 2(4), 329–340. https://doi.org/10.1016/j.oneear.2020.04.010.

Shrivastava, P., Zsolnai, L., Wasieleski, D., Stafford-Smith, M., Walker, T., Weber, O., Krosinsky, C., & Oram, D. (2019). Finance and Management for the Anthropocene. *Organization & Environment, 32*(1), 26–40. https://doi.org/10.1177/1086026619831451.

Steffen, W., Broadgate, W., Deutsch, L., Gaffney, O., & Ludwig, C. (2015). The Trajectory of the Anthropocene: The Great Acceleration. *The Anthropocene Review, 2,* 81–98. https://doi.org/10.1177/2053019614564785.

Steffen, W., Richardson, K., Rockström, J., Cornell, S. E., Fetzer, I., Bennett, E. M., Biggs, R., Carpenter, S. R., de Vries, W., de Wit, C. A., Folke, C., Gerten, D., Heinke, J., Mace, G. M., Persson, L. M., Ramanathan, V., Reyers, B., & Sörlin, S. (2015). Planetary Boundaries: Guiding human development on a changing planet. Science 347: 6223, 1259855, pp. 1–10. https://doi.org/10.1126/science.1259855.

Steffen, W., Rockström, J., Richardson, K., Lenton, T.M., Folke, C., Liverman, D., Summerhayes, C.P., Barnosky, A.D., Cornell, S.E., Crucifix, M., Donges, J.F., Fetzer, I., Lade, S.J., Scheffer, M., Winkelmann, R., & Schellnhuber, H.J. (2018). Trajectories of the Earth System in the Anthropocene. *Proceedings of the National Academy of Sciences* Aug 2018, *115* (33), 8252–8259. https://doi.org/10.1073/pnas.1810141115.

Tencati, A., Perrini, F., Hofstra, N., & Zsolnai, L. (2009). Engaging in Progressive Entrepreneurship. In L. Zsolnai & A. Tencati (Eds.), *The Future International Manager* (pp. 153–171). London: Palgrave Macmillan. https://doi.org/10.1057/9780230274068_9.

Tencati, A., & Zsolnai, L. (eds.) (2010). *The Collaborative Enterprise: Creating Values for a Sustainable World*. Bern: Peter Lang. https://doi.org/10.3726/978-3-0353-0005-5.

U.S. Army Heritage and Education Center. (2019). Who First Originated the Term VUCA (Volatility, Uncertainty, Complexity, Ambiguity)? https://usawc.libanswers.com/friendly.php?slug=faq/84869. Accessed 20 July 2020.

Wallis, J. (2020). Obstacles to Sustainable Change in Business Practice. In L. Zsolnai & M. Thompson (Eds.), *Responsible Research for Better Business: Creating Useful and Credible Knowledge for Business and Society* (pp. 211–233). London: Palgrave Macmillan. https://doi.org/10.1007/978-3-030-37810-3_10.

Wellbeing Economy Alliance. (2020). https://wellbeingeconomy.org/. Accessed 20 July 2020.

Zsolnai, L. (2014). Ethics Needs Spirituality. In L. Zsolnai (Ed.), *Beyond Self. Ethical and Spiritual Dimensions of Economics* (pp. 79–83). Bern: Peter Lang. https://doi.org/10.3726/978-3-0353-0596-8/20.

Zsolnai, L. (2015). Spirituality, Ethics and Sustainability. In L. Zsolnai (Ed.), *The Spiritual Dimension of Business Ethics and Sustainability* Management (pp. 3–11). New York: Springer.https://doi.org/10.1007/978-3-319-11677-8_1.

Index

A. Ócsai, *Ecologically Conscious Organizations*, Palgrave Studies in Sustainable Business In Association with Future Earth,
https://doi.org/10.1007/978-3-030-60918-4

The manufacturer's authorised representative in the EU is Springer Nature Customer Service Centre GmbH, Europaplatz 3, 69115 Heidelberg, Germany. If you have any concerns regarding our products, please contact ProductSafety@springernature.com

Printed and bound by CPI Group (UK) Ltd, Croydon, CR0 4YY
30/09/2026
02225698-0014